Computer Communications and Networks

Series editor

A.J. Sammes
Centre for Forensic Computing
Cranfield University, Shrivenham Campus
Swindon, UK

The **Computer Communications and Networks** series is a range of textbooks, monographs and handbooks. It sets out to provide students, researchers, and nonspecialists alike with a sure grounding in current knowledge, together with comprehensible access to the latest developments in computer communications and networking.

Emphasis is placed on clear and explanatory styles that support a tutorial approach, so that even the most complex of topics is presented in a lucid and intelligible manner.

More information about this series at http://www.springer.com/series/4198

Sudip Misra · Barun Kumar Saha
Sujata Pal

Opportunistic Mobile Networks

Advances and Applications

 Springer

Sudip Misra
Department of Computer Science
  and Engineering
Indian Institute of Technology Kharagpur
Kharagpur, West Bengal
India

Sujata Pal
School of Information Technology
Indian Institute of Technology Kharagpur
Kharagpur, West Bengal
India

Barun Kumar Saha
Department of Computer Science
  and Engineering
Indian Institute of Technology Kharagpur
Kharagpur, West Bengal
India

ISSN 1617-7975   ISSN 2197-8433   (electronic)
Computer Communications and Networks
ISBN 978-3-319-29029-4   ISBN 978-3-319-29031-7   (eBook)
DOI 10.1007/978-3-319-29031-7

Library of Congress Control Number: 2015960776

Printed on acid-free paper

This Springer imprint is published by SpringerNature
The registered company is Springer International Publishing AG Switzerland

Dedicated to my family

Sudip Misra

Dedicated to my parents

Barun Kumar Saha

Dedicated to my son, Soham

Sujata Pal

Endorsement

An excellent resource for both students and researchers interested in mobile opportunistic networking. A well balanced mix of theoretical and practical aspects.

Marco Conti, Institute of Informatics and Telematics (IIT), National Research Council of Italy (CNR), Italy

I've been working on opportunistic networking since a friend was caught up in a disaster zone after a tsunami and couldn't use his smart phone, as the net was down, yet there were many devices around the town which had wireless and could store, carry, and forward messages, but no protocols to implement this useful service. This book covers in clear and up-to-date style, the principles and practice of the main communications mechanisms to date, and offers helpful exercises and useful information on how to implement these systems in practice.

Jon Crowcroft, Fellow of IEEE, Fellow of ACM, Fellow of Royal Society (UK), University of Cambridge, UK

Foreword

In today's world, wireless connectivity acts as an enabling technology for a myriad of services and applications. One can hardly imagine a scenario where we were always plugging in Ethernet cables to our smartphones for electronic communication rather than using Wi-Fi! Wireless networks, in general, work in two modes—with and without any supporting infrastructure. Wi-Fi and cellular networks, which we are typically familiar with, are examples of infrastructure-based wireless networks. On the other hand, infrastructureless communication has also received significant focus in the past few decades largely in the form of mobile ad hoc networks (MANETs). One might be tempted to say that a MANET is much like the Internet sans the infrastructure. However, is deploying a MANET always feasible? What if there is no end-to-end communication path? How to develop protocols and support applications in such extreme, but likely, scenarios?

During my long tenure in research and academia, I have actively worked with different areas and aspects of wireless networks, such as mobile systems and communications, network security, sensor networks, network analytics, and quality of service. It, therefore, piqued my interest when, in the recent past, researchers looked upon Opportunistic Mobile Networks (OMNs), and its close variants, as a potential solution to the previously mentioned problem. In a layman's terms, it means that in OMNs, you do not necessarily need someone nearby you when you have created a message and want to have it delivered to the intended recipient. The capability of the nodes to buffer messages for a long time and transmit them at available contact opportunities gives OMNs an edge. In a sense, OMNs offer a truly best effort service—it attempts to deliver a message, but may not provide any related guarantee.

As the Editor-in-Chief of the IEEE Transactions on Mobile Computing, editorial board member of several other journals and IEEE Transactions, and chair of several conferences, I have come across a lot of cutting-edge research problems and solutions for OMNs. Based on these experiences, I personally feel that OMNs come with a great promise and wide usability. However, to realize its full potential, a detailed understanding of the fundamental issues of OMNs, as well as a vision for the future, is required. In this context, I am happy to say that the publication of the book *Opportunistic Mobile Networks: Advances and Applications* by Sudip, Barun, and Sujata could not have been more timely!

As a reader, you perhaps are already familiar with the authors of the book. I must say that the author trio has an impressive publication profile covering top conferences, journals, and IEEE transactions. Sudip is well known for his vast contributions to wireless sensor networks and allied areas, which are measured by not only in the number of published books and research articles, but also by patents, recognitions, and national and international awards. Barun, who has spent years in software industry, is highly experienced in the domain of network simulation, and known for his contributions toward the simulator community. Sujata has several years of teaching experience, and is a recipient of the prestigious TCS Research Fellowship and Schlumberger Fellowship. Therefore, a book by Sudip, Barun, and Sujata definitely does justice to the readers interested in OMNs and closely related areas.

I am, in particular, enthralled by the organization and focus of this book. The present book maintains a fine balance between theoretical and practical aspects. In fact, this book has something for everyone—be a student, researcher, teacher, or practitioner. Several other features of this book, such as largely cross-disciplinary approach, examples, illustrations and chapter highlights, and conceptual and programming exercises, are mentionworthy. The book begins with a concise overview of the evolution of OMNs, and gradually moves into greater technical depth. Although there are several treatments on routing protocols for OMNs available, the second chapter of the book does a nice job by presenting a few basic protocols and their subsequent enhancements together at the same place. The discussion about real-life traces in this chapter—in particular, the summary of characteristics of several traces—presents a nice perspective.

The third chapter of this book is a must-read for anyone who is serious about simulating OMNs. Although as researchers, we quite frequently implement our proposed protocols, simulate them, and gather results, we often tend to overlook some aspects of software engineering. I am impressed to find that a portion of this chapter is devoted toward testing of routing protocols developed with the simulator. There is no doubt that writing test cases is a necessity when you are concerned with serious development rather than a hobby.

The latter parts of this book deal with advanced topics, such as recent advances in sensor networks, communication in post-disaster scenarios, situation awareness, human intelligence, emotions, heterogeneity and their possible effects upon OMNs, and so on. A particular topic of my personal interest—use of social networks as modern sensor networks—is discussed in the fourth chapter. The third part of the book presents a gentle introduction to game theory and its applications toward cooperative communications. Subsequently, an evolutionary game theory-based cooperation scheme for OMNs is discussed. Among others, the last part of the book touches upon historical efforts toward representing dynamic networks as graphs. The final chapters of the book attempt to familiarize the reader with the current state of the art. It is interesting to see that there already is a handsome number of patents on OMNs and closely related topics.

To summarize, this book does a good job to bring together several dimensions of OMNs at a single place. The concluding part of the book discusses about some interesting applications, such as traffic off-loading, health care, and gaming. Since I have been involved with smartphone-based application and protocol development for a long time, these aspects, along with other topics discussed throughout the book, seem particularly promising to me. It is no doubt that smartphone-based computing and communication has become ubiquitous. But—as I raised the question raised at the beginning—would the existing infrastructure be able to support an exponential traffic growth in the future? The answer seems to be positive. OMNs can help in enabling communication not only in the absence of network infrastructure, but also without end-to-end communication paths. In other words, you would not require to search for an Wi-Fi access point while traveling with your smartphone. In this context, the present book sheds light on various related practical issues. I am sure this book would be useful for a diverse spectrum of readers out there!

Davis, California   Prasant Mohapatra
November 2015

Preface

As children, many of us were glued to the science fiction stories in comic books or television shows. Arguably, computer-generated imagery (CGI) in the recent sci-fi shows or movies have improved hugely as compared to the 1980s. Along with that, in the last few decades, sensing and computing technologies have progressed by leaps and bounds. Today, we have sensors around us almost everywhere—be inside a car, in our house, or pretty much anywhere else. Computing devices, on the other hand, are getting smaller and smaller in size. Those awesome gadgets, such as smart watches and smart glasses, from the sci-fi stories are no more just elements of fantasy, but are reality now! In other words, the terms "pervasive computing," "ubiquitous computing," and "wearable computers" were perhaps never more justified as they are today.

Together with advances in computing technologies, the need for communication among the devices, especially in wireless mode, is also rising. Our familiar Internet model has worked great so far. But what if some of the fundamental assumptions of the Internet model do not hold true? Moreover, when would such situations arise? And most importantly, would we still be able to communicate in such scenarios? The Delay-Tolerant Networking (DTN) architecture was proposed in the last decade to mitigate such situations where communication based on the Internet model fail. Ever since its inception, Delay-Tolerant Networks (DTNs) and its variants, such as Opportunistic Mobile Networks (OMNs), have hogged the limelight in the relevant research community to a large extent. In this book, we would talk about some of the fundamental aspects of DTNs and OMNs, their various applications, and breakthroughs made in this domain in the recent years.

Why Another Book on OMNs?

About 2 years back, we began toying with the idea of writing a book on OMNs. Our motivation and objective were rather simple. We wanted to bring the latest developments in OMNs to the readers in a concise manner. For example, as we shall see in the second part of this book, in contrast to the previously existing network communication models, human beings and their behaviors (such as movement patterns) are being increasingly intertwined with OMNs and related areas. Moreover, while cooperation in OMNs is a well-studied subject, we would observe in Part III of the book that an OMN by itself may move to equilibrium even in the absence of any incentive. In general, we wanted to talk about two specific topics—human aspects together with networking, and emerging forms of cooperation—among others in the context of OMNs.

However, we did not want to come up with yet another "reference book" on this subject. On one hand, our aim was not to make the treatment too shallow so that the reader misses the big picture. On the other hand, it should not be so much detailed that a novice reader shies away. Thus, in a way, this book is a quest for such a "just approach." Moreover, a narrow focus only on OMNs may not be a well-justified approach. There are various other closely related areas (for example, participatory sensing) that have emerged almost concurrently with OMNs. Much like one should be familiar with current affairs, it is desirable that a reader is equipped with such cross-disciplinary knowledge, which can lead toward further evolution and novel applications of OMNs. Furthermore, since network simulation is a quintessential tool for students and researchers alike, we wanted to have a specific focus on this aspect alongside the main matter. The present book, *Opportunistic Mobile Networks: Advances and Applications*, is a humble outcome of the above detailed thought process.

How to Use This Book?

This book can be used in different ways for different purposes.

As a Textbook

This can be used as a textbook for an introductory course on OMNs or an advanced course on communication networks that include DTNs/OMNs. Part I of the book is especially suitable for this purpose. Other parts of the book can be used to complement a similar course on OMNs with intermediate depth. This includes, but not limited to, post-disaster mobility and communication models, and heterogeneous networks. The remainder of the book would be suitable for a little advanced study. Although this is not supposed to be a book on Psychology, readers from that domain hoping to device online mechanisms for emotion detection might find a portion of this book interesting. In fact, the relevant bibliography would point to further resources in that area. Moreover, as we shall discuss below, this book is organized in a manner that both teachers and students would find it useful.

As a Reference Book

This book covers topics on some of the recent advances in OMNs as well as wireless sensor and ad hoc networks. Parts II–IV, in particular, are crafted to increase the appetite of a curious researcher. We hope that the huge volume of references presented herein would act as pointers for further advanced studies. In fact, several chapters of this book include discussions on some advanced topics, such as graph-based representation of DTNs, network equilibrium of cooperative and exploitative strategies, heterogeneous routing, and effects of emotions on communications. We hope that a reader with research mindset would find such content useful.

Who Is This Book For?

This book is intended to cater to the taste of a diverse audience including students, instructors, researchers, and practitioners. In general, this book is for anyone who wishes to learn about the recent advances in OMNs from a theoretical, as well as practical, point of view.

For Students

This book has been written with special care for the students. In particular, they would find Part I of this book particularly useful, where we present the general characteristics of OMNs, a detailed overview of contemporary routing protocols, and a chapter on simulating OMNs. Once comfortable, we encourage them to explore the remainder of this book. We have presented ample *illustrations* in this book to clearly convey the concerned ideas. Moreover, we have been careful to present relevant *examples* inside the chapters wherever appropriate. Such examples are a mean to better illustrate the underlying concepts. At the end of each chapter, there is a list of *review terms*. Much like the flash cards, we hope that these review terms would be helpful to quickly recollect the key concepts. Of course, there are *exercises* at the end of the chapters. Additionally, there are also accompanying *programming exercises* working on which would help in mastering the different techniques of network simulation.

Although we have tried to keep the language and flow of this book as simple as possible (but no more simpler, as Einstein once suggested), a previous exposure on the following topics would be helpful:

- Familiarity with wireless networks and basic concepts of mobile ad hoc networks.
- Familiarity with Java for working on the programming exercises.

For Instructors

Instructors would find this book useful to teach both undergraduate and post-graduate students. The book can be used to teach a course specifically on OMNs or as a part of a broader networking course covering OMNs. Part I of this book presents introductory concepts, which would be useful for the first-time students of this subject. The remaining parts of the book deal with slightly advanced topics that can be taught subsequently. Several examples and exercises are included in this book together with number of numerical problems that would be helpful for examination and evaluation purpose.

For Researchers

Researchers working in the domain of OMNs or other related areas would find this book a useful companion. This book not only present an exposure of a spectrum of contemporary research areas, but also discuss some of the specific problems in greater depth. Moreover, the book has a little cross-disciplinary flavor where we discuss certain aspects from artificial intelligence and psychology. A researcher might find those topics interesting. Moreover, this book contains around 300 bibliographic references to the current literature, which one might find useful.

For Practitioners

We hope that to a certain extent at least this book would be useful to practitioners. In particular, the final chapter of this book presents a tour of the existing technical specifications (requests for comments or RFCs) related to OMNs. The latter are essential if one wishes to develop real-life applications. Moreover, we also survey a sample of existing patents, which could be useful and motivating for future innovative ideas.

Organization of This Book

This book is divided into four parts as discussed below:

Part I: Introduction

The first part of this book presents a general introduction on the subject matter. Chapter 1 looks back at the origins of DTNs and its subsequent evolution over time giving rise to OMNs, Pocket Switched Networks (PSNs), and Mission-Oriented Opportunistic Networks (MOONs). The characteristics of DTNs are contrasted with existing conventional networks. This chapter, subsequently, presents a flavor of different research areas in DTNs. However, this is rather the tip of the iceberg, and some of these topics are discussed in greater depth in the remainder of this book.

Chapter 2 presents an overview of the contemporary routing protocols for DTNs and OMNs. This chapter also introduces the different performance evaluation metrics, which are essential to measure the performance efficiency of any networking scenario. Based on these metrics, some general insights into the broad class of routing protocols are provided. This chapter also presents a discussion on real-life traces that are increasingly being used in network simulations. Finally, this chapter concludes with a review of some of the existing applications of DTNs.

A quick introduction to the Opportunistic Network Environment (ONE) simulator is presented in Chap. 3. Specifically, this chapter guides the reader on setting up a simulation project using NetBeans and use it for application debugging. Subsequently, it describes how real-life traces can be incorporated into simulations. A key highlight of this chapter is a detailed discussion on development of a new routing protocol. We present a simple example with detailed code walk-through and insightful instructions. This chapter also introduces the reader to version controlling, which is essential in any real-life software development. In particular, basic usage of Git, a well-known distributed version control software, is discussed here. The latter portion of this chapter discusses unit testing of the protocols developed with the network simulator, and some of the best practices that one can follow while performing simulations.

Part II: Human Aspects in Opportunistic Mobile Networks

The second part of this book deals with an interesting topic—human aspects in the content of communication networks. Chapter 4 presents a quick tour of different forms of wireless sensor networks that have evolved in the recent past. Subsequently, their applications in disaster monitoring is discussed. In this context, we also discuss about some of the popular mobility models, as well as mechanisms of communication, in the aftermath of disasters. This chapter, then, presents an overview of agent-based systems, their applications, and typical representations of intelligence. This chapter concludes with a discussion of the effects of intelligence-induced mobility in a MOON formed in a post-disaster scenario.

Continuing the same thread, Chap. 5 deals with another fundamental aspect of human beings—emotion. This chapter begins with an overview of different theories and computational models of human emotions that has been proposed in Psychology and related domains. Next, different contemporary techniques of emotion detection and their applications are discussed. Various innovative techniques for quantifying emotions, for example, using smartphones and social media, are discussed. This chapter, and the second part of this book, concludes with a discussion on the effects of emotions on communication aspects in MOONs by considering a previously described computational model of emotion.

Part III: Cooperation in Opportunistic Mobile Networks

In distributed networks, communication involves more than a source node creating a message and a destination node receiving it. The intermediate nodes in such networks play a crucial role in relaying messages from one to another along a potential shortest path. It is, therefore, of utmost importance that the nodes cooperate among themselves for efficient operations. The third part of this book addresses these issues.

Chapter 6 introduces the reader to classical and evolutionary game theory. Game theory has been successfully used to define a collective set of strategies among a group of rationally behaving nodes. Subsequently, its application to different types of networks is presented. In particular, this chapter discusses about a particular type of game known as the Rock-Scissors-Paper (RSP) game. In the latter portion of this chapter, a set of strategies for cooperation (or lack thereof) among the nodes based on the RSP game are discussed. Mathematical representation of the game formulation is presented. This chapter concludes with an analysis of the cooperative strategies and a discussion on their relationship with one another.

Chapter 7 discusses several schemes of internode cooperation proposed for DTNs/OMNs. In particular, we look at different approaches for inducing cooperation based on incentives, which involve providing credits to (or updating reputation scores) of nodes in an OMN depending upon their message replication behaviors. Subsequently, we also look at a few game theoretic schemes for cooperation. Additionally, some of the approaches that do not fall in either of the previous two categories are explored. The latter portion of this chapter presents a detailed discussion on DISCUSS, a distributed cooperation enforcement scheme for OMNs. This is followed by an in-depth look at its operation together with its theoretical characterization. Finally, this chapter closes with a quick study of feasibility and efficiency of DISCUSS in practice.

Part IV: Advanced Topics

The final part of this book deals with relatively advanced topics. In contrast to the remainder of the book, where network homogeneity was assumed, Chap. 8 presents a detailed discussion on the origins of heterogeneity in a network and its impacts. Subsequently, aspects of heterogeneity that are seen in the context of DTNs are discussed. In particular, heterogeneity arising due to diverse contact patterns among the nodes is also considered. This chapter then describes different ways of representing a DTN with a graph—a typical approach in computer networks. However, unlike the traditional graph theoretic approaches, DTNs require a special treatment. Using the concept of time-varying graphs, the notion of communication index is presented in this chapter. Finally, different ways of mitigating the adverse effects of heterogeneity are discussed here.

Chapter 9 presents a comprehensive look at the current reality of OMNs. In particular, it looks at a sample of quantified volume of research efforts on DTNs/OMNs in the recent years. This is followed by a discussion on RFCs, their categories, and overview of some of the RFCs related to our area of interest. A review of a sample of related inventions and corresponding patents is also presented. This chapter then outlines some of the promising avenues and directions along which OMNs can evolve in the future. This chapter ends with a discussion on a set of prospective projective topics that an interested reader can undertake.

Finally, Chap. 10 concludes the book. Throughout this book, several topics related to OMNs and closely allied areas are discussed. This final chapter presents all such discussions in perspective and looks at the big picture emerging out of it.

Organization of the Chapters

The chapters in this book are organized in the following way. A chapter begins with a general *introduction* on the larger set of topics to be discussed in the remainder. This introduction also presents an overview of the organization of the corresponding chapter. Subsequently, concerned subject matters are discussed. In general, the chapters in Parts II and III usually begin with simple aspects; deeper analysis of related problems are presented in the latter sections. Toward the end, we present *summary* of the chapters where the previous discussions are put in context. Followed by this, a set of *review terms* are presented, which curates the key terminologies used in a chapter. Next, a set of *exercises* are presented. We present here both numerical as well as descriptive type of problems, which should be useful to evaluate one's learning. Finally, a chapter ends with a set of *programming exercises*, where we list a variety of problems to be solved using the ONE simulator. We very much encourage the reader to attempt these problems, especially if he/she is a beginner to network simulations.

Conventions Used in This Book

The following convention is used in this book. Mathematical notations are displayed in slightly italicized text, for example, $S = (l, \theta, \Delta t_f, \Delta t_p)$. Vectors are denoted with bold faces, for example, **e**.

File names or other strings are displayed in fixed width teletype fonts, for example, `File > New Project`. Directories are indicated with a front-slash at the end of their names, such as `test/`. Commands, class names, and other (inline) code inside the text are displayed in small fixed width teletype fonts such as `StationaryMovement`. Source code listings are displayed inside the boxes with line numbers shown outside, for example:

Listing 1: The getMessagesWithCopiesLeft() method

```
115  protected List<Message> getMessagesWithCopiesLeft() {
116      List<Message> list = new ArrayList<Message>();
117
118      for (Message m : getMessageCollection()) {
119          Integer nrofCopies = ↩
                 (Integer)m.getProperty(MSG_COUNT_PROPERTY);
120          assert nrofCopies != null : "SnW message " + m + " didn't ↩
                 have " +
121              "nrof copies property!";
122          if (nrofCopies > 1) {
123              list.add(m);
124          }
125      }
126
127      return list;
128  }
```

In listing of Java codes, the keywords are displayed in blue and string constants in purple, whereas class and method names are depicted with bold faces. In some cases, the line numbers are omitted if not relevant, for example, while displaying the contents of the simulation settings files as shown below:

```
Scenario.simulateConnections = true

## Movement model settings
# KAIST movement trace
Group.movementModel = ExternalMovement
ExternalMovement.file = my_scenarios/KAIST_92n_movement_trace.txt
```

Additionally, listing of commands executed in a terminal, and their output, are also shown inside a box without line numbers. The terminal prompt is indicated with a $ symbol, as shown below:

```
$ git commit -m "Adding source files"
$
```

Buttons (in software screens) are shown in an oval box with the text inside, for example, Next . Note that images and code snippets are available in color in the electronic version of the book. Insights or additional information on the concerned topics are presented inside a gray box as shown below.

> Due to intermittent connectivity, contemporary global information about the OMNs is often not available to the nodes. Therefore, while routing, a node typically attempts to make an optimal message forwarding decision based on locally available information.

Supplementary Resources

This book comes with the following supplementary resources:

- **Solutions Manual**: This contains solutions to most of the exercises provided at the end of each chapter.
- **Source Code**: Solutions to selected programming exercises are presented in this book, which are to be used together with the ONE simulator.

The solutions manual can be obtained from website of the publisher, Springer. Readers can access the following GitHub repository to download the above-mentioned source code: `https://github.com/barun-saha/one-simulator`

Acknowledgments

Writing a book is a project by itself. As with all real-life projects, publication of a book, too, involves efforts from a large team of people apart from the authors.

This book would not have materialized without the continuous support from Springer. In particular, we thank Mr. Wayne Wheeler, Senior Editor, Computer Science, Springer and Mr. Simon Rees, Associate Editor, Computer Science, Springer, for their faith in the idea of this book. Behind the scenes, they took care of all the relevant formalities, thereby allowing us to concentrate solely on preparing the manuscript.

While, as authors, our duty ended with submission of the manuscript, the role of the Editorial team, Springer, just began with that. We thank the team members for their painstaking efforts in typesetting and proofreading this manuscript and translating it into reality.

We express our gratitude to IEEE, IET, ACM, and Springer for allowing us to incorporate our previously published works in this book. In particular, we are thankful to Mr. Frank Pepe, Director, Pricing & Strategic Analysis, IEEE, and Ms. M.E. Brennan, IEEE, for their kind cooperation and permission. We also thank Ms. Alyssa Howell, Journals Permission Officer, IET, for granting us a similar privilege.

This list would be incomplete without thanking the user community of the ONE simulator. The members of this vibrant community have always been helpful, and patiently provided solutions to the various problems faced by users. On various occasions, we had fruitful discussions and exchange of ideas with different members of this community. In this context, we are also thankful to the community which has made Barun Saha's blog on Delay Tolerant Networks a success, and which serves as an additional platform for knowledge dissemination and interaction.

Acknowledging people directly or indirectly involved with this book is easy. However, apart from them, there are many other people in our lives whose support and inspiration have been instrumental in undertaking and successfully completing this process.

Sudip thanks his parents, Prof. J.C. Misra and Mrs. Shorasi Misra, his wife, Satamita, and children, Devadeep and Devadrita, for their continuous support and encouragement throughout this project.

Barun thanks his parents, Mr. Bimal C. Saha and Mrs. Kamala Saha, who have always been a source of motivation for him. All these months, they regularly inquired about the progress of the book, which helped Barun to stay focused on the writing track. He is also thankful to his brother, Mr. Arun K. Saha, and sister-in-law, Mrs. Shaonly Saha, for inspiring him. And, of course, Barun's sweet little nephews, Akash and Arush, who have been a constant source of joy and encouragement!

Sujata expresses her appreciation to her son Soham and huband Anup whom she missed a lot during the preparation of the manuscript. They are her constant sources of inspiration and strength. Nothing would have been possible without the support of her parents and parents-in-law. Sujata's sisters, Sumita, Kabita, and Shukla, brother Prabir, and other family members have been the pillars of strength in all her endeavors. Sujata thanks them from the core of her heart.

Kharagpur, India   Sudip Misra
November 2015   Barun Kumar Saha
                                                                             Sujata Pal

Contents

Part III   Cooperation in Opportunistic Mobile Networks

Acronyms

ADU	Application Data Unit
API	Application Programming Interface
ARPANET	The Advanced Research Projects Agency Network
CA	Central Authority
CBR	Contact-Based Routing
CCDF	Complementary Cumulative Distribution Function
CDF	Cumulative Distribution Function
D2D	Device-to-Device (communication)
DF	Delegation Forwarding
DISCUSS	Distributed Information-Based Cooperation Ushering Scheme
DISIDE	Distributed Strategy Identification Scheme
DTN	Delay Tolerant Network
EBR	Encounter-Based Routing
EGT	Evolutionary Game Theory
ESN	Environmental Sensor Network
ESS	Evolutionary Stable Strategy
ICT	Inter-Contact Time
IDE	Integrated Development Environment
IoT	Internet of Things
IP	Internet Protocol
IPN	Interplanetary Internet
MANET	Mobile Ad Hoc Network
MOON	Mission-Oriented Opportunistic Network
NSS	Neutrally Stable Strategy
OMN	Opportunistic Mobile Network
ONE	Opportunistic Network Environment
OSPF	Open Shortest Path First
PAD	Pleasure-Arousal-Dominance
PDA	Personal Digital Assistant
PDF	Probability Density Function
PRoPHET	Probabilistic Routing Using History of Encounters and Transitivity

PSN	Pocket Switched Network
PTU	Protocol Translation Unit
RAPID	Resource Allocation Problem for Intentional DTN
RFC	Request For Comment
RFID	Radio Frequency Identification
RIP	Routing Information Protocol
RWP	Random Waypoint
SD-OMN	Strategy Defined-Opportunistic Mobile Network
SMS	Short Message Service
SnW	Spray and Wait
TA	Trusted Authority
TCP	Transmission Control Protocol
TLW	Truncated Levy Walk
TTL	Time-to-Live
TVG	Time-Varying Graph
WSN	Wireless Sensor Network

Part I
Introduction

Chapter 1
Origins and Characteristics

The ARPANET project, sponsored by the Advanced Research Projects Agency Network, USA, in the 1960s was aimed at interconnecting computers at different research institutes for better work processing. It is well known that the TCP/IP protocol suite evolved from ARPANET, which serves as the backbone of modern Internet. However, the world of computing itself has undergone a massive change in the past five decades. Devices are getting smaller and smaller in size, and exhibiting significant computing power. While the first computers occupied a large room, modern smart phones nicely fit in our pockets. It is not just phones that are getting smart, but wrist watches, glasses and other wearable devices, too. And with such technical advances, connectivity among the devices has progressed as well. Today, we live in a world where access to the Internet highway ceases to be a luxury, and becomes more of a basic need.

Continued research in the domain of the computer networks have given rise to various other dimensions beyond the Internet, for example, mobile ad hoc networks (MANETs) and wireless sensor networks (WSNs). Although MANETs are different from the Internet, they still retain flavors of the Internet to a greater extent. The game, however, drastically changes when one tries to fit the good old Internet model in *challenged* environments. Such challenges arise due to multiple reasons including, but not limited to, harsh geography and very sparse connectivity because of which, certain assumptions of the Internet does not hold good any further. To provide connectivity in such environments and under such conditions, the delay (or disruption)-tolerant networks (DTNs) have evolved.

In this chapter, we briefly look at the origins and evolution of DTNs. We summarize the characteristics where the traditional Internet model of communication fails. Subsequently, we look at the characteristics of DTNs and underlying challenges. Specifically, we look at how routing in DTNs and conventional networks fundamentally differs. Next, we discuss about different variations of DTNs that have evolved over the years. This is followed by a review some of the contemporary research areas in DTNs. Finally, we conclude this chapter with a discussion on network simulation—a powerful tool for any network protocol designer.

© Springer International Publishing Switzerland 2016
S. Misra et al., *Opportunistic Mobile Networks*,
Computer Communications and Networks, DOI 10.1007/978-3-319-29031-7_1

1.1   Delay Tolerant Networks

Let us begin by talking about the Internet "model". The well-known TCP/IP protocol stack essentially powers the Internet atop which various applications, such as email and World Wide Web, are running. The ubiquitous Internet, however, fails to work in the environments that exhibit one or more of the following characteristics [1][1]:

- Intermittent connectivity among the nodes in the network, which invalidates the assumption of the existence of end-to-end paths. Such connectivity could be scheduled, for example, a satellite in the Mars' orbit communicating with the Earth station only at certain times or, could be opportunistic, for example, two smart phones carried by the humans in an urban scenario communicate when they come in one another's transmission range.
- High or variable latencies in the message deliveries as an effect of intermittent connectivity among the nodes.
- Asymmetric connections or link bandwidths—such constraints make an interactive protocol like TCP to fail.
- High error rates along the transmission media calls for multiple retransmissions, which may not be feasible given the limited availability of communication opportunities.

To contrast the so characterized *challenged networks* with the Internet, let us consider some quantitative examples. While the round-trip time (RTT) in the Internet is about a few hundred milliseconds [2], the speed of light pushes the propagation delay between Earth and Mars in the range of 3–20 min [3] depending on the positions of the planets. Moreover, the speed of transmission directly between Earth and Mars is only about 500 bps–32 Kbps, whereas a typical dial-up modem provides a data rate of about 56 Kbps [4]. In case of underwater acoustic networks, too, the transmission speeds are limited to around 38 Kbps [5].

To cope with such characteristics, and provide interoperability among heterogeneous networks, the DTN architecture was proposed [6]. DTN act as an "overlay" to make the diverse networking architectures inter-operate. Thus, DTN enables to connect the terrestrial Internet with the deep-space network. While traditional Internet use store-and-forward strategy, DTNs provide a paradigm shift by using the store-carry-and-forward strategy. Typical examples of DTNs include, but not limited to interplanetary Internet (IPN) [3, 7] and underwater acoustic sensor networks [8].

---

[1]Portions of this chapter are reproduced with kind permission from Springer Science+Business Media: Next-Generation Wireless Technologies, Cooperation in Delay Tolerant Networks, 2013, 15–35, Sudip Misra, Sujata Pal, Barun Kumar Saha.

1.1.1 Evolution

The evolution of DTNs began with the efforts of Cerf et al. [7] to define an IPN architecture. The choice of extending the terrestrial Internet for interplanetary communications was evident, given its tremendous success. However, several environmental challenges, as mentioned previously, prevented the direct adaptation of TCP/IP-based Internet for the purpose, which called for to define a new architecture. In the proposed architecture, the IPN was divided into multiple regions. Data among the nodes were transferred as "bundles" using a store-and-forward approach. In particular, IPN had three types of nodes: bundle agents (source and sinks of the bundles), IPN Relays, and IPN Gateways (connecting two or more IPN regions).

Fall [6] generalized the concepts presented in [7] into a delay-tolerant networking architecture to provide interoperability among diverse and challenged networks, where one or more assumptions of the Internet may not hold true. The DTN architecture so proposed served as an overlay on the top of the transport layers of the underlying networks. The concepts of regions and gateways from IPN were extended for the DTN as well. The DTN architecture identified two different types of nodes—with and without persistent storage. In addition, the author also discussed about different services available in, and issues related to, such networks.

Over the years, researchers have explored the idea of DTNs to give rise to multiple specialized sub-domains. Khabbaz et al. [9], however, observed that while DTN was originally meant as an "overlay" architecture [6], in many cases in the existing literature, the term DTN has been used to refer to the individual challenged networks as well. The authors differentiated the two categories as "Delay-Tolerant Networking" and "Intermittently Connected Networks". Two prominent subclasses of such challenged networks are the opportunistic mobile networks (OMNs) [10] and pocket switched networks (PSNs) [11]. An OMN[2] essentially captures the idea that, unlike the traditional networks, communication links do not always exist in such a network. Rather, mobility of the nodes provides them with opportunities to communicate with the other nodes in the network. On a similar note, the concept of PSNs was proposed. A PSN is formed among the portable devices, such as smart phones and PDAs, carried by humans. In this book, unless otherwise stated, by "DTNs" we refer to such challenged networks that exhibit characteristics enumerated at the beginning of this section.

---

[2]We prefer the term "Opportunistic Mobile Networks" to "Mobile Opportunistic Networks" although technically, they refer to the same thing. We note that the nodes in OMNs are inherently mobile—it is mobility that, in fact, makes communication possible. "Opportunistic" is a qualifier that characterizes such kind of communication.

The DTN literature, in fact, is rich in nomenclature. Apart from the names mentioned here, the reader likely would come across various terminologies, such as opportunistic networks, OppNets, mobile opportunistic networks, human-centric DTNs, and so on. As mentioned here, although terminologies vary, they typically represent the same basic characteristics. We, however, ask the reader not to be intimated by such variety of terminologies, but rather focus on the fundamental concepts. In particular, when we talk about OMNs in the remainder of the book, unless otherwise stated, we assume that no global network connectivity is available.

1.1.2 Characteristics and Challenges

The inherent characteristics of the DTNs ensure that the protocols from the Internet, wireless networks or MANETs fail to work with such networks. This implies that the network dynamics, for example, routing and security, be addressed from a different point of view. To illustrate, consider a popular routing protocol for the MANETs, the dynamic source routing (DSR) [12]. One of the fundamental assumptions of the protocol is the availability of at least one end-to-end communication path between any source and destination node pairs. In [13], multiple such paths are considered as backups, in case the primary path fails. Unfortunately, such possibilities are rare in DTNs.

One of the salient differences observed in DTNs is how a message is routed. Conventional networks typically employ the *store-and-forward* routing. For example, let us consider a traditional router with multiple interfaces. When a packet arrives along an input line, it is at first stored in the input port's buffer until it can be forwarded to a suitable output port.

Routing in DTNs, on the other hand, usually follows the *store-carry-and-forward* paradigm. In DTNs, when a node receives a message from another node, the former stores the message in its buffer. The concerned node can possibly move with the message stored in its buffer. In other words, a node *carries* a message while it moves. Finally, when the node comes in contact with another node—which could take a considerable amount of time—the former forwards the node to the latter with the hope that the latter can deliver the message to its corresponding destination. Figure 1.1 illustrates the difference in two routing paradigms.

Furthermore, in DTNs, we typically consider a *message* unlike packets in the traditional networks. This can be traced back to [6, 14]. Fall [6] presented DTN as a "message-oriented reliable overlay architecture" residing over the transport layers of the interconnected networks. Traditionally, the term "packet" is used for data units at the network layer. The focus of DTNs has been on "reliable message routing" rather than "best-effort packet switching" [6]. Request for comment (RFC) 4838 [14] provides further insight into this as quoted below:

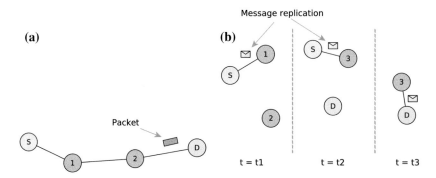

Fig. 1.1 Routing in **a** conventional networks via end-to-end paths and **b** DTNs using message replications. In the figure, "S" indicates the source node of the packet/message and "D" denotes its destination

In general, applications should attempt to include enough information in an ADU so that it may be treated as an independent unit of work by the network and receiver(s). The goal is to minimize synchronous interchanges between applications that are separated by a network characterized by long and possibly highly variable delays. A single file transfer request message, for example, might include authentication information, file location information, and requested file operation (thus "bundling" this information together).

Additionally, in DTNs, we talk about message *replication* rather than forwarding. Replication refers to the process where a transmitting node sends a *copy* of the message to the receiving node and itself retains the original message. Therefore, multiple copies of a message can exist in the network. Such replication is used to enhance the chances of message delivery given the scarcity of communication opportunities in DTNs.

To summarize, DTNs were originally conceived to exchange messages among the nodes of (possibly heterogeneous) networks. Therefore, the contemporary literature in DTNs/OMNs often talks about messages rather than packets. In the remaining of this book, we would mostly use the term "message." Although on few occasions the term "packet" also finds it use, the difference, if any, should be clear from the context.

In general, the following points summarize some of the basic characteristics of DTNs/OMNs:

- Hop-by-hop transfer of a message in the absence of end-to-end communication path.
- Storing a message, carrying it along often for a long time, and transmitting during a subsequent contact with another node.
- Replication of a message rather than forwarding it.
- Experiences high message delivery latency, typically in the order of hours.

1.2   Mission-Oriented Opportunistic Networks

Pocket switched networks (PSNs) [11]—another member of the DTN family—are formed among the mobile devices (for example, smart phones and PDAs) carried by humans. These devices make use of the global connectivity, such as Wi-Fi access point, when available. However, in the absence of any such network infrastructure, the devices engage in opportunistic communications with one another, enabled by Bluetooth and/or Wi-Fi interfaces. Two nodes in a PSN communicate when they come within the transmission range of one another. The data transfer among the nodes follow the store-carry-and-forward mechanism as in DTNs. Moreover, the mobility of the human owners carrying their devices help in the dissemination of messages across the concerned geography.

In this context, one may wonder about the significance of the nomenclature of *pocket* switched networks.[3] Two reasons can be cited for this [15]. First, the devices in PSNs—for example, smart phones—are typically carried in users trouser pockets. Second, by the virtue of a user's mobility, such a device is carried to a "pocket of wireless connectivity," and thereby, enabling multiple devices to communicate.

Modern generation smart phones, which are equipped with one or more sensors (for example, accelerometer and gyroscope), are increasingly becoming part of the everyday life. Therefore, the integration of humans in mission-oriented networks [16] with PSNs is inevitable. This hybridization of PSNs with mission-oriented networks induces interest in a new class of networks—*mission-oriented opportunistic networks* (MOONs) [17], which are primarily characterized by:

- **Mission objectives**: MOONs are often associated with some mission objectives, for example, providing prompt communication facilities in the aftermath of a large-scale disaster.
- **Opportunistic contacts**: Whenever two or more nodes come in the communication range of each other, they exchange available information. In case no communication opportunity is available at that moment, nodes store the messages in their buffer (for possibly long time).
- **Human–network interplay**: MOONs are not just networks where automated devices communicate among themselves. Rather, the presence of humans—the owners of the devices—as part of the network, adds a new dimension. In other words, the humans are not merely the end users who send and receive messages, and thus, are expected to affect the network dynamics.

While the areas of psychology and social sciences study the factors affecting the different aspects related to humans, the possible impact of such human factors on the network dynamics, and thus, mission objectives has been largely overlooked. The only human characteristic that reflects significantly in networking research is human

---

[3]The difference between PSNs and OMNs is rather implicit and hardly found in the literature. It may, however, be noted that PSNs inherently involves human users as the carrier of their devices. OMNs, on the other hand, do not make such assumption, but rather focus on *opportunistic* communications. Once again, we ask the reader not to be intimidated with all these terminologies.

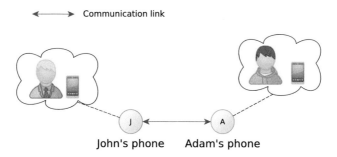

Fig. 1.2 Communication in a MOON. In MOONs, the communication end points are not just mechanical nodes. Rather, they are the portable devices carried by humans. Consequently, various human aspects could influence the network communication mechanism

mobility. This motivates one to study the effects, if any, of other human aspects, for example, emotion and intelligence, on the communication networks. The motivation is further strengthened by the fact that, in today's age, humans spend considerable time with their mobile devices giving rise to a form of *human–device relationship*. Therefore, in MOONs, the *nodes* are not mere mechanical devices, but those devices, together with their human owners, give rise to a *supernode*-like entity as shown in Fig. 1.2. This is in sharp contrast with the conventional point of view of networking, where nodes refer to devices, and humans are just end users.

> MOONs have been inspired by multiple evolving research areas in the domain of wireless networks and psychology including PSNs, human-centric sensing, mission-oriented sensor networks and effects of human emotions on their actions. The latter categories would be explored in the subsequent chapters.

1.3 Research Areas in OMNs

Different paradigms of communication networks—whether the Internet, or cellular networks, or OMNs—are conceptualized with the primary objective of supporting communication in scenarios where it was otherwise nonexistent or suboptimal. However, with the passage of time, providing a "vanilla" communication service is no more sufficient—support for other features is also required. DTNs/OMNs are no exceptions in this regard. Although originated as a mean to enable communication under challenged and heterogeneous network conditions, the past decade and a half has witnessed researchers focusing their efforts along different dimensions of this domain. Some of the commonly observed research trends in the area of DTNs/OMNs are shown in Fig. 1.3. In this chapter, we present an overview of some of these topics.

Fig. 1.3 Different research
trends found in the area of
DTNs

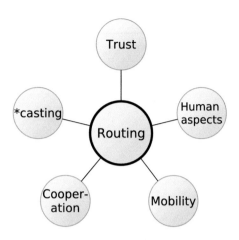

An exhaustive listing of, and discussion on, all the related areas is out of scope of
this book. However, a few of advanced topics would be discussed in greater depth in
the remainder of the book.

- **Routing protocols**: Undoubtedly, design of *the* most efficient routing protocol
  for DTNs/OMNs still remains an open problem. The key challenge of routing
  is to ensure delivery of as many number of messages as possible using different
  available information, if at all, under diverse scenarios. For example, the Epidemic
  [18] protocol is based on simple flooding. Whereas, the PRoPHET [19, 20] protocol
  makes routing decision based on the chances of meetings among the nodes. In the
  recent past, several routing protocols have been proposed that offer interesting
  performance profile. A detailed discussion on some of the contemporary routing
  protocols for DTNs/OMNs would be presented in Chap. 2.
- **Human mobility and analysis**: As noted earlier, mobility is rather a driving force
  for OMNs. It is mobility of the nodes that bring them in contact in the otherwise
  disconnected world. Therefore, a deep interest in studying different aspects of
  mobility is a natural consequence. Numerous real-life experiments, for example,
  [11, 21, 22] have been conducted by providing mobile devices to the human users.
  The data so obtained have been used to draw inferences related to mobility patterns,
  connection dynamics and others.
- **Cooperation**: Cooperation remains a critical issue in OMNs due to the intermittent
  connectivity among the nodes. Lack of cooperation among the nodes can affect
  the network performance resulting in less number of delivered messages.
- **Trust and security**: Determining trusted nodes for message delivery is often
  related to the aspects of cooperation. Trust, however, is different from cooperation since the former deals with malicious nodes while the latter with selfish
  nodes. Trifunovic et al. [23], for example, presented approaches for establishing
  social trust in OMNs. Chen et al. [24], on the other hand, considered dynamic trust
  management in DTNs in the presence of selfish as well as malicious nodes. Ayday

and Fekri [25] presented an iterative mechanism to thwart Byzantine attacks in DTNs. When concerned with real life, however, routing and security should not be parallel objectives, but go hand in hand.

- **Privacy and anonymity**: Respecting privacy and anonymity in communication may not be a privilege, but a necessity in many scenarios. For example, consider intelligence exchanged via an OMN, or voicing concerns in hostile environments. Lack of provision for anonymous communication could lead toward potential harm. Needless to say, this is an active research area in OMNs.
- ***casting**: By *casting we refer to the large family of message delivery schemes including multicasting, broadcasting, anycasting [26], manycasting [27], geocasting [28], and interestcasting [29]. To recollect, unicasting refers to sending a message to a single destination node. Multicasting, on the other hand, sends the same message to multiple destination nodes who are the members of a given multicast group. Anycasting refers to transmitting a message to any node at the least. Whereas, the objective of manycasting is to send a message to K nodes in the network. On the other hand, interestcasting [29] refers to dissemination of information to other users sharing similar interest. Since unicasting itself is difficult in OMNs/DTNs, multicasting and other variations become further challenging in such networks. RelayCast [30] is a popular example of scalable multicasting in DTNs based on the two-hop relaying [31] protocol.[4]

It may be noted from Fig. 1.3 that multiple aspects are eventually related to the routing issue. For example, whether or not a node would cooperate is a different decision than how it should route a message. Again, trust factor introduces an additional parameter to the underlying routing mechanism—whether or not a message should be forwarded to a given node. In the following, we briefly look into some of these areas.

1.3.1 Cooperation

A cooperative node is one that helps in delivery of message created by other nodes by receiving and replicating them (messages). A node that does not cooperate is said to be non-cooperative, in general. Their behavior can, however, span a wide spectrum, as we briefly discuss here.

Figure 1.4 illustrates a typical communication scenario in the presence of cooperative, selfish, and malicious nodes. In Fig. 1.4a, a cooperative intermediate node comes in contact with a source node at a time instant t_1. This intermediate node receives a message from the source node, and forwards that to the corresponding destination node at a later time instant t_2. On the other hand, in Fig. 1.4b, the intermediate node is selfish—although it receives the message from the source node, it

---

[4]In two-hop routing, the source node transmits a message copy to say, n, nodes. These nodes, however, cannot further replicate the message, but only directly deliver to the concerned destination node. As a consequence, any delivered message travels two hops at most.

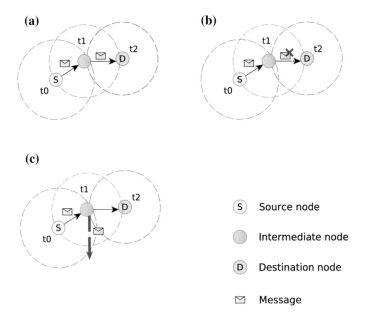

Fig. 1.4 Illustration of communication in the presence of **a** cooperative, **b** selfish, and **c** malicious nodes

does not forward the same to the destination. Such lack of cooperation, however, could possibly be accounted to several factors other than selfishness as we shall see next. Finally, the intermediate node in Fig. 1.4c is malicious—it intentionally drops the message.

It may be noted that, "cooperation in a network" can indicate numerous higher level phenomena. In practice, one must define the network dimensions where cooperation plays a role. Cooperation among the nodes in a network could be with regards to various aspects, including, but not limited to:

• **Buffer space**: A node in a DTN, which forwards it own messages to other nodes, must also accept and store the other nodes messages into its buffer. The Bundle protocol [32] provides the facility of custody transfer—a reliable mechanism for transferring bundles. In custody transfer, a DTN node accepting a bundle (the custodian) from another node cannot drop the bundle either until it has been forwarded to some node, or its time-to-live (TTL)[5] is expired.

• **Communication opportunities**: A node in a DTN not only carries message(s) from other nodes, but also forwards them without discrimination when suitable

---

[5]The reader may recollect that IP headers have a similar TTL field, which indicates the maximum number of network hops that a packet can travel across. In DTNs, however, TTL is expressed in time, and is usually large. For example, if the TTL of a message is t hours, the message is valid for that duration since its generation. Beyond that time interval, the validity of the message expires, and any node storing that message in their buffer is free to drop it.

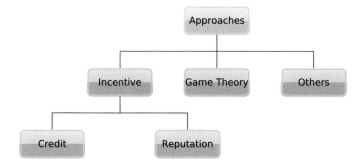

Fig. 1.5 Different approaches for cooperation enforcement in DTNs

contact opportunities are available. This is important, since, as discussed earlier, lack of connectivity among the nodes in a DTN is the norm rather than exception. Of course, a node forwards messages created by itself with higher priority.

- **Energy aspects**: A node with critical energy considerations may utilize the current communication opportunity with another node to seek cooperation in forwarding the messages. Differently, a node may increase its transmission power to reach the distant nodes [33].

Figure 1.5 shows some of the popular approaches taken for enforcing cooperation in DTNs/OMNs. The two major categories are incentive- and game theory-based approaches. Misra et al. [34], on the other hand, presented a scheme for cooperation in strategy defined OMNs based on evolutionary theory. We will discuss in details on these different approaches in OMNs in Part III of this book.

1.3.2   Human Mobility

Several real-life experiments (for example, [11, 21, 22]) have been conducted till date to study different characteristics of the OMNs. In these experiments, participating users were given some mobile devices—iMotes or phones—installed with software to record the sightings of nearby devices. These are small and low-power devices, which one could carry in their pockets. Typical data recorded in these cases include the contacts with the other devices, the contact durations, and the locations of the individual devices.

One of the well-explored features of these real-life traces is the inter-contact duration in the network. A *contact* is defined as an event that starts when two nodes come within the transmission range of one another, and continues until they move away from each other's transmission circle. The time duration for which two nodes stay in contact is termed as the *contact time* (or contact duration). Similarly, the *inter-contact time* (ICT) is defined as the time duration between two successive

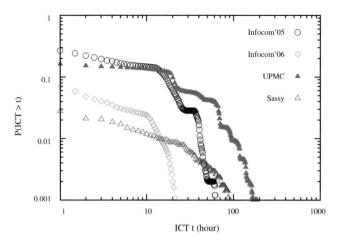

Fig. 1.6 Distribution of inter-contact time for some real-life connection traces. Both the axes of the plot are shown in logarithmic scale

contacts. Hui et al. [11] noted that the aggregated[6] ICT among the nodes in a network tend to resemble a heavy-tailed power law distribution, as shown in Fig. 1.6. Such revelations were of particular interest since for a long time, the mobility models usually considered in different types of networks were either the random waypoint (RWP) mobility model, or its variants, which results in exponential ICT distribution. Karagiannis et al. [35], on the other hand, observed that such a behavior is exhibited until a characteristic time typically of the order of 12 h. Beyond that characteristic time, however, the ICT exhibits an exponential decay. The authors further note that such a dichotomy, indeed, helps in achieving finite delivery times in OMNs.

Analysis of large number of GPS traces show that the human movement pattern can be modeled with the truncated Levy walk model (TLW) [22]. TLW consists of a sequence of steps, where each step is represented with a tuple $S = (l, \theta, \Delta t_f, \Delta t_p)$. Here, $l > 0$ denotes the length of the flight; $\theta \in [0°, 360°]$ denotes the direction; the flight time is denoted by $\Delta t_f > 0$; and $\Delta t_p \geq 0$ denotes the pause time. The distributions of the flight length and the pause time, $p(l)$ and $\psi(\Delta t_p)$, respectively, follow the Levy distribution: $p(l) \sim 1/l^{(1+\alpha)}$ and $\psi(\Delta t_p) \sim 1/\Delta t_p^{(1+\beta)}$, where $0 < \alpha, \beta < 2$. In [22], the authors, based on the experimental data, modeled the flight time as $\Delta t_f = k l^{(1-\rho)}$, where k and ρ are constants; $0 \leq \rho \leq 1$. Thus, every time a traveler generates a tuple S, so that $0 < l \leq l_{max}$, and waits for time $0 \leq \Delta t_p \leq \Delta t_{p,max}$ after reaching the destination. The step time, $\Delta t_s = \Delta t_f + \Delta t_p$, gives the time spent by a traveler for a given step.

---

[6]Aggregated ICT refers to the consideration of all ICTs in a network—between any pair of nodes—as a whole. Alternatively, one can also consider the pairwise ICTs, i.e., the ICTs for particular pairs of nodes (i, j) in the network, $\forall i, j, i \neq j$. We will discuss these aspects in further details in Chap. 8.

Fig. 1.7 Characteristics of the truncated Levy Walk model. **a** Diffusion of the nodes increase with decreasing α, and **b** ICT distribution is heavy-tailed

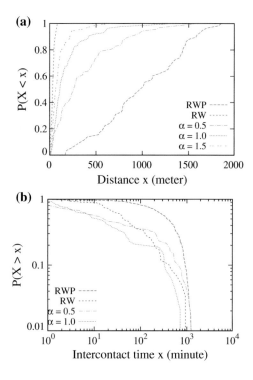

Figure 1.7 illustrates some characteristics of the TLW model and compares them with that of the RWP and random walk (RW) mobility models. Figure 1.7a shows that as α decreases, the diffusion of the nodes moving with TLW increases. In other words, with lower α, nodes travel more. It can be observed that RWP is the most diffusive mobility model in this context.

On the other hand, Fig. 1.7b shows that the ICT distribution under TLW exhibits heavy-tail characteristics, unlike the RWP and RW mobility models. Note that Fig. 1.7b uses a log-log plot—both the axes of the plot are in logarithmic scale.

1.3.3 Privacy and Anonymity

Replication of messages is undoubtedly one of the key aspects that brings in high success rate in OMNs. However, when we move beyond merely theoretical interest and look at OMNs with a more practical perspective, the aspects of privacy-assured and anonymous communication gain much importance. Since a message in OMNs consist of several metadata and pass through various nodes in the network, several key questions arise here. Can a message reveal my location information? Can one profile my interests and communication patterns? Can one convey some crucial information without revealing his/her identity? Unless such issues are addressed, users may not be largely interested in adapting OMNs. Unsurprisingly, these topics are receiving active focus by the research community [29, 36–38].

One of the major concerns of users is about their location privacy—revealing one's location while using an application or service. Parris and Henderson [36] conducted a survey on 80 Facebook users to understand the extent up to which one is ready to share his/her location. The authors found that many users (49 %) were largely comfortable in sharing their locations with their friends in Facebook. However, as high as 23 % of the users were not enthusiastic about sharing their locations, in general, with anyone. Interestingly, only 19 % of the users were ready to share their location with anyone. As a whole, these numbers indicate that most users are concerned about their location privacy to some extent at least. Motivated by this, Parris and Henderson [36] studied the message delivery performance in OMNs while taking into account the aforementioned privacy profile. It was found that different privacy models induced a large variation in the number of messages delivered. However, such deviations were much less pronounced in case of message delivery delays.

Parris and Henderson [37] also investigated the privacy aspects in case of "source routing" involving social network information. Since certain groups of people ("friends") are expected to encounter one another frequently, it makes sense for the source node to specify a list of other nodes (friends) that a message can pass through (replication). However, transmitting a list of friends in clear text is unsafe in the presence of any eavesdropper. On the other hand, encrypting such a list would render it unusable to the intermediate nodes. To overcome these challenges, Parris and Henderson [37] considered different schemes for obfuscating the friends list. In the first scheme, termed as statisticulated social network routing (SSNR), randomly chosen friends are removed from the friend list and/or non-friends are added to it. Therefore, inferring whether or not a given node is a friend of a source node becomes difficult. In the second scheme, termed as obfuscated social network routing (OSNR), the authors considered obfuscating the friends list using a Bloom filter. Finally, a hybrid of the two schemes was considered. Experimental results revealed that obfuscating friends list to a moderate degree can maintain about 90 % message delivery ratio as compared to the baseline social network routing scheme.

Shi et al. [39] considered the case of *strong sender anonymity*, where any node but the source can correctly identify the origin of any message with minuscule probability. Such sender anonymity is said to be *weak* if the recipient of a message can identify its source. Similarly, receiver anonymity is defined as the inability of other nodes to identify the destination of a message. To ensure anonymous communication in DTNs, Shi et al. [39] proposed ARDEN, which is motivated by the onion routing protocol [40]. In particular, the nodes in the network are dynamically partitioned into different groups. Similar to the onion routing, a source node in the DTN encrypts its message using keys of multiple groups. As the message passes through (members of) different groups, the corresponding layer of encryption is removed; the destination node is capable of completely decrypting the message. Thus, with a few more additional hops along the message delivery path, ARDEN is able to ensure anonymous communication. Shi et al. observed that, in the presence of adversaries, ARDEN can outperform onion routing in terms of both message delivery performance as well as computational overhead.

1.3.4 *Congestion*

Congestion mitigation in DTNs is particularly interesting (and difficult) due to its divergent characteristics from the traditional Internet. In particular, unlike the Internet, high latency in DTNs makes propagation of congestion indicator through the network impractical. Similar to other networks, research on congestion in DTNs have focused on two approaches—congestion avoidance and control.

Thompson et al. [41] noted that congestion in DTNs is a network-wide effect and the challenge is to detect it accurately. Global congestion detection algorithms are not suitable because of high delay and low reliability in DTNs. Therefore, the authors advocated the use of autonomous algorithms, which can respond to spatial and temporal fluctuations in congestion. The authors further noted that in highly dynamic environments like DTNs, popular message based replication cannot adapt to dynamic congestion conditions. Therefore the authors proposed a node-based message replication scheme. Each node locally maintains an approximated index of the global congestion value, CV, as the number of message drops to the number of message replications. The global message drops is estimated using node's local drops; the global replication count is approximated by the number of replicas a node receives. At each network sample period, the value of CV is updated as, $CV = 0.9 \times CV_{sample} + 0.1 \times CV$. The drop and replication counts are reset at every sample. Whenever the value of CV changes, the message replication limit is updated. The proposed scheme was found to result in improved performance when used with Epidemic [18], PRoPHET [19], and SnW [42].

Radenkovic and Grundy [43] aimed to improve network performance by offloading traffic from the congested areas of a DTN to the relatively less congested regions. The authors considered several heuristics such as delay, buffer space availability, and congestion rate of the nodes and their ego networks. In the resulting congestion-aware opportunistic forwarding algorithm, the nodes attempt to forward messages to those nodes with lower delay and lower congestion rates.

Lakkakorpi et al. [44], on the other hand, proposed a method for avoiding congestion, wherein the nodes advertise their buffer occupancy, and nodes with higher occupancy are avoided while forwarding messages. In particular, if B be the total buffer size, then the advertised buffer size is evaluated as $B_a = T_C \times B - B_o$, where B_o indicates the occupied buffer space, and T_C is the congestion threshold (safety margin). The authors proposed to dynamically adapt T_C based on the number of message drops.

Leela-amornsin and Esaki [45] addressed the looping problem caused by simple buffer management policies, and proposed a solution by keeping the deleted message headers in junk entry. The authors also proposed a credit-based scheme for heuristically deleting messages when the buffer is congested in order to retain high delivery rate with low number of message replicas. Apart from these, some other notable contributions in this field include token-based congestion avoidance scheme [46], a utility function-based congestion control mechanism [47], and social metrics-based active congestion prevention [48].

1.4 Network Simulation

Network simulation is one of the simplest ways to study the performance of one or more protocols under different scenarios, and subsequently have a comparative performance analysis. A wide range of network simulators exist today catering to different types of networks. The Network Simulator version 2, or popularly known as NS-2,[7] has been in use since a long time. NS-2 supports a wide range of networks such as MANETs, satellite networks, and ZigBee. Moreover, some specific application using NS-2 with focus on certain aspects have been developed over the years, for example, [49, 50]. NS-3,[8] on the other hand, is a relatively new network simulator, but steadily gaining momentum. Both NS-2 and NS-3 have modules to simulate DTNs[9]

Ns2web [50] is a web-based front-end to the NS-2 simulator. With Ns2web, users can directly jump into simulating different network scenarios without going through the installation steps of NS-2, which often is not straight forward. Apart from the basic networking modules of NS-2, Ns2web supports simulation of WiMax, Bluetooth, and wireless sensor networks. Moreover, Ns2web provides a GUI interface for defining the network scenario and generating the corresponding simulation scrip, which can be customized later on. To help users with analysis, Ns2web also provides tools for evaluating different performance measurement metrics and plotting them. Ns2web is available for use at http://vlssit.iitkgp.ernet.in/ns2web/ns2web/.

In this book, we would, however, particularly focus upon the opportunistic network environment (ONE) simulator [51] developed for simulating and evaluating DTN protocols. Since its inception, the ONE simulator has become largely popular in the research community and is being widely used. Apart from providing implementations of several well-known routing protocols and mobility models, the ONE simulator makes integration of real-life traces (such as, mobility and connection information) into simulation a simple and easy task. Moreover, contributions from other researchers (in form of implementation of their own protocols) are also available.

---

[7]http://www.isi.edu/nsnam/ns/.

[8]https://www.nsnam.org/.

[9]For example, http://www.netlab.tkk.fi/tutkimus/dtn/ns/. Also, a list of implementations of different components of DTN can be found at https://sites.google.com/site/dtnresgroup/home/code.

Perhaps one of the most useful features of the ONE simulator is its reporting framework. Readers familiar with NS-2 might compare the reports with trace files generated with NS-2. However, unlike trace files, reports generated by the ONE simulator do not necessarily contain only raw data. It is possible to capture detailed or aggregate information based on a simulation and output them into a report file. A collection of several reports is available with the simulator, which helps in obtaining commonly used performance metrics, for example, delivery ratio of messages and delay distribution. New reports can be created by implementing one or more interfaces.

Developed using Java, the ONE simulator can be used in any environment that supports Java, and, thus, one does not need to worry about related software dependencies. Several tutorials and documentations are available online to get started with the simulator.[10] Besides, the ONE simulator has a large and useful community. We would look into the ONE simulator at a greater depth in Chap. 3, where we would discuss about development and testing of protocols developed with the ONE simulator.

It may, however, be noted that the ONE simulator presently supports network layer and above. In other words, if one wishes to explore link layer or physical layer properties in DTN, he/she might consider other alternative simulators. Moreover, in the recent times, real-life experiments using real hardware, test beds, or smart phones are gaining momentum. However, this does not obviate the use of network simulators. For instance, a simulation-based performance evaluation would be of much useful before one actually plans to conduct a real-life experiment.

1.5 Summary

The ubiquitous Internet model fails to work in scenarios where one or more of its fundamental assumptions fail to hold true. In such scenarios, DTNs and its variants, such as OMNs and PSNs, can be leveraged. However, the inherent characteristics of such networks render routing a challenging task. This calls for cooperation among the nodes in terms of the message forwarding behavior. Although the nodes can be motivated for cooperation by providing some sort of incentives, the nodes themselves may judge to find cooperation a suitable strategy. On the other hand, while the traditional networking viewpoint focused only the actions of independent devices, MOONs add a new dimension by acknowledging the presence of shumans in PSNs. Thus, different human aspects—for example, intelligence and emotions—can affect the dynamics of a MOON.

---

[10]Interested readers might go through the blog of one of the co-authors available at http://delay-tolerant-networks.blogspot.com/.

In this chapter, we reviewed the evolution of DTNs, its characteristics and contemporary research trends. In the next chapter we would retrospect some popular routing protocols and applications in this domain. In the latter parts of the book, we would delve into details on some specific topics in DTNs/OMNs.

1.6 Review Terms

- Challenged networks
- Delay Tolerant Networks
- Pocket Switched Networks
- Store-and-forwarding
- Mission-Oriented Opportunistic Networks
- Anycasting
- Geocasting
- Inter-contact time
- Location privacy
- Receiver anonymity

- Interplanetary Internet
- Opportunistic Mobile Networks
- Message replication
- Store-carry-and-forward routing
- Human–network interplay
- Manycasting
- Contact
- Truncated Levy walk
- Sender anonymity
- Onion routing

1.7 Exercises

1.1 Discuss some of the fundamental differences between the Internet and DTN.

1.2 Suppose that an autonomous vehicle has been dropped on the surface of Mars from a satellite, which itself is orbiting Mars. The vehicle can communicate with the satellite and the satellite can communicate with a station on the Earth, but they are never in the line of sight altogether. Suggest a mechanism for data transmission from the vehicle to the Earth station.

1.3 Why underwater communications use acoustic waves in contrast to electromagnetic waves used terrestrially?

1.4 How message replication is different from message forwarding?

1.5 What is meant by the "store-carry-and-forward" mechanism of message transmission?

1.6 How the aspects of "cooperation" are different from "trust" and "reputation"?

1.7 Why human mobility is a fundamental aspect in PSNs/OMNs?

1.8 What is a buffer management scheme? Give example of some such typical schemes in the context of DTNs.

1.9 Why is congestion bad for a network? Discuss some congestion avoidance and control schemes for DTNs.

1.8 Programming Exercises

1.10 Familiarize yourself with the ONE simulator, which is widely used in research related to DTNs/OMNs. In particular, take a look at the `default_settings.txt` file, which contains several parameters and their values that altogether define a simulation scenario.

Chapter 2
Delay Tolerant Routing and Applications

The fundamental goal of any communication network is to have the data/packets/messages delivered to their corresponding destinations. With nontrivial network topologies having multiple paths from a given source node to a destination node, the data/packets/messages are required to be routed along an optimum (for example, least cost) path. In general, one may group the routing protocols into two broad categories—distance vector (DV) and link state (LS). DV-based protocols consider a *distance* metric to determine the best path between a pair of nodes. The *vector* in DV indicates the interface to which traffic should be forwarded so as to optimally reach the concerned destination. A typical example is the routing information protocol (RIP) used in the Internet, where the hop counts between routers act as the distance. The LS-based protocol, on the other hand, maintains a graph of the network and typically employs a shortest-path algorithm to determine the best route. Open shortest-path first (OSPF) is such an example used in the Internet.

Although similar routing algorithms have also been used in MANETs, such approaches become difficult in DTNs. This can be accounted to the different characteristics of DTNs noted in Chap. 1. The network topology of DTNs/OMNs is not only highly dynamic, but also exhibits high degree of network partitioning. In particular, the concept of end-to-end communication paths practically ceases to exist in such networks.

To overcome such issues and constraints, several schemes have been proposed in the literature to achieve efficient routing in DTNs/OMNs. In this chapter, we take a look at a broad category of such protocols. A common characteristic of all these routing protocols is that they are replication-based, i.e., they create multiple copies of a message in the OMN concerned. As we shall shortly see, the protocols attempt to restrict the extent of replication in different ways to achieve efficiency. We, then, present a set of commonly used metrics for measuring the performance of OMNs. Subsequently, we discuss about real-life traces that are increasingly being used with network simulations nowadays. Finally, we conclude this chapter by retrospecting some of the applications of DTNs.

© Springer International Publishing Switzerland 2016   23
S. Misra et al., *Opportunistic Mobile Networks*,
Computer Communications and Networks, DOI 10.1007/978-3-319-29031-7_2

2.1 Routing Protocols

Routing is a key aspect in any kind of communication network including DTNs/
OMNs. We, however, skip a detailed taxonomy of the routing protocols, and look at
a single aspect—the number of copies of a message that is forwarded as shown in
Fig. 2.1. Interested readers may look at [52–54] for a detailed treatment.

Single-copy routing is a rarity in OMNs, since intermittent connectivity results
in low delivery of the messages as well as high delivery latencies. A trivial example
is the direct delivery routing, where only a source node itself delivers its messages
to the corresponding destination node(s). Another example is first-contact routing,
wherein a node forwards a message to the first node that it comes in contact with,
and the process continues until the message reaches its corresponding destination.
In reality, these schemes are hardly used. For the sake of completeness, we note that
a few single-copy-based routing algorithms have been proposed for DTNs/OMNs
such as minimum estimated expected delay [55] and context aware routing [56].
However, in this book, we focus only on the multi-copy schemes.

The multi-copy routing schemes can be broadly classified into two types—limited
and unlimited. In limited multi-copy routing, a message is replicated for a fixed
number of times and forwarded to the different nodes in the network. A typical
example is the Spray and Wait [42] routing protocol.

In unlimited multi-copy routing protocols, for example, PRoPHET [19], there is
no fixed limit on the number of replications of a given message that occur in the
network. In particular, if N be the number of nodes in a DTN and K_i be the number
of replicas of the ith message, then $0 \leq K_i \leq N$. The extreme case is flooding, where
a given message is replicated and forwarded to all the nodes in the network. Often
some form of heuristic is used to decide upon the upper limit of K_i, i.e., how many
times a given message should be replicated. It has been shown that determining a
routing schedule even with complete information is an NP-hard problem [57], which
justifies the use of heuristics.

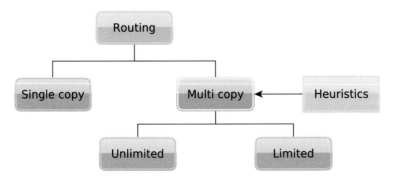

Fig. 2.1 Routing categories in DTNs. In practice, multiple replications of a message are created to
ensure better delivery chances

In the remainder of this section, we briefly look at some of the well-known routing protocols used in DTNs. The review provides insights about their basic characteristics operations. Furthermore, we also look at the different extensions suggested and changes made to those routing protocols in the literature.

2.1.1 Epidemic

Vahdat and Becker [18] proposed the epidemic routing protocol for ad hoc networks with high network partitions. The epidemic routing protocol (and similar protocols in distributed systems) is modeled after the spreading of epidemic diseases in real life. With this analogy, a node consisting of a message, m, may be considered as already *infected*. Now, when such a node comes in contact with another node not having that message, the former node transmits m to the latter and thereby, infects it. Eventually, after infecting multiple nodes in the network, the message m reaches to its destination. The goal of the epidemic routing protocol has been to maximize the chances of message delivery ratio and minimize latency while minimizing the aggregate resource consumption (for example, bandwidth and energy) in the network [18].

As illustrated in Fig. 2.2, the interaction between a pair of nodes—when within the transmission range of one another—using the Epidemic routing protocol involves two steps:

1. **Exchange of summary vectors**: Each node maintains an index of the messages (for example, unique message IDs) that it is carrying in its buffer. Let, V_X and V_Y, respectively, be the summary vectors of the two nodes X and Y. As shown in Fig. 2.2, the nodes exchange V_X and V_Y between themselves.
2. **Exchange of messages**: The node X computes $V'_X = V_Y \setminus V_X$, the set of messages that are carried by the node Y, but not present in the buffer of node X. Similarly, node Y computes $V'_Y = V_X \setminus V_Y$. If such a set computed by a node is not empty, it requests its peer to transmit the messages with the corresponding message IDs.

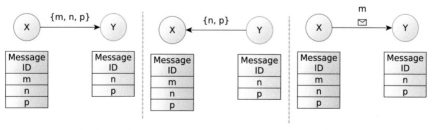

Exchange of summary vectors   Message exchange

Fig. 2.2 Message exchange between two nodes using the epidemic routing protocol

Example 2.1 With reference to Fig. 2.2, we have $V_X = \{m, n, p\}$ and $V_Y = \{n, p\}$. Therefore, $V'_X = \varnothing$ and $V'_Y = \{m\}$. So, node Y requests node X to send the message m. However, node X makes no such request to node Y.

In a similar way, the nodes X and Y subsequently replicate m to the other nodes in the network with which they come in contact. When one such node meets with the destination node, it delivers m to the latter.   □

The authors associated the following three properties [18] to each message generated in the network.

- **Message identifier**: This serves as the ID to uniquely identify a message in the network. The authors suggested that such an identifier can be obtained by concatenating the address of a node together with a local message ID value.
- **Hop count**: This field indicates the maximum number of time a given message can be exchanged using the aforementioned protocol. Messages with zero hop count can be removed from the buffer in the face of storage crisis depending upon the local buffer policies.
- **Acknowledgment request**: A node can optionally request for an acknowledgment message be sent by the destination node once a given message is delivered.

However, these properties are not just specifically used with Epidemic. In fact, these and others, such as source node identifier, destination identifier, TTL, message length, and so on, are some common metadata maintained with the messages by almost any routing protocol. Often other additional information are also maintained as demanded by specific protocols. Unlike traditional networks, acknowledgments are typically not used in DTNs/OMNs, since the round trip time of having a message delivered and subsequent receipt of its acknowledgment by the source node is often very large. However, if appropriate, acknowledgments can be used to "clean up" the network. Let us look at how such clean up can be useful in real life.

Since in epidemic routing a message is replicated to all possible nodes in a network, the overall storage requirements become high in such networks. Theoretically, if n messages are created in an OMN, the complexity of storage, and transmission, overhead becomes $O(n)$—linearly increasing with the number of messages. To overcome such effects, different recovery schemes, such as *immunity* and *vaccine*, and their effects have been considered by researchers [58–62]. The idea of these schemes arise from biology, wherein a person either has immunity to a given epidemic disease and does not get affected by the it, or is given a vaccine when affected by the same. The objective of these proposed schemes in communication networks is similar—to help propagate delivery information of messages to other nodes so that they (the other nodes) can stop replicating and storing those particular messages.

In the immunity scheme [58, 59], when a node delivers a message to its corresponding destination, the former node removes the message from its buffer. Moreover, it stores the ID of the delivered message with itself so that it does not accept the same message again from any other node in the future. In other words, the node is immune and could no more be *infected* by the *disease*. The stored ID (of the delivered message) is often referred to as *anti-packet*, similar to the concept of anti-body in biological systems.

A proposed variation of the immunity scheme is to propagate the anti-packet information to other nodes who are currently infected, i.e., contain a copy of the message that has been already delivered. In the vaccine scheme, on the other hand, such anti-packet information is propagated to all possible nodes with the intention of preventing any node receiving an already delivered message.

Zhang et al. [59] presented a performance modeling of the epidemic routing protocol together with different recovery schemes discussed above. The authors used a Markov model for the purpose and derived ordinary differential equations (ODEs) as the limiting process of such model considering natural scaling. On the other hand, Li et al. [60] considered a two-dimensional continuous-time Markov chain to represent the message delivery in the presence of social selfishness and studied the resulting effects on the routing performance.

Mundur et al. [61] considered epidemic routing in DTNs together with the immunity scheme. Similar to the summary vector list, the authors considered an "immunity" list maintained by each node, which stores the IDs of the messages already delivered. During an encounter, a pair of nodes exchange both the summary vector and the immunity list. This prevents a node from receiving a copy of a message again that it has delivered to the corresponding destination in the past.

Matsuda and Takine [62] considered the (p, q)-Epidemic routing protocol, wherein a message is probabilistically replicated by a node. In particular, a relay (non-source) node replicates a message to another relay node with probability p, whereas a source node replicates a message to a relay node with probability q. As such, the (p, q)-Epidemic protocol encompasses a large class of routing protocols:

- When $p = q = 0$, it degenerates to direct delivery routing protocol, where a source node itself delivers a message to the corresponding destination node (i.e., zero replication in the network).
- When $p = q = 1$, we get the original Epidemic routing protocol.
- When $p = 0$ and $q = 1$, (p, q)-Epidemic transforms into two-hop routing protocol.

Subsequently, the authors considered the dynamics of the network and presented a corresponding model for the performance of the (p, q)-Epidemic routing protocol together with the vaccine scheme.

2.1.2 Spray and Wait

Spyropoulos et al. [42] proposed the Spray and Wait (SnW) routing protocol, which imposes a maximum limit on the number of possible replications of a message. The SnW protocol consists of two phases:

1. **Spray phase**: In the spray phase, a particular message is spread to at most L different relay nodes. Such spraying begins with the source node, i.e., the node that created the message. Any node receiving such a message can, in turn, engage in spraying. The spraying process could be of different types, as we would soon discuss below.

2. **Wait phase**: If the destination node was not encountered with during the spray phase, the intermediate nodes carrying a copy of the message perform a direct delivery when they come in contact with the corresponding destination of the message.

It may be noted that multiple nodes carrying a copy of a message say, m, may come in contact with the destination node and forward the same. The destination node, however, is assumed to receive only a single copy of m—from the node that it first comes in contact with. When the other nodes try to forward the same message, the destination can "refuse" to receive them or just simply discard them.

There are different variations of the spraying technique of the messages. In the "source" spraying technique [42], the source node itself distributes L copies of a given message up to first L nodes that it comes in contact with which does not already have a copy of the message. Of course, if the destination node is encountered before that, the message is directly forwarded to it, and the routing of that particular message successfully terminates. Otherwise, any of those L nodes can directly deliver the message to the destination node if and when encountered.

The binary spraying technique [42] is slightly different. In this version, the source node initially has L copies of a given message. Now consider any node—whether the source or a relay—that has n copies of the message, $L \geq n > 1$. If one such node encounters another node having no copy of the message, the latter node is given $\lfloor n/2 \rfloor$ copies of the message, while the former itself retains $\lceil n/2 \rceil$ copies. It has been shown that the binary spraying technique is optimum and results in minimum expected delay among all Spray and Wait routing mechanisms.

Spyropoulos et al. [42] observed that SnW brings in the together the speed of Epidemic routing and the simplicity of direct delivery routing. As such, this can be viewed as a trade-off between single- and multi-copy routing protocols. The authors proved that the minimum value of L to achieve an expected delay does not depend on the size of the network or the transmission ranges of the nodes, but only on the number of nodes in the network. Subsequently, the authors also provided a method to estimate L when the network parameters are not known.

The authors further observed that in SnW, the message copies are sprayed as fast as possible to the nodes' neighborhood [63, 64]. Thus, if the mobility of such nodes is confined to a small locality (for example, movement of the students inside a university campus [65]), the chance of encountering the destination, and subsequent delivery of the message, is greatly reduced. So, although SnW is fast and resource friendly, constrained mobility imposes limitations to the protocol. Consequently, the authors [63, 64, 66] proposed the timer transitivity-based Spray and Focus (SnF) algorithm to overcome such adversity.

SnF [63] has a spray phase similar to SnW. However, instead of the wait phase, SnF introduced the "focus" phase, where a forwarding decision is based on a utility. The authors defined single-copy utility-based routing as follows. Let us consider that each node i in the network maintains a utility value, $U_i(j)$, for every other node j in the DTN, $\forall i, j$. Further, let us assume that a node X, which has a message destined for the node Z, comes in contact with another node Y, $Y \neq Z$. Then, X forwards the message to Y, if and only if their utility functions satisfy the following relationship:

$$U_Y(Z) > U_X(Z) + \theta, \tag{2.1}$$

where θ is some threshold.

The authors considered timer transitivity to define the utility functions. In particular, let $\tau_i(j)$ be the duration since the last encounter between the nodes i and j as maintained by the node i. Initially, $\tau_i(i) = 0$, $\tau_i(j) = \tau_j(i) = \infty$. Moreover, at every encounter between i and j, these counters are reset by the respective nodes, i.e., $\tau_i(j) = \tau_j(i) = 0$.

Let the distance between the nodes X and Y be d_{XY} at the time when they encountered. Further let, M be the mobility model followed by the nodes so that $t_M(d_{XY})$ be the time required, on an average, to move the distance d_{XY} as per the mobility model. Then, for any node $Z \neq Y$, if $\tau_X(Z) > \tau_Y(Z) + t_M(d_{XY})$, update the counter as $\tau_X(Z) = \tau_Y(Z) + t_M(d_{XY})$. The rationale here is that, due to the transitivity property, the node X had a more "recent" encounter with Z.

Based on this, the working of the SnF algorithm can be described as follows:

- Each node maintains a summary vector of the messages as described earlier.
- Each node maintains the last encounter timers as discussed above.
- Any node—either the source or an intermediate relay—with $1 < n \leq L$ copies of a message, sprays the message copies according to the binary spraying algorithm. However, if $n = 1$, the node forwards the message copy according to (2.1). As mentioned earlier, the last encounter timers are used to determine the values of these utility functions.

Example 2.2 Let us consider that the nodes A and B, moving with uniform velocity, came in contact with another node j at time instants $t = 1$ and $t = 3$, respectively. Further, let $t_M(d_{AB})$ be 1. Therefore, at an instant of time $t = 6$, we obtain $\tau_A(j) = 6 - 1 = 5$, $\tau_B(j) = 6 - 3 = 3$. So, $\tau_B(j) + t_M(d_{AB}) = 3 + 1 = 4 < \tau_A(j)$. Therefore, $\tau_A(j) = \tau_B(j) + t_M(d_{AB}) = 5$. □

The above example is meant for the purpose of simple illustration. A few aspects, however, should be kept in consideration in this context as discussed in [64]. Spyropoulos et al. suggested corrections in the (larger) timer value when two individual values differ by a large extent. However, in our example, we illustrated the procedure with small values. Moreover, the term $t_M(d_{XY})$ mentioned above is a property of the underlying mobility model M. So, unless the nodes are moving with constant velocity, certain related computations might be required. Interested readers should also look at [64] for a variation of the Spray and Focus algorithm.

Spray and Focus algorithm is not a single-copy routing protocol. Similar to SnW, messages are sprayed here, too. However, unlike the SnW algorithm, where a node with a single copy of a message directly delivers it to the corresponding destination, SnF uses forwarding (not replicating) of such messages. Whether or not such a message with a single copy should be forwarded to the other node is decided based upon their utility functions.

2.1.3 PRoPHET

Lindgren et al. [19] proposed the probabilistic routing protocol using history of encounters and transitivity (PRoPHET). PRoPHET is a greedy algorithm, where a node forwards a replica of a message to another node only if the latter node has a greater chance of encountering the destination of the message than the former itself. PRoPHET uses the concept of *delivery predictabilities* (DP), the likelihood of one node meeting with another node in the OMN. Every node running PRoPHET in the network maintains a table of such probabilities of meeting with every other known node. Moreover, the DPs are dynamic—if two nodes do not meet for a certain time period, their DP value for the other node decays appropriately. Therefore, a pair of frequently meeting nodes would have high values of DP for one another.

Equations (2.2) through (2.4) [19] related to updating delivery predictabilities on encounter, aging of predictabilities, and capturing transitive predictabilities, govern the functionality of this routing protocol:

$$P_{(a,b)} = P_{(a,b)\text{old}} + (1 - P_{(a,b)\text{old}}) \times P_{\text{init}} \qquad (2.2)$$

$$P_{(a,b)} = P_{(a,b)\text{old}} \times \gamma^k \qquad (2.3)$$

$$P_{(a,c)} = P_{(a,c)\text{old}} + (1 - P_{(a,c)\text{old}}) \times P_{(a,b)} \times P_{(b,c)} \times \beta \qquad (2.4)$$

Here, $P(A, B)$ and $P(A, B)_{\text{old}}$, respectively, indicate the current and previous delivery predictabilities of node A for another node B, $P(A, B) \in [0, 1]$. The parameter $P_{\text{init}} \in [0, 1]$ is an initialization constant. The delivery predictabilities are aged with time when a pair of nodes does not encounter for long. The parameter $\gamma \in [0, 1)$ is the aging constant, and k denotes the number of time units expired since the last update of this predictability. Such aging helps in eliminating stale information maintained by the nodes. The delivery predictability $P(A, B)$ acts as heuristic here.

The scaling parameter $\beta \in [0, 1]$ controls the extent to which transitivity should affect the delivery predictability. The transitivity phenomenon is explained as follows. Suppose that node X encounters node Y frequently and node Y, in turn, encounters node Z frequently. Therefore, any other node i, which has a message to send to node Z, may as well forward the message to node X. Although node X may not have frequent encounters with node Z, the transitivity property allows the messages to be transferred from X to Z via node Y. The proposed values of P_{init}, β, and γ, respectively, are 0.75, 0.25, and 0.998 [19].

Figure 2.3 illustrates the sequence of actions taken when two nodes A and B using PRoPHET come in contact with one another. After the exchange of summary vector of the messages and the delivery predictabilities, each node updates their own DP values. Subsequently, candidate messages, if any, are exchanged between them. In other words, node A (B) replicates messages to node B (A) if the latter has a higher likelihood to encounter the destination of the message—directly or via others—than A (B) itself.

Example 2.3 Let $P(A, B) = 0.5$, and the pair of nodes does not have any subsequent encounter. If a single time unit comprises 60 s, then after two time units, the updated delivery predictability would be $P(A, B) = P(A, B)_{old} \times \gamma^2 = 0.5 \times 0.998^2 = 0.498$. □

PRoPHET has inspired many subsequent protocols, and several improvements to PRoPHET itself have been proposed. The utility evaluation of the Spray and Focus protocol [63, 64] discussed earlier was inspired by similar transitivity. Huang et al. [67] proposed PRoPHET+, where, apart from delivery predictability of the nodes, other performance indicators, such as buffer capacity, power level, and popularity, are also considered. Such scalar values are combined together with appropriate weights to determine the *deliverability* of a node based on which the messages are replicated. On the other hand, Li and Das [68] proposed a trust-based mechanism for data forwarding, which was evaluated by integrating with PRoPHET. The proposed scheme was shown to be efficient against attacks.

Fig. 2.3 Sequence of interaction between two nodes using PRoPHET. Here, DP indicates the delivery predictabilities maintained by each node

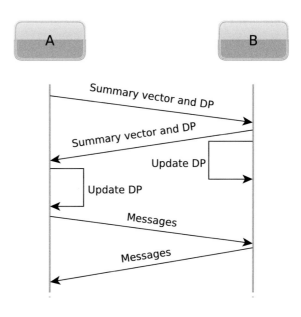

Grasic et al. [20], however, pointed out that the delivery predictabilities used in PRoPHET may not always be able to reflect the real-world dynamics. The authors discussed several scenarios where high values of delivery predictabilities were computed, but were undesirable. For example, in the parking lot problem [20], devices were apparently found to have frequent encounters in short time periods. In reality, the devices had multiple reconnection among themselves due to fluctuating Wi-Fi signals, which were treated as "new" encounters. Such observations do not correspond to typical human movement patterns scaling over hours or days. The authors observed that when $\beta > 0$, the DP evaluated by any node using (2.4) would always keep increasing.

To overcome such problems of PRoPHET, Grasic et al. [20] proposed an improved version of the protocol—PRoPHETv2. In this version, P_{init} is computed as

$$P_{\text{init}} = \begin{cases} P_{\text{max}} \times \Delta_j/\Delta, & 0 \leq \Delta_j \leq \Delta \\ P_{\text{max}}, & \text{otherwise,} \end{cases} \tag{2.5}$$

where Δ_j is the time since the last encounter with node j and Δ indicates the typical time interval, on an average, between two successive connections for a given network scenario. Moreover, the transitive predictability evaluation is modified as

$$P(a, c) = \max\{P(a, c)_{\text{old}}, \ P(a, b) \times P(b, c) \ \times \beta\} \tag{2.6}$$

Such an operation stabilizes unexpected amplification of the delivery predictabilities.

2.1.4 RAPID

Balasubramanian et al. [57] treated routing as a resource allocation problem in DTN and proposed the resource allocation problem for intentional DTN (RAPID) protocol. By *intentional*—in contrast to *incidental*—it is meant that the proposed protocol can explicitly optimize a routing metric of choice. To achieve this, RAPID used an in-band control channel to exchange various metadata including expected contact time with other nodes, list of messages delivered, and average size of past transfer events.

RAPID essentially defines a per-message[1] utility function. A message i is replicated in such a way that it locally optimizes the marginal utility $\delta U_i/s_i$, where U_i is the contribution of message to the concerned routing metric, and s_i is its size. RAPID is composed of three functional components:

---

[1] In [57], the authors considered routing of packets. For the sake of consistency, we use the term message here. The general problem of routing, however, remains the same irrespective of whether the data unit considered is a packet or a message.

1. **Selection algorithm**: When two nodes come in contact, at first they exchange their metadata. Subsequently, they directly deliver the messages, if any, destined for the other node in decreasing order of their respective utilities. Finally, the remaining messages—that are not already carried by the other node—are replicated in decreasing order of their marginal utility $\delta U_i / s_i$.
2. **Inference algorithm**: It is used to compute the utility of a message for a specific routing metric. For example, in order to minimize the average delivery latency of a message, the concerned utility can be defined as $U_i = T_i + R_i$, where T_i and R_i, respectively, denote the time duration since creation and expected remaining time for delivery of the message i. Similarly, utilities for other desired metrics can be defined as well.
3. **Control channel**: The control channel is used to exchange the previously described metadata. It was found that the overhead due to the control channel was quite less. Moreover, although the control channel cannot always provide most recent and accurate information, even such inaccuracies were significantly helpful.

Balasubramanian et al. discussed examples of three particular metrics that can be used with RAPID to evaluate the utility function U_i. The first metric aims to minimize delivery delay, $D(i)$, of a message i, on an average, in the network so that $U_i = -D(i)$. The second metric aims to maximize the chances of delivering a message i to its destination before its TTL, $L(i)$, expires. In other words, the probability that $L(i) - T(i)$ is greater than the expected delivery time is considered. The final metric targets to minimize the maximum delay experienced by the set of messages in a node's buffer. Once the parameters are plugged into U_i, the marginal utility can be evaluated, and the replication decision subsequently taken. The authors evaluated RAPID using simulations and test beds. The performance in both cases was found to be very close and better than other contemporary routing protocols.

It may be noted here that although both PRoPHET and RAPID use heuristics for efficient routing of the messages, their approaches differ. In case of PRoPHET, the heuristic is defined as per destination node, whereas in RAPID, per-message utility is considered.

Due to intermittent connectivity, contemporary global information about the OMNs is often not available to the nodes. Therefore, while routing, a node typically attempts to make an optimal message forwarding decision based on locally available information.

2.1.5 Bubble Rap

In real life, interactions among people can largely be represented in terms of interacting communities. With such knowledge, it is possible to determine, or predict, how a message from a person passes on to a different person. Hui et al. [69] exploited such social relationships among the people—and therefore, their devices—in OMNs, and thereupon developed a message forwarding algorithm termed as Bubble. Two social metrics—*centrality* and *community*—are combined together for taking message forwarding decisions.

The Bubble algorithm assumes that every people in the OMN belong to at least one community. Moreover, each node has two rankings associated with themselves. The first ranking is a local ranking, where a node is assigned a relative rank in its own community. The other one is the global ranking, which is valid across the entire OMN. Message forwarding using Bubble essentially involves climbing—or *bubbling*—up the hierarchical ranking tree. Initially, a message is replicated by the nodes using their relative global ranking until the message reaches to a node that belongs to the same community as the destination of the node. Once a member of the desired community receives the message, it uses local ranking of the other nodes in the community to further disseminate it until the message reaches to its corresponding destination node. Figure 2.4 illustrates this process. The authors used the term "bubble" to indicate the communities in an OMN.

Bubble uses the inherent mobility patterns of human carriers of the devices as a metric for forwarding data in OMNs. The forwarding paths are selected based on the correlated interaction between humans in a community. More popular people have more number of connections. In other words, they have high *centrality*, and are identified as hubs. A group of nodes in an OMN having common interest and social links form a community. Based on the community, each node has two different types of centrality as follows:

- **Local centrality**: A node sends the message according to local ranking within the local community.
- **Global centrality**: A node sends the message according to global ranking across the whole network (global community).

In graph theory and social network analysis, centrality is a measure of the *importance* of a given node in a graph. For example, in a star topology, the central node has the highest centrality, since it makes communication with all the other nodes possible. This particular measure is called the degree centrality, where the "degree" of a node is the number of links that a node has. Several other measures of centrality have been proposed in the literature, for example, those based on closeness and betweenness. Interested readers may take a look at [70–72] for further details.

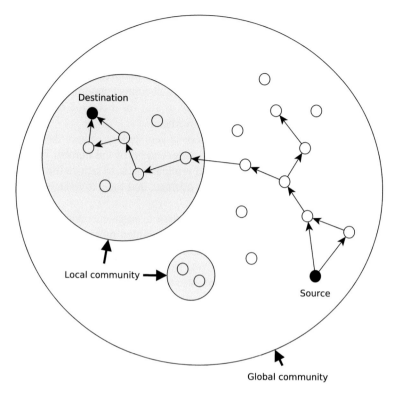

Fig. 2.4 An instance of message forwarding using Bubble

Bubble works well when the destination node's communities member belongs to high rank. It fails to work, when the destination node belongs to communities whose members have low global centrality values. In such situation, the nodes of global communities may not be able to reach the appropriate local communities of the destination node.

2.2 Routing Based on Encounter Statistics

In this section, we look at three routing protocols that take into consideration the number of encounters (contacts) among the nodes in an OMN. Their message replication decisions are optimized based on such encounter statistics. The simplicity of these routing protocols make them interesting candidates for study.

2.2.1 Encounter-Based Routing

Nelson et al. [73] proposed the encounter-based routing (EBR) protocol for DTNs. The authors noted that certain nodes (for example, ambulances in a post-disaster scenario) in a DTN might have more contact (or encounter) opportunities than other nodes. EBR essentially exploits this property to aim for increased delivery of the messages while maintaining a low routing overhead.

EBR introduces the concept of *quota-based* replications, where the maximum number of replications of any message is independent of the number of nodes in the network. The objective of quota-based replications is to reduce the number of replicas of messages in the network, on an average, and thus, to reduce the routing overhead.

In EBR, each node in the DTN maintains their average encounter rate with the other nodes in the network using exponentially weighted moving average method. In particular, each node maintains a variable, EV, to store the node's past rate of encounters. Additionally, the variable CWC stores the number of encounters with the node that happened in the current time window. Mathematically,

$$EV = \alpha \times CWC + (1 - \alpha) \times EV. \tag{2.7}$$

The novelty of EBR lies in the fact that EBR uses the value of the variable EV to determine how many replica(s) of a given message should be sent to the other node. Specifically, if nodes A and B come in contact, and node A has a message m with itself, then the number of replicas of m, $r(m)$, sent to B is given by

$$r(m) = m_A \times \frac{EV_B}{EV_A + EV_B}, \tag{2.8}$$

where m_A indicates the number of replicas of m carried by A. Subsequently, node A updates the value of m_A for the message m as

$$m_A = m_A - r(m). \tag{2.9}$$

The performance of the EBR protocol was evaluated under diverse scenarios. It was found that EBR resulted in improved message delivery ratio (up to about 40 %) as compared to the contemporary routing protocols, while concurrently yielding a better goodput. In other words, EBR was found to be resource friendly.

2.2.2 Contact-Based Routing in DTNs

Contact-based routing (CBR) [74] is a flooding-based protocol that reduces the average message storage time at a node, while maintaining similar or higher message delivery ratio as compared to other routing protocols. In CBR, on every connection establishment between a pair of nodes, each node exchanges information such as frequency of contacts with the other nodes. Based on these frequency of contacts, each node identifies the relatively better candidate nodes, and forwards their messages to the such nodes.

Every node running CBR maintains the frequency of contact with other nodes in a table F. For example, a tuple $\langle a, k \rangle \in F$ indicates that there has been k contacts with a node identified by the address a, $a \in N$, where N is the set of nodes in the OMN. Whenever a new node comes in contact for the first time, a corresponding record is inserted into F with $k = 0$. For subsequent contacts, the value of n is incremented by 1 each time.

The CBR protocol has two main components—contact history maintenance and candidate node selection. During its lifetime, when a node comes in contact with another node, it increments the contact count for the latter and updates F accordingly, as discussed above. Subsequently, for every message held in the buffer, a node determines the most appropriate node currently in contact, and forwards the corresponding message to the selected node. In case of CBR, the frequency of encounters of a node determines its underlying utility. Therefore, when a node is in contact with multiple nodes, it selects the one that has maximum encounter frequency, and replicates a message to that node. To represent mathematically, let C_i be the set of nodes that node i is in contact with at a given time instant; $|C_i| \geq 1$. Moreover, let us consider that node i has a message m, which is destined for j. Further, let $I(m, k)$ be an indicator function, which is 1 if node k contains the message m, and 0 otherwise. Let

$$c = \underset{x, I(m,x)=0}{\arg\max} \ F_x[j] \tag{2.10}$$

where F_x is the frequency table of node x. Then, node i replicates the message m to the candidate node c. In case multiple nodes have the same highest frequency of encounters, one such node is randomly chosen among them.

2.2.3 Delegation Forwarding

Erramilli et al. [75] proposed Delegation Forwarding (DF), a novel routing protocol for OMNs. The objective of DF is to reduce the overhead of message delivery in OMNs, which is achieved using a astonishingly simple algorithm.

In DF, every node is assigned a *quality* metric, which captures the likelihood of a node to deliver a given message to its corresponding destination. Thus, quality, $\mathbf{q}$, is a vector of real numbers for each node in the OMN. Additionally, every message created in the network is assigned a property called *threshold*. Initially, when a message m is generated by a source node say, s, its threshold, $\theta(m)$, is set to $\mathbf{q}[i]$, the current measure of quality of s for the node i, which is the destination of the message. Subsequently, when the source node comes in contact with another node say, j, it (node s) replicates the message m to j if $\mathbf{q}[j] > \theta(m)$. The threshold value of the message is then updated to $\mathbf{q}[j]$ by both s and j. Similarly, when node j comes in contact with k, the replication decision is based on the result of comparison between $\mathbf{q}[k]$ and $\theta(m)$. This process is repeated for each message held by the nodes in the OMN.

Essentially, each node running DF always tries to find the *best* forwarding candidate with respect to time. The threshold property assigned to each message helps in storing the past history. Therefore, threshold of such a message is a strictly monotonically increasing function of time. The authors suggested various metrics that can be used to measure the quality of a node such as, frequency of encounters, last time of contact with any node, destination-specific frequency of encounters, and destination-specific last contact time. The first two metrics are independent of destinations of the messages. Therefore, in such cases, quality is a scalar quantity rather than a vector.

Experimental results indicate that the message delivery overhead obtained using DF is remarkably less when compared to other routing protocols. However, such optimization comes with a tradeoff. Waiting for a better candidate node to replicate a message increases the message delivery latency, on an average. Nevertheless, when overhead is a critical issue—and higher delays can be tolerated—DF stands out as an excellent choice for routing in OMNs.

Table 2.1 presents a comprehensive summary of the different routing protocols for OMNs studied so far. Table 2.1 shows that most of the protocols except for Epidemic and SnW use some form of intelligence to make routing decisions. This is true because both Epidemic and SnW blindly replicate the messages in the OMN. However, SnW, as well as SnF and EBR, replicates only up to a fixed upper limit. As such, they have relatively lower delivery cost in terms of the number of message replicas. Among these routing protocols, Bubble uses social metrics for message forwarding.

Overhead of the routing protocols in terms of storage and transmission cost are also noted in the Table. In particular, if the OMN has n nodes, the storage overhead of PRoPHET becomes $O(n)$. This is due to the reason that a node running PRoPHET has to store the predictabilites of the other nodes in the OMN, which asymptotically reaches to $O(n)$. Similarly, the transmission overhead of PRoPHET is also $O(n)$, since the delivery predictabilities vector need to be exchanged between two nodes. SnW and Epidemic, however, have zero overhead in both the cases assuming that they do not exchange summary vectors of the messages. If such an exchange is involved, then the transmission cost runs into $O(m)$, where m is the number of messages created in the OMN. Alternatively, use of Bloom filters can reduce that to constant cost. Finally, the overhead of DF is either $O(n)$ or constant depending upon whether or not destination-specific utility values are used by the protocol.

Table 2.1 Comparative summary of different routing protocols

Protocol	Routing intelligence	Maximum replications	Social metrics	Routing storage overhead	Routing transmission overhead
Epidemic [18]	–	Unlimited	–	0^a	0
PRoPHET [19]	Delivery predictabilities	Unlimited	–	$O(n)$	$O(n)$
PRoPHET+ [67]	Delivery predictability, buffer capacity, popularity	Unlimited	–	$O(n)$	$O(n)$
Spray and Wait [42]	–	Limited (L)		0	0
Spray and Focus [63]	Timer transitivity	Limited (L)	–	$O(1)$	$O(1)$
RAPID [57]	Time duration since message creation and expected remaining time for delivery	Unlimited	–	$O(\max\{m, n\})^b$	$O(\max\{m, n\})$
BUBBLE [69]	Mobility pattern of the nodes	Unlimited	Local and global centrality	$O(1)$	$O(1)$
Encounter-based Routing [73]	Most encounters	Limited	–	$O(1)$	$O(1)$
Contact-based Routing [74]	Neighbor with most encounters	Unlimited	–	$O(n)$	$O(n)$
Delegation Forwarding [75]	Quality and threshold	Unlimited^c	–	$O(1)$ to $O(n)^d$	$O(1)$ to $O(n)$

[a] Ignoring storage and transmission of summary vectors

[b] Metadata exchanged includes expected meeting times with every node and message (packet) replication information

[c] Practically, however, it is very low

[d] $O(n)$ when destination-specific metrics are used. Here, n is the number of nodes in the OMN

2.3   Performance Indicators and Key Insights

In this section, we present some metrics typically used for the performance evaluation of an OMN. However, by no means this is an exhaustive list. The reader should also consider metrics suitable for his/her specific scenario, if relevant. Subsequently, we also compare the performances of some of the previously discussed routing protocols under a few scenarios.

2.3.1   Performance Evaluation Metrics

Let, M and M_d, respectively, be the set of messages generated and delivered in the network, $M_d \subseteq M$. Further, let t_i and t_i', respectively, be the time instants when a message $m_i \in M_d$ was created and delivered to its destination, $t_i' > t_i$. Note that due to the decentralized nature of OMNs, a given message can be "delivered" multiple times to its corresponding destination by different nodes in the OMN. However, by "delivery," we refer to the first such instance when a node receives a message destined for itself.

Moreover, let us consider that each message $m_i \in M$ has r_i replica in the OMN, $0 \le r_i \le N$, where N indicates the total number of nodes in the network. Based on these, we define the following metrics.[2]

- **Delivery ratio of the messages**: This metric indicates, on an average, the fraction of the messages created in the network that were successfully delivered to their corresponding destinations. The average delivery ratio of the messages is evaluated as $|M_d|/|M|$. This is one of the primary metrics used to evaluate the performance under a given scenario or using a particular protocol.
- **Mean message delivery latency (delay)**: This metric gives an indication of, on an average, the time required to deliver a message from its source to its corresponding destination. This is evaluated as $\sum_{i=1}^{|M_d|}(t_i'-t_i)/|M_d|$. Typically, the average delivery latency of a message in OMNs is high and could be even up to few thousand seconds. Therefore, one of the objectives usually considered is minimizing this latency.
- **Median message delivery latency (delay)**: Let the set of delivery latencies of the messages be $D = \{t_i' - t_i\}$, where $|D| = |M_d|$. Sort the elements of D in ascending order. The $(|D| + 1)/2$ th element of the sorted list gives the median message delivery latency.
- **Overhead ratio of message delivery**: Routing in OMNs is usually replication based. As such, multiple copies of a message exist in the network. This has repercussions on the storage capacities and energy levels of the nodes. The overhead ratio is computed as $(\sum_{i=1}^{|M|} r_i - |M_d|)/|M_d|$.

---

[2]An initial version of some of these definitions appeared in one of the authors blog at http://delay-tolerant-networks.blogspot.com/2014/03/commonly-used-metrics.html.

- **Average hop count**: Let h_i be the hop count for the delivered message $m_i \in M_d$. In other words, the message m_i has passed through h_i nodes before reaching to its destination. Then, the average hop count is determined as $\sum_{i=1}^{|M_d|} h_i / |M_d|$.

It may be noted that the above defined metrics are generic in nature and can be used in any scenario. However, when planning for performance evaluation under specialized metrics, the reader may consider using some additional metrics that are relevant and suitable for such scenarios. For example, if one is working toward detection of misbehaving nodes in an OMN, the degree and accuracy of detection would be metrics of interest in such a scenario.

> Delivery ratio of the messages, delivery latency, and overhead ratio are three tightly intertwined metrics. In practice, whenever one of them improves, another one deteriorates. For example, having large number of replicas of a message is helpful to improve its delivery chances. But, at the same time, it also incurs high overhead. The goal of an optimal routing protocol is to design such a scheme that maximizes the delivery ratio, and at the same time, minimizes the average delivery latency and overhead ratio as well. Such an objective is attempted to achieve by utilizing various heuristics, some of which were discussed in this chapter.

2.3.2 General Insights into Routing

While a comprehensive performance evaluation of different routing protocols under various scenarios is out of scope of this text, we look at the performance of a few such protocols under the influence of different buffer capacities.[3] All simulations in this book were performed using the Opportunistic Network Environment (ONE) simulator [51].

Figure 2.5 shows the delivery ratio of messages obtained when SnW was used with different number of copies of the messages (L) and the nodes had different storage capacities. It can be observed that buffer size strongly influenced the delivery ratio when the simulation durations were for 12 h. This is due to the reason that in the absence of any TTL expiry, increased buffer size enabled the nodes to exchange greater number of messages during the available communication opportunities. When the simulation duration was increased, the nodes in the network eventually met with one another, and delivered the messages.

---

[3]This section has been reproduced by permission of the Institution of Engineering &Technology.

B.K. Saha, S. Misra, "Effects of heterogeneity on the performance of pocket switched networks," *IET Networks*, 2013, vol. 3, no. 2, pp. 110–118.

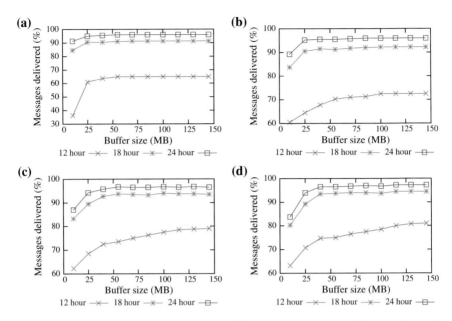

Fig. 2.5 Delivery percentage of the messages for different buffer sizes under SnW with **a** 8-, **b** 16-, **c** 24- and **d** 32-copies

The effect of the number of copies used by SnW on the delivery ratio of the messages is also evident. It could be seen that, as the number of copies increased, for a given buffer size, the delivery ratio also increased. The graphs indicate that unlimited buffer sizes may not help in achieving higher delivery of the messages.

Figure 2.5 also reveals that the rate of increase in the delivery ratio of the messages is faster for buffer sizes up to 40 MB. Although this threshold is specific to the parameters that were considered (message size and rate), this indicates that if more messages are to be delivered within a short time, the nodes should have a certain minimum buffer size.

Figure 2.6 shows the message delivery ratio with different buffer sizes when First contact [51, 76], PRoPHET [19], and Epidemic [18] routing protocols were used. In First-contact routing, a node forwarded (not replicated) any message it had to the first node that it came in contact with. Figure 2.6b, c indicate that with no fixed limit on message replication, larger buffer sizes enhanced the delivery ratio. This is due to the reason that during each communication opportunity, more number of nodes got a copy of a message, and, thereby, increased their respective chances of delivery.

Figures 2.5 and 2.6 provide some insights on how buffer size and message forwarding schemes are related:

- With a fixed limit on message replication, consideration of excessive buffer sizes is not useful. This is reflected by the performances of the SnW and the first-contact routing protocols.

Fig. 2.6 Effects of buffer
sizes on the message delivery
ratio using **a** First contact, **b**
PRoPHET and **c** Epidemic
routing

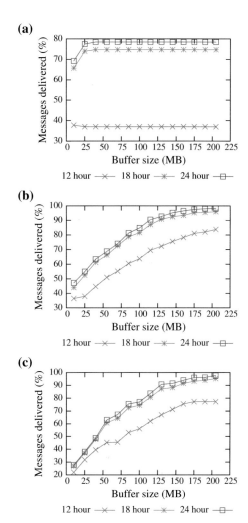

• With no fixed limit on the number of message replicas, larger buffers enhance
  the delivery ratio. This behavior, verified from the evaluation of the PRoPHET
  and Epidemic protocols, is due to the reason that during each communication
  opportunity, more number of nodes get a copy of a message.

Thus, it might be helpful to keep the above observations in mind while developing
a routing scheme for use in real life.

2.4 Real-Life Traces

Performance evaluation and model validation through simulations have been an integral part of research in the domain of computer networks. To facilitate the simulation of scenarios where the nodes are mobile, several synthetic mobility models (for example, Random Waypoint and Random Walk) have been proposed. When such models are integrated with the simulation tools, the nodes move according to the stochastic mobility model in the terrain concerned. Therefore, it is possible to add dynamism into the network scenario for example, a MANET with dynamically changing routes between a source and destination pair of nodes.

Although such synthetic models dominated the research domain for long, it was realized that these models, while simple, are idealistic and often do not reflect reality. For example, if one keeps track of the movement pattern of a human being, it would be far from random. Rather, a lot of similarity would possibly be observed in a person's daily movement patterns for example, traveling to the office, visiting a mall, and returning back to home. It may also be noted that in a network with mobile nodes—unlike networks with stationary nodes—the connectivity among the nodes is also governed by their movement. This is particularly true in OMNs, where the nodes typically have short communication ranges, for example, up to 10 m radius of communication for the Bluetooth interfaces. In fact, as noted in an earlier chapter, nodes' mobility is the driving force that enables communication in the otherwise disconnected network.

Kim et al. [77] extracted mobility statistics of the users from a large set of logs captured by several wireless access points over a time period of 13 months. The authors observed that the speed and pause time distribution of the users were lognormal. Moreover, contrary to the assumption of many mobility models, users are unlikely to move along all directions with equal probability. Rather, such mobility distribution is strongly related to the direction of the roads. Rhee et al. [22], on the other hand, analyzed human movement by recording the GPS-based locations of the users over a long period of time. Subsequent analysis revealed that human movement bears similarity with the truncated Levy Walk model, which is also observed in other animals.

> The CRAWDAD (http://crawdad.cs.dartmouth.edu/) repository has a huge collection of real-life traces belonging to different categories including, but not limited to, mobility logs, measurements related to Wi-Fi and WiMax, Bluetooth connectivity, GPS locations, and mobile phone activity records. As of October 2015, the CRAWDAD repository had a collection of 106 data sets, and 7717 registered users from across 111 countries. (Source: CRAWDAD news email list)

Table 2.2 summarizes a few characteristics of some of the data sets available in CRAWDAD. The "dataset" column indicates a contextual name for the concerned

Table 2.2 Summary of selected data sets available at CRAWDAD

Year	Dataset	Trace name	Trace type	Device	Network type	Transmission range (m)	No. of devices	Duration (days)
2005	MIT	Reality Mining [79]	Connection	Phone	Bluetooth, Cellular	10	94	246
2005	Cambridge/Haggle	Infocom'05 [80]	Connection	iMote	Bluetooth	30	41	4
2006	Cambridge/Haggle	Infocom'06 [80]	Connection	iMote	Bluetooth	30	78	4
2006	UPMC	Content [81]	Connection	iMote	Bluetooth	10, 20	54	~2 months
2007	Nottingham	Cattle [82]	Mobility	Phone	Bluetooth, GPS		7	2
2008	Unimi (University of Milano)	PMTR [83]	Connection	PMTR (Pocket Mobile Trace Recorder)	RTX-RTLP	10	44	19
2008	UMass	Diesel [78]	Connection, Mobility	HaCom Open Brick	IEEE 802.11b		30–40	
2009	UPMC	RollerNet [84]	Connection	iMote	Bluetooth	10	62	0.12
2009	NCSU	KAIST [85]	Mobility	T-mote	GPS receiver		4	
2011	St_Andrews	Sassy [86]	Connection	T-mote	Bluetooth	10	27	79
2014	Roma	Taxi [87]	Mobility	Tablet	GPS		320	~2 months

set of data, for example, name of the organization conducting the experiment, the place, and so on. Each such dataset consists of one or more traces collected from the corresponding experiment. The "Cambridge/Haggle" dataset has connection traces connected from different experiments, for example, during the INFOCOM conferences in 2005 and 2006. It can be noted from the Table that various kinds of devices have been used for the experiments based on the requirements. For example, in the UMass data set [78], computers fitted to buses were used. However, in other experiments, smaller devices were used since they were meant for carrying by human beings.

Availability of such a huge collection of data sets available at our disposal has altered the simulation dynamics in the related research domain. The use of real-life traces has now become the *de facto* approach toward simulation of OMNs and subsequent performance evaluation. As mentioned earlier, these traces capture a diverse category of information and can be used for appropriate purposes. A commonly observed pattern is the use of device connectivity information. Multiple data sets (for example, Infocom'06 [80] and Sassy [86]) capture information on the Bluetooth device sightings. For example, when a Bluetooth-enabled device B comes in the proximity of another similar device A, the two devices are said to have a "contact" (or encounter). Each device records the earliest time when one detected the presence of the other. The devices often record the contact duration as well—either the actual duration or the time when the other device was last seen. Therefore, from a simulation perspective, such information helps in determining when a link in the network should be up and for how long that should remain up.

As an example, let us consider the following records from the Infocom'06 [80] trace.

```
22   21   9259    9734    1   0
 1    4   12641   12768    9   372
 1    5   13750   14349    4   266
16   50   20534   20778    2   8958
```

Here, the first column indicates the address of the node which was recording. The second column indicates the address of the node with which the first came in contact. The third column indicates the time when the encounter began, and the fourth column indicates the time when the encounter was over. Therefore, the contact duration (time) for the pair of nodes in the first row was $9734 - 9259 = 475$ s.

In a similar way, one may use the GPS-based location traces (for example, [85]) of the nodes. In such a case, the nodes move as per the pre-specified coordinates in a given simulation terrain. The connection pattern among the nodes is derived from their underlying movement in the terrain. The following describes two popular ways—among others—the real-life traces can typically be used along with a network simulator:

- Make the nodes in the simulator move as per the mobility of the nodes captured via GPS logs. The connection among the nodes occur as dictated by such mobility.

- Ignore mobility of the nodes, but make them connect as per the connection traces available. As noted previously, such traces usually provide the precise time when a pair of nodes come in contact and how long they do stay in contact.

To close this section, one should, however, keep in mind that it may not be possible to use real-life traces in all kinds of scenarios. For example, when one is planning to simulate a post-disaster scenario, it would be difficult to take into consideration how would people communicate or move in a similar scenario in real-life. Therefore, the user should make judicious use of other simulation models in such cases. In Chap. 3, we shall look at how real-life traces can be imported into the ONE simulator and used in network simulations.

2.5 Applications

As noted earlier, DTNs are the suitable mode of communication in different challenged environments such as, deep-space networks and underwater sensor networks. Several implementations of DTN-related protocols are publicly available.[4] Some of the known implementations are the DTN2[5] project, Interplanetary Overlay Network (ION),[6] Postellation,[7] and IBR-DTN.[8] These implementations often come with sample applications for sending and receiving files, news delivery, and so on.

In this section, we discuss about some scenarios for DTN and DTN-like applications [88]. Some of these applications have been implemented and tested in real life. The following list is, however, by no means not exhaustive. A notable omission in this section is the Haggle project [89], which we will cover in details in Chap. 8.

2.5.1 DakNet

The DakNet project [90] has been developed at MIT Media Lab to provide low-cost and energy-efficient communication facilities. DakNet—more of an ad hoc network rather than a DTN—combines asynchronous services together with wireless mode of communication to connect remote villages with towns and cities. DakNet consists of kiosks, mobile access points (MAPs) equipped with portable storage devices and Internet access points or hubs. The MAPs are mounted on—and powered by— vehicles such as bus, motorcycle, bicycle and even ox cart. Whenever such a MAP comes within the proximity of a kiosk, data are transferred between them. Subse-

---

[4]A detailed list can be found at https://sites.google.com/site/dtnresgroup/home/code.

[5]https://sites.google.com/site/dtnresgroup/home/code/dtn2documentation.

[6]https://ion.ocp.ohiou.edu/.

[7]http://reeves.viagenie.ca/.

[8]https://trac.ibr.cs.tu-bs.de/project-cm-2012-ibrdtn.

quently, when a MAP comes near a hub (possibly in the town or city), data from the kiosks are synchronized with the Internet. The MAPs, therefore, essentially act as data mules. DakNet has been deployed in the villages of India and Cambodia.

2.5.2 Bytewalla

Ntareme et al. [91] developed Bytewalla—an Android application to enable delay-tolerant networking. Bytewalla implements the DTN v2 specification by the DTN research group, and can send/receive data bundles. A user can specify where to store the bundles as well as how much memory to be used. It has provision for sending emails via DTN, which is useful in rural areas with no Internet connectivity. A DTN server receiving such an email converts it into a bundle, and are forwarded to other smartphones using the application. On the other hand, a server extracts the email from a bundle before delivering it to a client.

2.5.3 DTWiki

Du and Brewer [92] implemented DTWiki—a Wiki system operating in DTNs with intermittent connectivity. The DTWiki application runs on the top of TierStore [93]—a distributed file system used to manage file replication, synchronization, and consistency. The backend of DTWiki is not a simple relational database, but consists of several components to ensure data consistency under intermittent connectivity. The data back end stores pages and their revisions with related metadata, user account information, attachments (for example, images and videos) contained in the pages, discussion pages for coordination among the users, facility for searching and indexing content, and mechanism for sharing. Such features help in providing robustness and scalability in the face of frequent network partitions and intermittent connectivity.

2.5.4 DT-Talkie

Islam et al. [94] developed DT-Talkie, a voice-based communication application much like the well-known walkie-talkies. At a high level, DT-Talkie consists of five modules. The first module, audio capture and play, is responsible for capturing the input voice message at the sender's side and storing the encoded voice message in the local file system. At the receiver's end, this module plays a received voice message. The MIME (Multipurpose Internet Mail Extensions) create and parse module creates multipart MIME message at the sender's side containing the encoded voice message and sender's information. At the receiver's end, this module is responsible for parsing the appropriate data from the received bundle. There is also a bundle send and receive module for creating, sending, and receiving a bundle; a GUI module to let the user interact with the application. Finally, the DTN daemon runs in the background providing necessary service for operations with bundles. Apart from one-to-one "conversations," DT-Talkie also allows group-based communication.

2.5.5 ZebraNet

The ZebraNet [95–97] project was aimed at monitoring the movement and activity of zebras at the Mpala Research Center, Kenya. Such wildlife movement spanned over a large terrain size and a long temporal scale. Although ZebraNet can be considered as a typical project on ad hoc WSNs, it exhibited several challenging requirements. For example, the size and weight of the collars could not be large so that they do not inhibit the movement of the zebras. On the other hand, with collars attached to the necks of zebras, a sparse network topology is created with lower chances of communication. Moreover, since the tracking devices, equipped with GPS sensors, were required to operate over long periods of time without human intervention, relying only on battery sources for power was not feasible. These factors led toward the design of custom-made hardware for the collars. However, from the perspective of network communications, a more stringent requirement was that the base station, which itself was mobile, should receive almost all the tracking data gathered—as close to 100 % as possible. The delay in obtaining all such data was not critical. In other words, the ZebraNet project exhibited delay-tolerant characteristics, where eventual delivery of messages was acceptable. The collars attached to the zebras would normally transfer data in peer-to-peer fashion. Subsequently, the base station received data from the zebras when they came within proximity.

Other related application include Web searching from buses [98], Twimight—a Twitter client for disaster mode [99], and so on. More similar applications are expected to come up and used in the future. In this context, it may be noted that Lindgren et al. [100, 101] have presented valuable insights gathered while deploying and testing DTN systems in real-life. Such anecdote of experiences would be of much value and use to the future developers.

2.6 Summary

Routing is one of the fundamental requirement in any kind of communication network including OMNs. In this chapter, we reviewed some of the popular routing protocols proposed for OMNs along with their variations and improvements. Unlike traditional networks, routing in OMNs involve repeated replications of a message in order to improve its chances of delivery. Some schemes impose an upper limit on replications, whereas others do not. We observed that most of the routing protocols in existence today use some form of intelligence for decision making. They range from as simple as total count of encounters to complex metrics, such as dynamic delivery predictability. Moreover, many routing protocols exploit the fact that social interactions among people exhibit some long-term patterns.

Subsequently, in this chapter, we formally defined a set of metrics commonly used for performance evaluation in OMNs. These metrics are generic in nature and can be used in any scenario. Some of the context-specific evaluation metrics would be discussed later in this book. Next, we discussed about the real-life traces and their characteristics. As we shall see in the following chapter, these traces can be used in network simulations. Finally, we closed this chapter by reviewing a few applications based on DTNs.

In the remainder of this book, we shall look at different aspects related to the OMNs, where different schemes are proposed and their effect on the network performance evaluated. This chapter, in fact, prepares the ground for introducing the reader to the latter chapters focused on specific research topics.

2.7   Review Terms

- Multi-copy routing
- First-contact routing
- Summary vectors
- Vaccine
- Binary spray
- Transitivity
- Community
- Message overhead ratio
- DakNet

- Direct delivery
- Flooding
- Immunity
- Anti-packet
- Delivery predictability
- Centrality
- Message delivery ratio
- Real-life traces
- Bytewalla

2.8   Exercises

2.1 Write down some metrics relevant to the evaluation of routing protocols. Should these metrics be maximized or minimized to obtain a better performance? Why routing overhead is typically considered in DTNs, but not in the traditional networks, for example, Internet and mobile ad hoc networks?

2.2 Consider the following contact pattern among a set of five nodes:

$\mathscr{C} = \{(A, B, 10), (A, C, 14), (C, D, 16), (B, D, 19), (D, E, 22)\}$,

where $(X, Y, t) \in \mathscr{C}$ indicates that nodes X and Y encountered one another at the time instant t. Further, let the set of messages created be $M = \{(m_1, A, 12), (m_2, B, 15), (m_3, C, 16)\}$, where $(m, X, t) \in M$ indicates that a message with ID m was created by the node X at the time instant t.

Which messages would be carried by the node E after time instant $t = 23$ when Epidemic and Spray and Wait routing protocols are used? Assume that each contact between a pair of nodes lasted for one time unit. Also, assume that SnW is used in binary mode with initial number of copies of a message as 4.

2.3 What are the advantages of SnF over SnW? What similarity does the Spray and Focus and PRoPHET protocols have?

2.4 Support or reject the following claim with justifications.

When the initial number of copies of a message is infinitely large, SnW degenerates into the Epidemic routing protocol. This is true irrespective of whether or not binary mode of spraying is used.

2.5 Assume that the nodes in a network move with a constant speed of 5 m/s. Suppose that the nodes A and j had their last encounter at $t = 1$ s, and the nodes B and j recently encountered at $t = 2$ s. Consider that at $t = 6$ s, nodes A and B come in contact, and the distance between them is 10 m. According to the Spray and Focus algorithm, what would be the value of $\tau_A(j)$ after this contact?

2.6 Assume that the delivery predictabilities between two pairs of nodes (using PRoPHET) just before the time instant t be $P(A, B) = 0.6$ and $P(B, C) = 0.7$, respectively. Further assume that $P(A, C)_{\text{old}}$ is 0.5. At time t the two nodes A and B come in contact. What would be the value of $P(A, C)$ after 120 s from the time instant t if the nodes A and B did not had any subsequent encounter? Assume that each time unit comprises of 60 s.

2.7 Consider some well-known algorithms for finding shortest path in a graph (for example, those proposed by Dijkstra and Floyd–Warshall). What is the challenge in their direct adaptation to OMNs?

2.8 What is a Bloom filter? How Bloom filters can be useful in the context of the vaccine scheme used with Epidemic (or other) routing protocol?

2.9 Suppose that the nodes in an OMN use SnW routing protocol. Let L be a constant that indicates the initial number of copies of any message created in the network. Prove that the overhead ratio in the OMN cannot be greater than L.

2.9 Programming Exercises

2.10 All routing protocols implemented in the ONE simulator are subclasses of the `ActiveRouter` class. However, the simulator also provides another class, `PassiveRouter`, which actually does no routing. Can you think of any scenario where such a nonfunctional router class would be useful?

2.11 The ONE simulator provides an implementation of the Epidemic routing protocol in the `EpidemicRouter` class. Complement the `EpidemicRouter` class with
the vaccine scheme as discussed in this chapter. In particular, use a `HashSet` as an
instance variable to store the ID of the messages delivered for the concerned node.
During any contact, make the two nodes in contact exchange their sets and update
accordingly.

2.12 The SnW routing protocol is interesting in the sense that it imposes a upper
limit on replication of any message. On the other hand, the efficiency of PRoPHET
is in the fact that it leverages the delivery predictability metric to identify better
candidates for message delivery.

Develop a new, hybrid routing algorithm, which combines the essence of SnW
and PRoPHET together. In other words, each message should have a maximum upper
limit, L, on its replication. However, unlike SnW, a message should not be blindly
replicated to another node. Rather, a node using the new routing algorithm should
replicate to a given node only if the latter has higher delivery predictability for the
destination of the concerned message.

Chapter 3
A Developer's Guide to the ONE Simulator

In Chap. 1, we briefly looked at the ONE simulator [51]. In this chapter, we would explore it in detail from the point of view of a developer. In a simple scenario, a developer's workflow is also simple—write code, test it, and repeat the cycle. However, when the required effort for coding becomes large, it is useful to have a more pragmatic approach, and take help of various tools—for example, an integrated development environment (IDE) and version control systems—in the process.

In this chapter, we look at how to work with the ONE simulator using the NetBeans IDE. Use of such an IDE often makes the workflow easier. Next, we discuss in detail a key topic—creating a new routing protocol using the ONE simulator. We illustrate this process with a simple example by considering a variation of the SnW protocol. A walkthrough of the code, together with insights, is provided. In real life, developing software is usually a complex process. In such scenarios, tracking back the code and debugging becomes easier using the graphical user interface (GUI) based IDEs. In this regard, we take a look at version control of source code using Git. Git operations can be performed from a terminal or from within NetBeans—we illustrate both approaches here. Later in this chapter, we cover the issue of testing a protocol developed with the ONE simulator. We discuss how to write a test case from scratch by using JUnit[1] and the framework provided by the ONE simulator. Finally, we conclude this chapter with a set of best practices based on our experiences with the ONE simulator.

3.1 Development with NetBeans

As mentioned in the beginning of this chapter, using an IDE, such as NetBeans, for development is both effective and productive. Features like autocompletion of code and automatic code generation can help save time from repetitive tasks. Moreover,

---

[1]http://junit.org/.

© Springer International Publishing Switzerland 2016
S. Misra et al., *Opportunistic Mobile Networks*,
Computer Communications and Networks, DOI 10.1007/978-3-319-29031-7_3

debugging an application becomes easier. However, when someone is learning a software for the first time, it is often recommended to begin with writing code in a simple text editor. This is helpful in learning the basics of the concerned technology. Once comfortable, the reader can move to using an IDE of his/her choice.

3.1.1 Setting Up a Project

In this section, we would look at how to work with the ONE simulator by creating a Java project using the NetBeans[2] IDE. It is assumed that the reader already has the NetBeans IDE available in his/her system. Moreover, the source code[3] of the ONE simulator is also assumed to be present in the reader's system.

Once NetBeans has been opened, go to the menu bar, click on `File > New Project` and select `Java Project with Existing Sources`, as shown in Fig. 3.1a. Then, click on the `Next` button.

In the following dialog box that appears, type in a name for the project (for example, "The_ONE") and (optionally) select a project folder, as shown in Fig. 3.1b. Note that the location of the project folder has been automatically generated by NetBeans based on the name of the project. Unless you are sure about what you are doing, it is recommended to keep the project folder location as it is. Again, click on the `Next` button.

In the next step, NetBeans would ask you to provide the location of the source code, as shown in Fig. 3.1c. Click on the `Add Folder` button, browse to the location where your source code is stored. We do not need to add any test package folder here. Once done, click on the `Finish` button. Now we have a project created in NetBeans!

> The source code of the ONE simulator has a `test/` directory. In the dialog shown in Fig. 3.1c, you can browse to and select that directory under the option `Test Package Folders`. This can also be done later by right clicking on the NetBeans project, selecting `Properties`, and then selecting `Sources` under the `Categories` option. Inclusion of the `test/` directory is necessary to run test cases, as we shall see later in this chapter.

However, a few more things need to be done before we can actually run any simulation from within NetBeans. If you look at the IDE now, you would see that several packages and source files are indicated to be in error (under the `Projects`

---

[2]https://netbeans.org/.

[3]The ONE simulator has been developed using Java 6. In the remainder of this chapter, all examples provided were executed using NetBeans running with OpenJDK 6. Minor changes might be required if Java 7 is used instead. The reader is suggested to consult appropriate documents/forums regarding this.

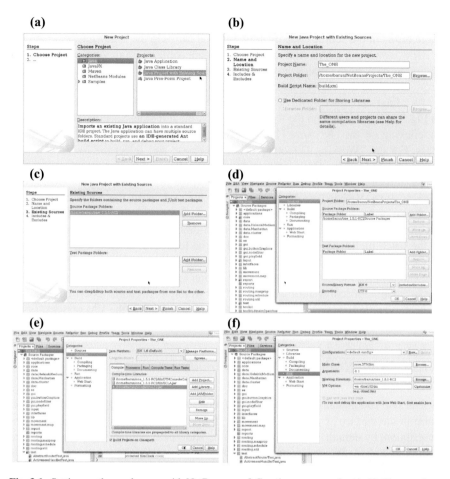

Fig. 3.1 Setting up the workspace with NetBeans. **a–d** Creating a new project in NetBeans using the source code of the ONE simulator, **e** addling library JAR files, **f** selecting the main class and working directory

tab), as shown in Fig. 3.1d. To fix this, right click on the project name, and then click on `Properties` displayed at the bottom of the context menu. A dialog box would appear, as shown in Fig. 3.1e.

Now, under the `Categories` section in the dialog box, click on `Libraries`. Next, under the `Compile` tab on the right-hand side, click on the ⟨ Add JAR/Folder ⟩ button. We need to add three JAR (Java Archive) files here.

The first two JAR files are located inside the `lib/` directory of the source code of the ONE simulator. These two files are named `DTNConsoleConnection.jar` and `ECLA.jar`. Then, click on the ⟨ OK ⟩ button. You would notice that all the errors (on the left side under the `Projects` tab), except for the `test` package, have now been resolved. The `test` package is still in error because NetBeans could not locate

JUnit. (Note: If your NetBeans installation is able to locate JUnit, there would be no more errors in the source code by now, and you can skip the following step.)

Now, download a JAR file[4] of JUnit and store somewhere in your system. Once again, go back to the dialog box shown in Fig. 3.1e, and under the `Compile` tab, browse and add the JAR file corresponding to JUnit. Then, click on the `OK` button.

As a final step, once again open up to the project properties dialog box, as shown in Fig. 3.1f. Now, click on the `Run` category on the left side. Specify the main class by clicking on the `Browse` button, and selecting `core.DTNSim` as the main class. For the "Working Directory" option, browse and select the location where the source code of the ONE simulator is located that was imported earlier (not the project directory of NetBeans). Now we are ready to run simulation by clicking on the `Run > Run Main Project` menu (or keyboard shortcut, the `F6` button). Also, as shown in the Fig. 3.1f, one can specify `-b` in the `Arguments` text field to run the simulation in batch mode. In case multiple simulations say, 5, are to be executed in the batch mode, type in `-b 5` instead in the text field. However, if one wishes to view the simulation in the GUI mode, keep the said text field empty. Optionally, one can also set the "VM Options." Any option specified here would be passed to the Java virtual machine during execution. For example, to enable assertions[5] during project execution, one needs to provide the `-ea` argument to the VM options.

3.1.2   Using Real-Life Traces in Simulations

Before one can actually run a simulation, one has to describe the underlying scenario in the `default_settings.txt` file. As the name suggests, the scenario description, by default, is read from this file. Additional settings files, however, can also be specified. While a complete description of the different settings used with a simulator is out of scope of this book,[6] in this chapter, we look at a particular aspect—the use of real-life traces.

It is possible to simulate a scenario with the ONE simulator using a real-life connection or mobility trace. Note that these two types are mutually exclusive for a set of nodes. If we use a connection trace for a set of nodes, we do not use a mobility trace together with it. The reverse also follows. In fact, when mobility traces are used, the set of connection patterns in the network are generated based on the transmission range(s) of the nodes. Moreover, the ONE simulator also allows importing a set of messages to be generated by the nodes in an OMN. This is in addition to the uniformly distributed message generator model that is provided by the simulator.

Listing 3.1 shows how to use real-life connection traces with the ONE simulator. Appropriate settings should be specified in the `default_settings.txt` file. As depicted in Listing 3.1, the movement model of the nodes in the network is set to

---

[4]https://search.maven.org/#searchgav|1|g:%22junit%22%20AND%20a:%22junit%22.

[5]By default, assertions are disabled in Java.

[6]Such descriptions are made available by many users in the simulator's community.

`StationaryMovement`. In other words, the nodes do not move in the simulation. Moreover, we ask the simulator not to simulate connections among the nodes. This is due to the reason that we are already using predefined connection traces. So, the simulator should create connections among the nodes exactly as our trace specifies and not in any other way. We also need to specify the location of the trace file for one of the events. However, it should be noted that a raw trace file may not suitable for use as it is. A preprocessing step would likely be required to convert the real traces into a format that is recognized by the ONE simulator. In particular, if GPS-based mobility traces are used where locations are represented with their latitude and longitude, they need to be converted into 2D Cartesian coordinates before using with a simulator. The reader is advised to consult related documents in this regard.

Listing 3.1 Settings for using real-life connection traces with the ONE simulator

```
## Scenario settings
Scenario.name = infocom06_router-%%Group.router%%
Scenario.simulateConnections = false
Group.movementModel = StationaryMovement

## Message creation parameters
# How many event generators
Events.nrof = 2

## Trace information
Events1.class = ExternalEventsQueue
Events1.filePath = my_scenarios/infocom06-78n-24hr.txt

# Events2 is for message generation events
```

The example shown above is for using connection traces. Real-life mobility traces can be used in a similar fashion, as shown in Listing 3.2.

Listing 3.2 Using real-life movement traces with the ONE simulator

```
Scenario.simulateConnections = true

## Movement model settings
# KAIST movement trace
Group.movementModel = ExternalMovement
ExternalMovement.file = my_scenarios/KAIST_92n_movement_trace.txt
```

The external movement (mobility) trace file should be formatted in a specific way for use in the simulation. This is documented in, and quoted below from, the `ExternalMovementReader` class.

First line of the file should be the offset header. Syntax of the header should be:

`minTime maxTime minX maxX minY maxY minZ maxZ`

Last two values (Z-axis) are ignored at the moment but can be present in the file.

Following lines' syntax should be:

`time id xPos yPos`

The first header line defines the valid terrain dimension and simulation duration to the simulator. Subsequent lines specify the positions of the nodes at different time instants. The external connection traces, too, have a similar format, as shown below.

```
timestamp1 CONN node1 node2 up
timestamp2 CONN node1 node2 down
```

The file is sorted based on time stamps of connection up/down events. In case there are multiple Event classes for a given scenario, their relative order is not important. For example, the first event can point to a real-life connection trace and second to a message generator event or vice versa. However, the path to the trace file should be correctly specified, and the contents of the trace file should be in appropriate format as required by the corresponding Event class.

3.1.3   Debugging with NetBeans

One of the advantages of using an IDE is that it often makes debugging an application easier. We would now briefly look at the debugging process using NetBeans.

The fundamental primitive in debugging is a break point. One or more break points can be inserted at different lines (in the same file or different files) of the source code. Subsequently, when the application is executed in debug mode, the execution would pause when the control reaches any predefined break point. Break points can be classified into two types:

- **Unconditional**: If the type of a break point is unconditional, the application, when executed in the debugging mode, would always pause when the control reaches that break point.
- **Conditional**: In case of conditional break points, one or more condition(s) are specified using zero or more relational operator. The application, when executed in the debugging mode, would pause only when the control reaches that break point and the predefined condition(s) is(are) satisfied.

The latter type of break points are useful when the possible causes of bug-prone behavior of an application have been narrowed down. As a hypothetical example, let us consider that message replication was found to fail when there are more than 5 messages in a node's buffer. In that case, we would be interested to investigate only those scenarios where a node has more than 5 messages in its buffer. Let us now look at how we use NetBeans for debugging.

Let us consider the SprayAndWaitRouter class, which contains the implementation of the Spray and Wait routing protocol. This class has the following method (line numbers correspond to the actual line numbers from the source file):

Listing 3.3 The getMessagesWithCopiesLeft() method

```
115  protected List<Message> getMessagesWithCopiesLeft() {
116      List<Message> list = new ArrayList<Message>();
117
118      for (Message m : getMessageCollection()) {
119          Integer nrofCopies = ↩
                 (Integer)m.getProperty(MSG_COUNT_PROPERTY);
120          assert nrofCopies != null :"SnW message" + m + "didn't ↩
                 have" +
121              "nrof copies property!";
122          if (nrofCopies > 1) {
123              list.add(m);
124          }
125      }
126
127      return list;
128  }
```

Line number 118 presents a situation similar to the one we talked above. To set a break point, open the SprayAndWaitRouter.java file in NetBeans from the project that we created earlier in this section. Then, scroll down near to the specified line number. On the outer grayish margin to the left of the code display area (where the line numbers are shown), right click on the number 118. This would create a break point at that line number, and the background color of the corresponding line of code would change to red. To turn this break point into a conditional one, right click on the break point, then select Breakpoint > Properties. Then, select the Condition checkbox, and type in the following condition in the corresponding text field: getMessageCollection().size()> 5, as shown in Fig. 3.2a. Now anytime the control pauses at this break point while debugging, you would be sure that the concerned node has more than 5 messages in its buffer. Figure 3.2b shows such a scenario where the execution paused when a node had 6 messages in its buffer. One might also observe the lower part of figure, where, under the Variables tab, a list of local variables and their current values are displayed. Note that to execute an application in the debug mode in NetBeans, the Ctrl+F5 buttons should be pressed (instead of F6, which is for normal execution).

Of course, setting break point(s) is not the goal of debugging—it is just a tool. Once the control pauses at a break point, one can *watch* one or more variables, continue debugging (until the next break point, if any, is replaced), or resume normal execution. While continuing debugging, one might *step over* the code or *step into* the code. These are general terminologies in the domain of debugging and not specific to NetBeans.

(a)

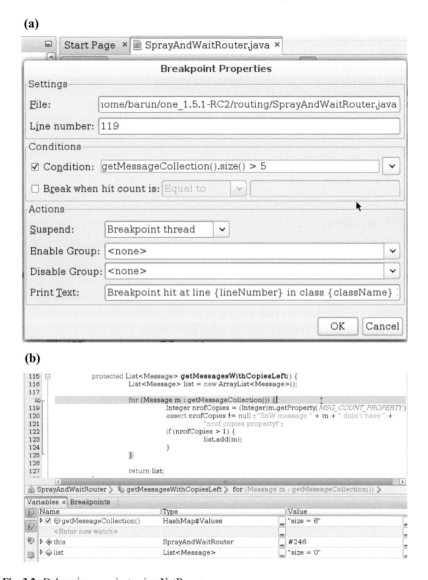

(b)

Fig. 3.2 Debugging a project using NetBeans

3.2   Developing a New Routing Protocol

In this section, we will look at the development of a new routing protocol from scratch in the ONE simulator. Our goal in this is very modest. We are not aiming to propose an efficient routing algorithm here, but to merely demonstrate the best practices in implementing a routing mechanism. In particular, in this section, we consider a simple variation of the Spray and Wait routing protocol termed as, Spray and Wait with Utility (SnWU). As the name suggests, we wish to augment the classical SnW with a notion of utility. In SnWU, the utility $U(i, j)$ of a node i for any other node j is given by the total number of times these two nodes came in contact. Similar to SnW, in SnWU, too, a message is not replicated if its current number of copies is less than two. Additionally, any node i replicates a message to another node j only if the utility of the latter node, $U(j, d)$, is greater than that of the former, $U(i, d)$, where d is the destination of the message, $i \neq j \neq d$. The reader may look at [102–104] for examples of similar other routing protocols that improve SnW with different utility functions.

3.2.1 The Roadmap

Before beginning with the implementation of any routing algorithm, one should identify whether or not it represents any similarity with one of the existing routing protocols in the ONE simulator. In our case, it actually does. We only need to adapt a certain behavior of the existing SnW protocol. In particular, we need to alter the logic for message replication by considering another factor, the utility values of the nodes. Now, this gives us an useful insight—the replication process should not only consider the messages, but also the other node in contact in order to compare the utilities.

It is useful to carefully study the implementations of Epidemic, SnW, and PRoPHET routing protocols available with the ONE simulator. In general, almost any new routing algorithm can be developed by using, or combining, the code of these protocols.

With these aspects in mind, we can begin with the implementation of our routing protocol. In particular, we would create a new routing class say, `SprayAndWait-UtilityRouter`, by extending the existing `SprayAndWaitRouter` class. Note that one can also extend the `ActiveRouter` class instead. However, in our scenario, taking the former approach has an advantage—it allows us reusability of functionality, which is an important aspect of Object Oriented Programming. In particular, the following two redundancies in parent and subclasses can be avoided in this case.

1. The `SprayAndWaitRouter` class has several member variables to store states, for example, namespace, initial number of copies of a message, and mode of operation. By extending it to create the `SprayAndWaitUtilityRouter`, we do not need to bother about such state maintenance.
2. Additionally, the `SprayAndWaitRouter` class has different methods for implementing several behaviors of the protocol that are common to `SprayAndWaitUtilityRouter` as well. Thus, using our approach, we do not need to duplicate those behaviors.

In general, wherever possible, common behaviors should be encapsulated in a
parent class, and specialized behaviors should be obtained by extending it. However,
in certain scenarios it might be difficult to extend an existing routing protocol. For
example, it might be so required that we want to bypass the behavior of the parent,
but still want to retain that of the grandparent class. Or maybe there is no existing
protocol whose implementation has similarity with the proposed one. In such cases,
one may simply extend the `ActiveRouter` class to create the new routing protocol.

3.2.2 Implementation Details

In the following, we discuss the implementation of the `SprayAndWaitUtility-`
`Router` class. Inline comments within the code are provided for contextual details.
The Java package import statements are not shown here, which is why the line
numbering does not begin with 1. It is advisable that the reader understands the basic
functionality demonstrated here, and then attempts to write the code on his/her own.

Listing 3.4 shows the `SprayAndWaitUtilityRouter` class, its two con-
structors, and the `replicate()` method. The first constructor takes a `Settings`
argument and passes it on to its superclass. The other constructor is a copy constructor, which takes an argument of the same class type. The `replicate()` method is
invoked repeatedly during a simulation setup to create the router module for other relevant nodes (`DTNHost`) in the OMN. It is essential that the `replicate()` method
is overridden with proper types. Otherwise, the behavior of the `replicate()`
method from the parent class would be retained, which would result in the nodes
having a different router type than the desired one.

Listing 3.4 The main methods

```
24  public class SprayAndWaitUtilityRouter extends ↩
        SprayAndWaitRouter {
25      // For storing per node utility values
26      // The key would be address of a node, and the value would be
27      // the number of contacts
28      protected HashMap<Integer, Integer> utilities;
29
30      public SprayAndWaitUtilityRouter(Settings s) {
31          super(s);
32          utilities = new HashMap<Integer, Integer>();
33      }
34
35      public SprayAndWaitUtilityRouter(SprayAndWaitRouter r) {
36          super(r);
37          utilities = new HashMap<Integer, Integer>();
38      }
39
40      @Override
41      public SprayAndWaitUtilityRouter replicate() {
42          // Create a new router of this type
43          return new SprayAndWaitUtilityRouter(this);
44      }
```

Note that in line number 28, we declared a data structure, `utilities`. This is a `HashMap` that stores key–value pairs. We would use this data structure to manage the utility values of the nodes. In particular, the utility of a node is monotonically increasing—it stays the same or increases, but never decreases. Such increments occur only when two nodes come in contact, and, therefore, we need to handle the event when a connection with another node is established.

Listing 3.5 shows the implementation of the above discussed details. In particular, in line number 50, we check whether or not the current event corresponds to connection establishment. Given the symmetric nature of our utility, we could also handle the connection termination events with same effect. As annotated within the code, on such a connection change event, a node retrieves the previous utility value, if any, for the other node it is in contact with, increments the value by 1, and updates the entry for the concerned node in the `utilities` data structure.

Listing 3.5 Updating utility values

```
46    @Override
47    public void changedConnection(Connection con) {
48        super.changedConnection(con);
49
50        if (con.isUp()) {
51            // Increase the number of contacts with the other node
52            int nContacts = 0;
53            DTNHost otherNode = con.getOtherNode(getHost());
54            int otherAddress = otherNode.getAddress();
55
56            if (utilities.containsKey(otherAddress)) {
57                // Already had some contact(s)s with this node
58                // Get the current utility value for the node
59                nContacts = utilities.get(otherAddress);
60            }
61
62            // Update the hash table with the new utility value
63            // for the concerned node
64            utilities.put(otherAddress, nContacts + 1);
65        }
66    }
```

Listing 3.6 shows the `update()` method. This is perhaps the most important method of a routing class where the replication logic resides. In particular, the first `if` block of the method verifies whether or not there is an ongoing transmission. If so, the method does nothing because simultaneous transmissions are not possible. The following `if` block invokes delivery of messages, if any, to the other node for which the latter is the destination.

Listing 3.6 Disabling default functionality of SnW

```
68   @Override
69   public void update() {
70       super.update();
71
72       if (!canStartTransfer() || isTransferring()) {
73           return; // nothing to transfer or is currently ←
                 transferring
74       }
75
76       // try messages that could be delivered to final recipient
77       if (exchangeDeliverableMessages() != null) {
78           return;
79       }
80
81       tryOtherMessages();
82   }
83
84   @Override
85   protected List<Message> getMessagesWithCopiesLeft() {
86       // This is required to override the functionality of the←
                 superclass
87       // where a message is replicated just if nrofCopies > 1
88       return new ArrayList<Message>();
89   }
```

An important method here is tryOtherMessages(), into which we would incorporate the replication decision of SnWU. We would look into this shortly. But before that, a vital issue deserves attention.

A reader familiar with the implementation of SnW would realize that the update() method shown above introduces a potential bug. The problem arises with an invocation of the same method of the parent class in line 68. Note that we cannot disable it—otherwise, the entire update sequence from ActiveRouter class would be ignored! On the other hand, presence of this method call will make SnWU act as SnW by replicating a message based solely on its current number of copies. To mitigate this issue, we also override the getMessagesWithCopiesLeft method from the SprayAndWaitRouter class by returning an empty list. The reader is advised to verify that this, indeed, prevents any replication dictated by SprayAndWaitRouter.

It is finally time to look at the tryOtherMessages method, which is borrowed from the implementation of ProphetRouter. Listing 3.7 shows the relevant code. In general, it is possible that a node is in contact with multiple nodes at a given instant of time. The nested for loops in Listing 3.7 help in comparing per-message utility between a pair of nodes. If the other node in contact has a greater utility for a given message, the current node marks it for replication to that node (line number 123). Finally, in line number 132, transmission actually begin.

Listing 3.7 Utility-based message replication in SnWU

```
91    protected Tuple<Message, Connection> tryOtherMessages() {
92        List<Tuple<Message, Connection>> messages =
93            new ArrayList<Tuple<Message, Connection>>();
94
95        Collection<Message> msgCollection = ←
              getMessageCollection();
96
97        /**
98         * For all connected hosts collect all messages that ←
               have a
99         * higher utility
100        */
101       for (Connection con : getConnections()) {
102           // Get reference to the router of the other node
103           DTNHost other = con.getOtherNode(getHost());
104           SprayAndWaitUtilityRouter othRouter = ←
                  (SprayAndWaitUtilityRouter)
105               other.getRouter();
106
107           if (othRouter.isTransferring()) {
108               continue; // skip hosts that are transferring
109           }
110
111           for (Message m : msgCollection) {
112               if (othRouter.hasMessage(m.getId())) {
113                   continue; // skip messages that the other one←
                         has
114               }
115
116               int destination = m.getTo().getAddress();
117
118               if (getCopiesLeft(m) > 1
119                       && othRouter.getUtility(destination)
120                       > getUtility(destination)
121                       ) {
122                   // the other node has higher utility
123                   messages.add(new Tuple<Message, ←
                         Connection>(m,con));
124               }
125           }
126       }
127
128       if (messages.isEmpty()) {
129           return null;
130       }
131
132       return tryMessagesForConnected(messages); // try to send←
              messages
133   }
```

Listing 3.8 shows the method for computing utility values for a given node. In general, it is helpful to encapsulate such computations within a method rather than hard-coding at relevant place(s). It keeps the source code clean and readable.

Listing 3.8 Computation of utility values

```
135    protected int getUtility(int address) {
136        int utility = 0;
137
138        if (utilities.containsKey(address)) {
139            utility = utilities.get(address);
140        }
141
142        return utility;
143    }
144
145    protected int getCopiesLeft(Message m) {
146        Integer nrofCopies = ↩
                (Integer)m.getProperty(MSG_COUNT_PROPERTY);
147        assert nrofCopies != null : "SnWU message" + m + "didn't↩
                have" +
148            "nrof copies property!";
149
150        return nrofCopies;
151    }
```

Thus, we have looked into the implementation of an entire algorithm. An important lesson obtained herein is that an algorithm, by itself, can seem very simple. However, when we are trying to implement it, several considerations should be kept in mind. We hope that, based on the examples provided in this section, the reader would feel confident in developing a new routing protocol.

3.3 Version Control

In this section, we discuss about one of the vital aspects of software development—version controlling the source code. Once again, note that this is not something specific to NetBeans or any other IDE, although many of the popular IDEs have support for version control. Here, we would look at some basic usage of Git, one of the most popular distributed version control systems in use today. A detailed coverage of the topic, however, is beyond the scope of this book. Interested readers are suggested to consult various resources on Git available on the Internet. An extensive documentation[7] on using Git with NetBeans is also available online, which the interested readers should refer to.

Simply stated, a version control software maintains a repository of the source code, and maintains the changes made to the source code at each successive *commit*. For each commit, a revision number is maintained. Git assigns a SHA-1 checksum to each commit. Additionally, it also maintains several other metadata, such as by whom and when the commit operation to the repository was performed.

---

[7]https://netbeans.org/kb/docs/ide/git.html.

Before proceeding further, it might be good place here to recall two fundamental guidelines for version control.

1. Any code that does not compile successfully should not be put under version control.
2. Executable binary files should not be put under version control.

The first principle says that if the source code could not be compiled successfully, it is not yet ready to be marked as a new version. In other words, the source code should not have any syntax error, for example, unmatched parentheses. However, any code that is under version control may not rule out runtime exceptions. This is precisely why extensive testing of a software is required.

To understand why it is not a good idea to commit code having compilation error(s), let us consider an example. Suppose that two developers are working on the same code in parallel. Now, let us consider that the first developer commits his/her changes. The second user "pulls in" (*checkout*) the changes made by the first user, and makes his/her own changes based on it. However, since the first user has left the code in an erroneous state, the second user cannot test his/her own changes.

The second guideline suggests not to add executable binary files under version control. To understand why, let us consider a project in C/C++ or any other compiled language (in contrast to interpreted languages). Based on the given source code, one can generate the corresponding binary executable file. Therefore, maintaining different revisions of that executable file is redundant and useless. Java, however, is not an entirely compiled language, and it does not generate a binary executable file similar to C/C++. However, each Java class is compiled to generate intermediate byte codes (`*.class` files), which should be avoided from version control.

In fact, it is not only executable files, but several other files that could (or should) be skipped from version control. For example, automatically generated backup copies of files by text editors and files containing password. On the other hand, binary files such as, images, are often put under version control. However, note that this is not a strict guideline—whether or not to version control an executable may eventually depend on the type of project and environment.[8] With Git, one can add the patterns of all the file names (for example, `*.class`) to be excluded from version control in the `.gitignore` file.

Let us now have an hands-on session with NetBeans and git. To begin with, go to NetBeans, right click on the (previously created) project "The_ONE" in the `Projects` tab. In the context menu that appears, navigate to `Git > Initialize Git repository`. A dialog box would be displayed asking for the path of the project directory, as shown in Fig. 3.3. Note that here you should select the directory inside which the source code of the ONE simulator is located. Once selected, click on the [OK] button. A new Git repository would be initialized at the specified location.

---

[8]For an interesting discussion, refer to http://programmers.stackexchange.com/questions/110518/binaries-in-source-control.

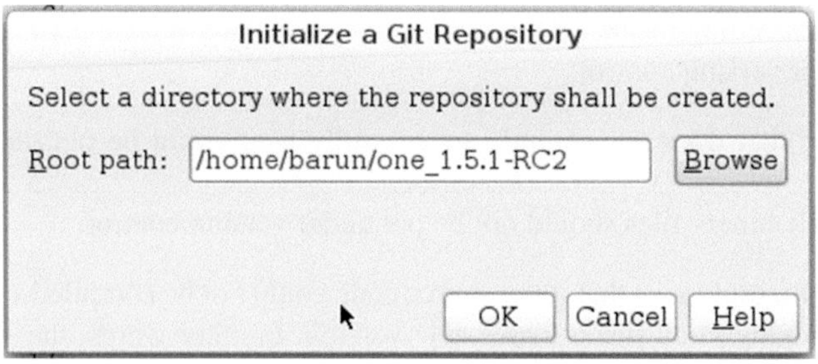

```
Initialize a Git Repository

Select a directory where the repository shall be created.

Root path:  /home/barun/one_1.5.1-RC2          Browse



                              OK   Cancel   Help
```

Fig. 3.3 Initializing a Git repository with NetBeans

Now, open a terminal[9] and switch to the location of your source code directory using cd. To view the status of any Git repository, type in the command git status in the terminal and press Enter. The first few lines of the output of the command executed are shown below.

```
$ git status
# On branch master
#
# Initial commit
#
# Untracked files:
# (use "git add <file>..." to include in what will be committed)
#
# HISTORY.txt
# LICENSE.txt
```

One might have noticed the string "Untracked files" in the output above. All files listed under this section are currently untracked by Git. In other words, these files are not yet under version control. As indicated in the output, to put some of these files under version control, one should execute the command git add <file_name1> <file_name2> in the terminal. However, if all files present inside the current directory are to be added, one should execute git add . in the terminal. Note the . (dot) after add that indicates the current directory.

Instead of performing git add . from the terminal, one can do that from within NetBeans. To do this, close all the source files, if any, currently opened in NetBeans (belonging to the corresponding project). Then, click on Team > Git > Add to add all the files.

Note that by now we have merely *added* files into the Git repository—the changes have not been yet *committed*. To verify this, once again go to the terminal, and execute git status the first few lines of the output are shown below.

---

[9]Windows users can use a command prompt for this purpose. Another alternative is to install Cygwin. Of course, whether you are using Linux or Windows, a Git client must be installed in your system.

```
$ git status
# On branch master
#
# Initial commit
#
# Changes to be committed:
# (use "git rm --cached <file>..." to unstage)
#
# new file: HISTORY.txt
# new file: LICENSE.txt
```

Also note that, if any of these files are modified after adding them to the repository, those files need to be added again. Otherwise, the recent changes would not be tracked in the current commit. Let us now commit the changes and view the status of repository again.

```
$ git commit -m "Adding source files"
$
$ git log
commit 6a42dbc9396d158311559c9782e6d4391a436673
Author: Barun Saha <barun.saha04@gmail.com>
Date: Wed Apr 8 00:21:45 2015 +0530

    Adding source files
$
$ git status
# On branch master
nothing to commit (working directory clean)
```

Observe that while committing our changes, we added a message using a -m option to git. The corresponding message is associated with that particular commit operation. Such a commit message briefly states what changes were made at a particular commit operation upon the source code. The commit message should be descriptive enough to understand the changes made. In other words, very generic messages, for example, "Removed file," should be avoided. The reader might point out that our commit message in the example sounds generic. This is, however, not a problematic scenario since this was our first commit operation with the sole purpose of adding the entire source code into the version control. As a final note to the readers, always add a brief yet meaningful commit message.

A history of the commits to a repository can be viewed by issuing the `git log` command. In this case, we had only a single commit. Note that our previously used commit message is also displayed.

Continuing with the example above, when we executed `git status` immediately after committing to the repository, git reported that there is no more file available with any change.

Next, let us make a simple change and look at how Git behaves. Let us modify the `default_settings.txt` file by adding the current date, as highlighted in Fig. 3.4, and then save the file.

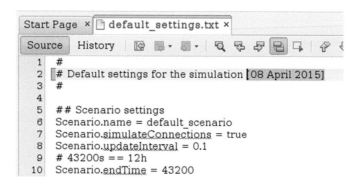

Fig. 3.4 Changing a file that is already under version control

```
$ git status
# On branch master
# Changes to be committed:
# (use"Git reset HEAD <file>..." to unstage)
#
# modified: default_settings.txt
#
$
$ git add default_settings.txt
$ git commit -m "Added current date to default_settings.txt"
$ git log
commit 09555e6bd4eb87bfc2a1c703c5572e75b8c07e10
Author: Barun Saha <barun.saha04@gmail.com>
Date: Wed Apr 8 00:35:15 2015 +0530

    Added current date to default_settings.txt

commit 6a42dbc9396d158311559c9782e6d4391a436673
Author: Barun Saha <barun.saha04@gmail.com>
Date: Wed Apr 8 00:21:45 2015 +0530

    Adding source files
```

As shown above in the output of a series of command executions, git recognized that changes have been made recently to the default_settings.txt file. However, those changes are not yet committed. Before we can commit, we added the modified file (default_settings.txt) using git add. Subsequently, we committed our changes. The output of the final git log command confirms that we have made two commits to the repository until now.

The operations shown in the previous examples, which were performed from within a terminal, can be executed entirely from NetBeans. It may be observed in Fig. 3.4, when a file is locally modified (but not yet committed), NetBeans highlights the file name with green color. Also, any addition of line(s) that has(have) been made to the file, is(are) highlighted with a light blue color.

Before concluding this section, a few words may be mentioned here for readers who choose to compile the source code and version control it from both within NetBeans and a terminal. After creating a Java project when one executes the project, NetBeans internally launches a build procedure. All the source code and dependency libraries are copied to the NetBeans project directory (which was specified during creating the project), compiled there, and a JAR file is generated. The output JAR file can be independently executed. The point to note here is that compilation is not performed in the project's source code directory. Thus, no `class` files are generated in that directory, which is unlike the case when one executes the `compile.bat` file to compile the source code. Therefore, in such a scenario, one might consider excluding the `*.class` files from Git.

3.4 Testing Protocol Development

Earlier, in this chapter, we briefly discussed about the debugging process with specific focus on NetBeans. It may be noted that the primary goal of debugging is not checking whether or not an application behaves correctly. Debugging is involved when someone knows that there is one more bug in the software, but does not know what/where they are. In other words, debugging is the process of identifying where or why a software fails. On the other hand, testing is the process that unravels whether or not a software (its intended behavior to be more precise) fails when subjected to different possible scenarios or environments.

One may ask—why should one bother about a *formal* testing procedure? More specifically, why should one write test cases? For example, if I am writing a program to add two numbers, can't I just provide two numbers as input, verify the output, and thus, conclude on the correctness of the program?

One may realize that even such simple manual testing has an underlying systematic approach—one needs to know what the input and corresponding output should be. Such a process of manually testing the code every time there is a change may be suitable for very small programs, but becomes tedious for a large software. In general, writing test cases for a formal testing procedure are helpful in many ways.

- Test cases help in automating the testing procedure, which, otherwise, would have to be done manually. Thus, when test cases are used, one would not require to run the simulations, generate reports, and verify whether desired results are obtained.
- Test cases, once written, can be used forever. Thus, whoever uses the code in whichever environment, successful execution of the test cases would give him/her the confidence that the code is working fine.
- Whenever any changes are made to the existing code, available test cases can be executed again to verify that he original behavior of the code is still retained. Of course, if the intention code change is to alter the original behavior, then the test cases should be modified accordingly.

Therefore, it is wise to invest some time to write test cases for the module developed—it might save a lot of time for repetitive tasks in the future.

3.4.1 An Overview of JUnit

While there are several variants of testing, in this chapter we would focus on unit testing. The goal of unit testing is to check whether the actual behavior of a piece of code matches with its intended behavior. The scope of testing is usually limited, for example, testing the methods of a class or a single class as a whole. JUnit is one of the popular unit testing frameworks used with Java.

3.4.1.1 An Example

Listing 3.9 shows an example of a simple test case implemented using the JUnit framework. Note that our test case class, `SimpleTest`, extends[10] the `TestCase` class of JUnit. The `SimpleTest` contains a member, `status`, of type boolean.

The test case overrides two methods from its parent class, `TestCase`. These two are special methods—the `setUp()` method is invoked *before* execution of all other method, whereas the `tearDown()` method is invoked *after* execution of all other methods. In testing terminology, they have a special name, *test fixture*. To quote from the JUnit Wiki[11]:

> A test fixture is a fixed state of a set of objects used as a baseline for running tests. The purpose of a test fixture is to ensure that there is a well known and fixed environment in which tests are run so that results are repeatable.

There are two other methods in this class—`testTrue()` and `testFalse()`—which actually performs the desired testing. These methods aim to create environments wherein certain conditions are checked for their validity. Note that each test method has a similar signature—`public void testXYZ()`. The test methods are `public`, have return type `void`, their names begin with `test`, and they do not take any argument. Any method not adhering to these specifications would not be considered as a test method. We would follow this convention of method naming (in contrast to annotations) in the remainder of this chapter.

Listing 3.9 A simple test scenario

```
package test.routing;

import junit.framework.TestCase;

public class SimpleTest extends TestCase {
```

---

[10]An alternative to this approach is to use the `@org.junit.Test` annotation for each test method.
[11]https://github.com/junit-team/junit/wiki/Test-fixtures.

```
 6
 7    private boolean status;
 8
 9    @Override
10    protected void setUp() throws Exception {
11        super.setUp();
12        status = true;
13    }
14
15    public void testTrue() {
16        assertTrue(status);
17    }
18
19    public void testFalse() {
20        assertFalse(status);
21    }
22
23    @Override
24    protected void tearDown() throws Exception {
25        super.tearDown();
26    }
27 }
```

3.4.1.2   Testing Primitives

JUnit provides only a few assert statements that are sufficient to test all possible conditions.

- **assertTrue** and **assertFalse**: These assertions allow us to check whether a given boolean expression is true or false, respectively. For example, `assertTrue (true)` evaluates to `true`.
- **assertEquals**: This assertion checks the equality of values. For example, `assert Equals(5, 6)` evaluates to `false`.
- **assertNull** and **assertNotNull**: These two assertions are used to check whether a given object reference is null or not, respectively.
- **assertSame** and **assertNotSame**: This pair of assertions check whether or not two variables refer to the same object.

The reader by now must have realized that in Listing 3.9, the `testTrue()` method would always evaluate to `true`, whereas the `testFalse()` method would always evaluate to `false`.

To execute a test, rebuild the project in NetBeans. Then, right click on the test class file, and click on `Test File` displayed in the context menu. The result of execution of Listing 3.9 is shown in Fig. 3.5a. As expected, our test failed.

Next, let us change the assertion inside the `testFalse()` method from `assert False(status)` to `assertTrue(status)`. On rebuilding the project and running the test again, both the test methods pass, as shown in Fig. 3.5b.

(a)

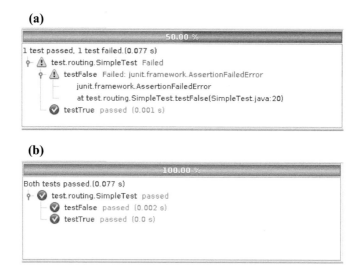

(b)

Fig. 3.5 Execution of a JUnit test case in NetBeans. A (**a**) failed, and a (**b**) successful scenario

Note that JUnit assumes that the test methods can be executed in any arbitrary order. Therefore, it is essential that each test method is written **independently** of the others. If, for example, there is a test method testB() that depends on the state changes made by another method testA(), then the method testB() should be rewritten to make it independent of testA(). Having said that, JUnit also provides a feature wherein the precedence order of the test methods can be specified.

> After a test case has been modified, it should be manually built by selecting the Build Project option from the Run menu in the menu bar of NetBeans. Otherwise, the changes made would not be reflected.

3.4.2 Testing with ONE

In this section, we would focus on testing routing protocols implemented in the ONE simulator. Note that the test/ directory already contains a collection of JUnit test cases to verify various functionality of the ONE simulator including the implementation Epidemic and PRoPHET routing protocols. Interested readers are suggested to look at those examples. Here, we would look at how to write a set of JUnit test cases from scratch to validate the working of the SnW routing protocol.

3.4.2.1 Important Classes

There are two important classes, namely `AbstractRouterTest` and `Message Checker`, that needs to be considered while writing test cases for routing protocols. As the name suggests, `AbstractRouterTest` is an abstract class,[12] which extends the `TestCase` class of the JUnit framework. Although it is not strictly required, it is recommended that while writing test cases for a router, one extends the `AbstractRouterTest`, which already provides some utility methods and instance variables. For example, in the `AbstractRouterTest`, there are seven `DTNHost` instance variables identified by h0 through h6. Therefore, any subclass of the `AbstractRouterTest` class can directly use these instance variables.

The `MessageChecker` class plays a crucial role. In fact, most of the testing would involve invoking methods from `MessageChecker`. Note that, the `AbstractRouterTest` also provides an instance variable of type `Message Checker`, which we would be using in this regard. One may realize that the routing process is equivalent to the transfer of (or lack thereof) a message from one node to another when certain conditions hold true. Therefore, assertions, coupled with information about a message (or its state) provided by the `MessageChecker` class, are sufficient to verify a routing algorithm.

The primary component of the `MessageChecker` class is a queue[13] of events of type `MessageCheckerEvent`. A `MessageCheckerEvent` event encapsulates the state of a message by capturing information such as, which is the concerned message, the *from* node, the *to* node, and so on. Whenever a `MessageChecker Event` event is removed from the queue, the `MessageChecker` object stores all the relevant information about that event. These are the different states against which assertions are to be checked.

A `MessageCheckerEvent` can be of one of the following types:

- TYPE_CREATE: This indicates that the last event was creation of a new message. In this case, the *from* and *to* nodes, respectively, indicate the source and destination of the message.
- TYPE_START: This type of event corresponds to the beginning of transfer of a message from one node to another.
- TYPE_RELAY: This indicates that a message transfer from one node to another was complete.
- TYPE_ABORT: If an ongoing message transfer is aborted, an event of this type is added to the event queue.
- TYPE_DELETE: This event is fired when a node deletes a message from its buffer.
- TYPE_NONE: This represents an invalid event type.

With these information, we are now ready to write a test case for the SnW routing protocol.

---

[12]An abstract class is one which cannot be instantiated, i.e., no objects of the type of the abstract class can be created.

[13]Technically, it is an `ArrayList` with FIFO operations.

It may be noted here that the JUnit test cases can be executed from the terminal. However, to execute any test case for the ONE simulator, all of them must be compiled at first, as shown below:

```
$ javac -extdirs lib/ -cp .:/home/barun/ONE/junit-4.1.jar test/↩
      *.java
```

Following a successful compilation, any test case contained inside the `test` package (say, `SprayAndWaitRouterTest`) can be executed in the following way. Note that the absolute (and correct) path of the location of the JUnit JAR file must be specified.

```
$ java -cp .:/path/to/junit-4.1.jar org.junit.runner.JUnitCore ↩
      test.SprayAndWaitRouterTest
```

We would, however, execute the test cases directly from within NetBeans. Note that a minor change[14] is required inside the `createHost()` method in `TestUtils.java` inside the `test` package. The code depicted in Listing 3.10 after line number 112 should be added to the original file.

Listing 3.10 Required change in TestUtils.java

```
112 NetworkInterface ni = new TestInterface(settings);
113 // Add the following line
114 ni.setGroupSettings(settings);
115 //
```

3.4.2.2   Testing the Spray and Wait Routing Protocol

Listing 3.11 shows the skeleton of the `SprayAndWaitRouterTest` class to which we would add test methods for testing the routing protocol. In general, we would add routing protocol specific settings and other relevant settings inside the `setUp()` method, which would be invoked before execution of all the test cases. The `tearDown()` method is optional. However, if the memory requirement by the test cases is very high, one can assign `null` to the objects so that JUnit can request garbage collection.[15] As a good practice, we would set the instance variables of the test class to `null` inside this method.

The `SprayAndWaitRouterTest` class has four instance variables—three of type `SprayAndWaitRouter` and the other of type `Message`. This corresponds to our plan to test the routing protocol by considering three nodes (hence three routers) and a single message. Note that instead of declaring them as instance variables, we could have declared them as local variables for different methods as well. However, we would be performing the following two tasks involving these variables repeatedly:

---

[14]This is required for version 1.5.1 RC2. The behavior with other versions was not tested.

[15]http://junit.org/faq.html#atests_18.

- Create an instance of `Message`, and
- Get the routers of the individual nodes (`DTNHost`).

As discussed earlier, it is, therefore, useful to move the above two operations inside the `setUp()` method.

Aside from the instance variables, the `SprayAndWaitRouterTest` class also has two class variables. The `NROF_COPIES` indicate the initial number of copies that a newly created message has, whereas `BINARY_MODE` indicates whether or not the `SprayAndWaitRouter` protocol works in binary mode. By default, we assume the binary mode to be true.

Listing 3.11 Skeleton of the SprayAndWaitRouterTest class

```
package test;

import core.Message;
import routing.MessageRouter;
import routing.SprayAndWaitRouter;

public class SprayAndWaitRouterTest extends AbstractRouterTest {

    private static int NROF_COPIES = 4;
    private static boolean BINARY_MODE = true;

    private SprayAndWaitRouter r0;
    private SprayAndWaitRouter r1;
    private SprayAndWaitRouter r2;
    private Message m1;

    private static final String MSG_COUNT_PROPERTY =
            SprayAndWaitRouter.MSG_COUNT_PROPERTY;

    /**
     * Set up test case fixture before execution of each test.
     *
     * @throws Exception
     */
    @Override
    public void setUp() throws Exception {
    }

    /**
     * Tear down test case fixture after each test case is ←
        executed.
     *
     * @throws Exception
     */
    @Override
    protected void tearDown() throws Exception {
    }
}
```

Before proceeding further, let us identify what all aspects we are going to validate by writing test cases. In general, what we would be testing? In the context of the SnW routing protocol, we would be testing the following functionality (requirements):

- **R1**: A node replicates a message to another node in contact when $L > 1$, where L is the number of copies of the message.
- **R2**: A node does not replicate a message to another node in contact when $L = 1$.
- **R3**: A node directly delivers a message if it is in contact with the destination node.
- **R4**: In the binary mode, both the nodes have $L/2$ copies of the replicated message each after the message has been replicated.
- **R5**: In the nonbinary mode, the replicating node and the receiving node have $L - 1$ and 1, respectively, after the message has been replicated.

Accordingly, we would be implementing the following three test methods:

- `testDirectDelivery()`: This method would validate the implementations of the requirement R3.
- `testRouting()`: This method would validate the implementations of the requirements R1, R2 and R4.
- `testNonBinaryReplication()`: This method would validate the implementations of the requirement R5.

Let us now look at the individual test methods.

At first, let us consider the `setUp()` method, as shown in Listing 3.12. Line numbers 8 through 15 initialize various settings for the `SprayAndWaitRouter` class. Next, a new `Message` is instantiated, but not yet created by any host. Finally, we retrieve the routers of the three hosts considered in our scenario.

Listing 3.12 The setUp() method

```
/**
 * Set up test case fixture before execution of each test.
 *
 * @throws Exception
 */
@Override
public void setUp() throws Exception {
    ts.setNameSpace(null);
    ts.putSetting(MessageRouter.B_SIZE_S, "" + BUFFER_SIZE);
    ts.putSetting(SprayAndWaitRouter.SPRAYANDWAIT_NS + "." +
            SprayAndWaitRouter.NROF_COPIES , NROF_COPIES + "");
    ts.putSetting(SprayAndWaitRouter.SPRAYANDWAIT_NS + "."
            + SprayAndWaitRouter.BINARY_MODE, "" + BINARY_MODE);
    setRouterProto(new SprayAndWaitRouter(ts));
    super.setUp();

    // This same message is used by every test case
    m1 = new Message(h0, h3, msgId1, 1);

    r0 = (SprayAndWaitRouter) h0.getRouter();
    r1 = (SprayAndWaitRouter) h1.getRouter();
    r2 = (SprayAndWaitRouter) h2.getRouter();
}
```

The `testDirectDelivery()` method is shown in Listing 3.13. The previously referenced message, `m1`, is created by the host `h0` in line number 6. Subsequently, we verify that precisely a single message have been created until now by invoking the `checkCreates()` method with an argument 1.

In line numbers 13–14, a connection is initiated between hosts `h0` and `h3`. Observe that here, we are not concerned about the underlying mobility models of the nodes. We are only concerned with whether or not a pair of nodes is in contact (and at what time). Subsequently, we advance the world by 1 s, i.e., the simulation clock advances by a second. The `advanceWorld()` method is shown in Listing 3.14.

Listing 3.13 A method to test direct delivery of messages with SnW

```
/**
 * Test direct delivery of a message between two nodes.
 */
public void testDirectDelivery() {
    // Host h0 creates the message m1
    h0.createNewMessage(m1);
    // Test that a single message has been created
    checkCreates(1);

    updateAllNodes();

    // Contact between h0 and h3 -- source & destinations of m1
    h0.forceConnection(h3, ←↩
        h0.getInterfaces().get(0).getInterfaceType(),
            true);

    advanceWorld(1);

    assertTrue(mc.next());
    // The last event was start of message transfer
    assertEquals(mc.getLastType(), mc.TYPE_START);
    // Verify the ID of the message
    assertEquals(mc.getLastMsg().getId(), msgId1);
    // Transfer was from host
    assertEquals(mc.getLastFrom(), h0);
    // Transfer was to host
    assertEquals(mc.getLastTo(), h3);

    // Clock advance is necessary to check the relay event
    advanceWorld(1);

    assertTrue(mc.next());
    // The message transfer has been completed
    assertEquals(mc.getLastType(), mc.TYPE_RELAY);
    assertEquals(mc.getLastFrom(), h0);
    assertEquals(mc.getLastTo(), h3);
    // This was the "first delivery" of m1 to h3
    assertTrue(mc.getLastFirstDelivery());
}
```

Line numbers 18 through 26 verify that a message transfer was initiated between the two nodes in contact. Here, `mc` is an instance variable from the `Abstract RouterTest` class. When the `mc.next()` method is invoked, an event from the events queue, if any, is removed. Similarly, line numbers 31 through 37 verify that the relay of the message to the host h3 was complete. Moreover, since h3 is the destination of the message m1, and h3 has not received m1 earlier, the aforementioned message relay corresponds to its "first delivery."

Listing 3.14 The advanceWorld() method

```
private void advanceWorld(int seconds) {
    clock.advance(seconds);
    updateAllNodes();
}
```

Next, Listing 3.15 shows the code for testing message replication with SnW when used in the binary mode. Recall that our test methods should be independent. So, the host h0 has to create the message m1 afresh. Note that any node other than h0 could have created the message as well. However, we plan to simulate a scenario, as described by line numbers 11 through 15 in the Listing, where host h0 acts as the source of the message m1. Also note that, unlike normal simulations, we do not require to create a bunch of messages while testing. Depending upon what we plan to test, a single message or just a few messages would suffice to test the logic of any routing algorithm.

Although Listing 3.15 might seem to be a lot of code, essentially we are doing only a four things:

1. In line numbers 24–50, hosts h0 and h1 come in contact, and the former replicates the message m1 to the latter. Since m1 initially had NROF_COPIES and the binary mode of replication was used, after the message relay is over, both h0 and h1 would have m1 in their respective buffers, but the message count property of m1 in each case would now be NROF_COPIES / 2.
2. Similarly, in line numbers 55–85, a replication of m1 takes place from h1 to h2. At the end of replication, h1 and h2 each would have NROF_COPIES / 4 copies of m1.
3. Lines numbers 89–100 show that hosts h1 and h4 come in contact. However, since the value of NROF_COPIES / 4 equals to 1, host h1 would not replicate m1 to h4.
4. In the final lines of codes, we verify that a direct delivery of m1 takes place from h1 to h3.

Listing 3.15 Testing message replication with SnW (in binary mode)

```
/**
 * Test the SprayAndWait routing implementation (binary mode).
 */
public void testRouting() {
    h0.createNewMessage(m1);
    checkCreates(1);
```

```
7
8      updateAllNodes();
9
10     /*
11      * Contact scenario:
12      * Time = t1, contact: h0 - h1 (h0 replicates m1 to h1)
13      * Time = t2, contact: h1 - h2 (h1 replicates to h2)
14      * Time = t3, contact: h1 - h4 (no replication)
15      * Time = t4, contact: h1 - h3 (h1 delivers m1 to h3)
16      */
17
18     // Initially h1 has NROF_COPIES copies of m1
19     assertEquals(m1.getProperty(MSG_COUNT_PROPERTY),
20             NROF_COPIES);
21
22     // t1
23     // Contact between h0 and h1
24     h0.forceConnection(h1, ↩
             h0.getInterfaces().get(0).getInterfaceType(),
25             true);
26     advanceWorld(1);
27
28     // Message replication from h0 to h1
29     assertTrue(mc.next());
30     assertEquals(mc.getLastType(), mc.TYPE_START);
31
32     advanceWorld(1);
33
34     assertTrue(mc.next());
35     assertEquals(mc.getLastType(), mc.TYPE_RELAY);
36     assertEquals(mc.getLastFrom(), h0);
37     assertEquals(mc.getLastTo(), h1);
38
39     // Contact between h0 and h1 terminates
40     h0.forceConnection(h1, ↩
             h0.getInterfaces().get(0).getInterfaceType(),
41             false);
42     advanceWorld(1);
43
44     // Now both h0 and h1 each have a single copy of m1
45     Message m2 = (Message) ↩
             r0.getMessageCollection().toArray()[0];
46     assertEquals(m2.getProperty(MSG_COUNT_PROPERTY),
47             NROF_COPIES / 2);
48     m2 = (Message) r1.getMessageCollection().toArray()[0];
49     assertEquals(m2.getProperty(MSG_COUNT_PROPERTY),
50             NROF_COPIES / 2);
51
52
53     // t2
54     // Contact between h1 and h2
55     h1.forceConnection(h2, ↩
             h1.getInterfaces().get(0).getInterfaceType(),
56             true);
57     advanceWorld(1);
```

```
58
59    // Message replication from h0 to h1
60    assertTrue(mc.next());
61    assertEquals(mc.getLastType(), mc.TYPE_START);
62
63    advanceWorld(1);
64
65    assertTrue(mc.next());
66    assertEquals(mc.getLastType(), mc.TYPE_RELAY);
67    assertEquals(mc.getLastFrom(), h1);
68    assertEquals(mc.getLastTo(), h2);
69
70    // Contact between h0 and h1 terminates
71    h1.forceConnection(h2, ↵
          h1.getInterfaces().get(0).getInterfaceType(),
72            false);
73    advanceWorld(1);
74
75    // Now both h0 and h1 each have a single copy of m1
76    m2 = (Message) r1.getMessageCollection().toArray()[0];
77    assertEquals(m2.getProperty(MSG_COUNT_PROPERTY),
78            NROF_COPIES / 4);
79    m2 = (Message) r2.getMessageCollection().toArray()[0];
80    assertEquals(m2.getProperty(MSG_COUNT_PROPERTY),
81            NROF_COPIES / 4);
82
83    h1.forceConnection(h2, ↵
          h1.getInterfaces().get(0).getInterfaceType(),
84            false);
85    advanceWorld(1);
86
87
88    // t3
89    // Contact between h1 and h4
90    h1.forceConnection(h4, ↵
          h1.getInterfaces().get(0).getInterfaceType(),
91            true);
92    advanceWorld(1);
93
94    // There is no message transfer event since h1 has only
95    // a single copy of m1
96    assertFalse(mc.next());
97
98    h1.forceConnection(h4, ↵
          h1.getInterfaces().get(0).getInterfaceType(),
99            false);
100   advanceWorld(1);
101
102
103   // t4
104   // Contact between h1 and h3
105   h1.forceConnection(h3, ↵
          h3.getInterfaces().get(0).getInterfaceType(),
106            true);
107   advanceWorld(1);
```

```
108
109     // Message replication from h1 to h3
110     assertTrue(mc.next());
111     assertEquals(mc.getLastType(), mc.TYPE_START);
112
113     advanceWorld(1);
114
115     assertTrue(mc.next());
116     assertEquals(mc.getLastType(), mc.TYPE_RELAY);
117     assertEquals(mc.getLastFrom(), h1);
118     assertEquals(mc.getLastTo(), h3);
119     assertTrue(mc.getLastFirstDelivery());
120
121     // Contact between h0 and h1 terminates
122     h1.forceConnection(h3, ↵
            h1.getInterfaces().get(0).getInterfaceType(),
123         false);
124     advanceWorld(1);
125 }
```

Listing 3.16 shows a similar testing of the replication scenario when SnW is used in nonbinary mode. A major difference with our previous Listing is shown here from line numbers 7 through 14. Recall that, in the `setUp()` method, we had specified SnW to work in binary mode. Therefore, to unset the binary mode, we reinitialize the settings of SnW once again inside the `testNonBinaryReplication()` method.

Listing 3.16 Testing message replication with SnW (in nonbinary mode)

```
1  /**
2   * Test the replication in nonbinary mode.
3   */
4  public void testNonBinaryReplication() throws Exception {
5      // Since the implementation of setUp() above defaults to ↵
           binary mode,
6      // we need to assign those settings once again here
7      ts.setNameSpace(null);
8      ts.putSetting(MessageRouter.B_SIZE_S,"" + BUFFER_SIZE);
9      ts.putSetting(SprayAndWaitRouter.SPRAYANDWAIT_NS + "." +
10          SprayAndWaitRouter.NROF_COPIES , NROF_COPIES + "");
11     ts.putSetting(SprayAndWaitRouter.SPRAYANDWAIT_NS + "."
12          + SprayAndWaitRouter.BINARY_MODE, "false");
13     setRouterProto(new SprayAndWaitRouter(ts));
14     super.setUp();
15
16     h0.createNewMessage(m1);
17     checkCreates(1);
18
19     assertEquals(m1.getProperty(MSG_COUNT_PROPERTY),
20          NROF_COPIES);
21
22     // Contact between h0 and h1
23     h0.forceConnection(h1, ↵
           h0.getInterfaces().get(0).getInterfaceType(),
```

```
24          true);
25      advanceWorld(1);
26
27      // Message replication from h0 to h1
28      assertTrue(mc.next());
29      assertEquals(mc.getLastType(), mc.TYPE_START);
30
31      advanceWorld(1);
32
33      assertTrue(mc.next());
34      assertEquals(mc.getLastType(), mc.TYPE_RELAY);
35      assertEquals(mc.getLastFrom(), h0);
36      assertEquals(mc.getLastTo(), h1);
37
38      // Contact between h0 and h1 terminates
39      h0.forceConnection(h1, ↩
            h0.getInterfaces().get(0).getInterfaceType(),
40              false);
41      advanceWorld(1);
42
43      // Now h0 has three copies of m1; h1 has one copy
44      r0 = (SprayAndWaitRouter) h0.getRouter();
45      r1 = (SprayAndWaitRouter) h1.getRouter();
46
47      Message m2 = (Message) ↩
            r0.getMessageCollection().toArray()[0];
48      assertEquals(m2.getProperty(MSG_COUNT_PROPERTY),
49              NROF_COPIES - 1);
50      m2 = (Message) r1.getMessageCollection().toArray()[0];
51      assertEquals(m2.getProperty(MSG_COUNT_PROPERTY), 1);
52  }
```

Finally, Listing 3.17 shows the tearDown() method, where we set the object references to null. Note that, in this specific test case, this method is optional as we consume only a meager amount of memory.

Listing 3.17 The tearDown() method

```
1  /**
2   * Tear down test case fixture after each test case is executed.
3   *
4   * @throws Exception
5   */
6  @Override
7  protected void tearDown() throws Exception {
8      super.tearDown();
9
10     // http://junit.org/faq.html#atests_18
11     // In case you have memory limitations, free up the resources
12     m1 = null;
13     r0 = null;
14     r1 = null;
15     r2 = null;
16 }
```

With this, we wrap up our discussion on writing test cases to test protocols developed with the ONE simulator. Here, we discussed how to test the SnW routing protocol. The reader is encouraged to go through the code Listings in this section, and convince himself/herself that the various assertions, indeed, hold true. The ONE simulator comes with a lot of test cases for verifying various other features (such as mobility and connections). An interested reader should consult them as well.

3.5 Best Practices

We conclude this chapter with a set of guidelines/tips that one might keep in mind while working with the ONE simulator. These are not claimed as universal rules; one should employ them if found appropriate.

- Wherever possible, try to use primitive types [105] instead of objects. For example, although an `ArrayList` is flexible, an array is cheaper in terms of memory consumption.
- When implementing a new routing protocol, try to find out if the proposed protocol can be achieved by specializing some behavior(s) of one of the existing routing protocols. If it seems feasible, create the new routing protocol as a subclass of the existing ones, and override the relevant method(s). If such inheritance seems infeasible, create the new routing as a subclass of the `ActiveRouter` class.
- If the proposed protocol has multiple settings, some of which can be enabled while others are disabled during a simulation, consider using Boolean flags. Relevant code can be wrapped around by the respective flags.
- When several variations of a single protocol are considered, it might be useful to create a base class (subclass of the `ActiveRouter` class) and then implement each variation as a subclass of the base class.
- When same data needs to be collected repeatedly, it is perhaps better to write a `Report` to output the required data. Moreover, if such data is to be used for plotting, it would be wise to format the output data in a way that is understood by the plotting software used. This would help in eliminating any intermediate preprocessing step.
- Often it might be required to print some aggregate statistics at the end of a simulation. One may write a report by implementing the `UpdateListener` interface. The final moment of simulation can be obtained as shown in Listing 18. Here, it is assumed that the update interval is 1 s so that the current time is checked at every second.

Listing 3.18 The updated() method from UpdateListener interface

```
public void updated(DTNHost hosts) {
    int curTime = SimClock.getIntTime();
    if (curTime % simulationDuration == 0) {
        // This is executed at the last second of simulation
    }
}
```

- Moreover, it might be required to collect some information from all the nodes in the simulation, and then print an aggregate statistic. Note that this could not be achieved if we attempt to print the relevant information for each node. An alternative approach is suggested. Since any simulation would have at least a single node, the node address 0 is valid. Thus, check if the address of the current node is 0, and if so, gather (and aggregate) statistics from all other nodes. The list of nodes can be obtained from the argument to the `updated()` method discussed above.
- Linux users might create a directory say, `my_scenarios`, and store all the simulation scenario files there. Subsequently, create a symlink by the name `default_settings.txt`, and make it point to the desired scenario file. This eliminates the need to overwrite the `default_settings.txt` file every time a different simulation is to be executed. Also, there is no need to specify the name(s) of any extra settings files.

3.6 Summary

This chapter presented a developer's toolbox for working with the ONE simulator. Experienced developers would realize that writing code is but a single step of the development process. When the scope of a project grows large, one should definitely consider putting the code base under version control even if a single person is working on it. At the same time, use of an IDE for development increases productivity by providing various features like automatic code generation and error highlighting. However, a general word of caution may be sounded here. Use of an IDE is no substitute of learning a language or system, for example, the ONE simulator. One should have sufficient understanding of the simulator before he/she starts working with an IDE.

We also looked at some basic operations using Git, such as initializing a repository and committing to it. Modern IDEs are integrated with different version control systems so that changes can be tracked (and added) from within the IDE itself. Additionally, several third-party tools are available for GUI-based visualization of the changes in a source code repository.

We also had a detailed look at the process of testing a protocol developed using the ONE simulator. The `AbstractRouterTest` and `MessageChecker` are the two primary classes to be considered while testing a routing protocol. Finally, we provided a set of best practices for working with the ONE simulator. These tips were assimilated based on our experience with the simulator. A reader might find them useful in the process of development.

3.7 Review Terms

- Breakpoints
- Conditional breakpoints
- Version control
- Unit test
- Test fixture
- JUnit
- Step over
- Step into
- Commit
- Test case
- Assertion

3.8 Exercises

3.1 In Listing 3.13, why a call to the `advanceWorld(1)` method is required in line number 29? What would happen if that call is omitted? What would happen if the time argument is changed to 10?

3.2 Explain the difference between *step over* and *step into* in the process of debugging code.

3.3 In the domain of version control, *branching* is an quintessential process. A branch represents a parallel line of development. Write the command to create a branch using Git. How would you switch to the newly created branch so that changes are made to the new branch and not to the original branch? Note that in Git, the default branch is called *master*.

3.4 Why is unit testing a better approach than performing actual network simulations?

3.5 Consider that you are writing unit test cases using JUnit for a set of routing protocols. Explain why would you have your test cases extend the `AbstractRouter Test` class rather than using JUnit annotations?

3.9 Programming Exercises

3.6 Simulate and compare the performances of SnW and SnWU using real-life traces and synthetic mobility models.

3.7 In Listing 3.6, what would happen if the `getMessagesWithCopies Left()` method returned `null`?

3.8 Modify the `getUtility()` method in the `SprayAndWaitUtility Router` class so that it degenerates to SnW. What else would you need to modify?

3.9 Redefine utility for a node as the product of number of contacts and logarithm of the total contact duration with that node. Update the `SprayAndWaitUtility Router` class accordingly.

3.10 A test suite is a collection of test cases. Create a test suite using JUnit.

3.11 In PRoPHET, a host replicates a message to another node if the delivery predictability of the latter host for the destination of the message is higher than the former node. Write an assertion to verify this.

3.12 Write unit test cases to verify the functionality of the `SprayAndWait UtilityRouter` class.

Part II
Human Aspects in Opportunistic Mobile Networks

Chapter 4
Emerging Sensing Paradigms and Intelligence in Networks

In Chap. 1, we noted that the last two decades have witnessed efforts to take network communication capacities beyond the earth, which resulted in the proposal of IPNs [7]. IPN was the precursor of DTN. Since then, variants of DTNs such as, PSNs and OMNs, have evolved. Concurrently, the field of Wireless Sensor Networks (WSNs) has also made interesting progress giving rise to different paradigms, such as human-centric sensing. Now, whether we talk about OMNs or human-centric sensing, some aspects are common to both—human users, their mobility, and actions. Seen in a different way, the notion that a communication network is made up of merely some devices is perhaps becoming blurred.

The recent advances in WSNs and related fields have a different implication, too. These domains are no more exclusive-only-for-scientists. Rather, interesting applications are emerging enabling end users to take an active part in such operations. For example, a modern smartphone, which comes equipped with a lot many sensors, can help track the total distance that one walks daily, and can provide health recommendations based on that. In this chapter, we take a look at such emerging areas of WSNs along with mission aspects in WSNs. Subsequently, we devote a larger focus on studying some of the popular and diverse applications of WSNs in the modern world. Interestingly, whenever we talk about ad hoc networks or WSNs, a fertile ground for applications is related to disaster scenarios. Thus, we look in detail at mobility models and communication aspects pertaining to pre- and post-disaster scenarios. Subsequently, we review the applications of agent-based systems in such scenarios.

The latter portion of this chapter deals with the notion of intelligence with specific focus on situation awareness. Next, we formally define a MOON with a set of different parameters. Finally, we conclude this chapter by studying the effects that intuitive intelligence-based decision-making on the mission prospects in MOONs.

© Springer International Publishing Switzerland 2016
S. Misra et al., *Opportunistic Mobile Networks*,
Computer Communications and Networks, DOI 10.1007/978-3-319-29031-7_4

4.1   Emerging Paradigms of Sensor Networks

WSNs have emerged as a popular research field, and find its applications across various domains, for example, security, geology, and health care. The rapid miniaturization and portability of the devices have given rise to the scenario, where the sensors are wore or carried by human beings. This has added a new dimension to the research in WSNs. In this section, we take a brief look at the human-centric sensing paradigm. Moreover, we look into how mission aspects have been considered together with the WSNs.

4.1.1   Human-Centric Sensing

Modern generation portable devices (for example, smartphones and PDAs), which are often equipped with one or more sensors (for example, accelerometer and gyroscope), are increasingly becoming part of the everyday life. The proliferation of such mobile devices in the modern society has opened up several new possibilities and research opportunities. In the recent years, the research space of wireless sensor networks (WSNs) has forked into different directions giving rise to new areas, variously identified as participatory sensing [106], people-centric sensing [107, 108], and opportunistic sensing [107].

Burke et al. [106] proposed participatory sensing, where the mobile devices carried by human beings act as sensor nodes, and engage in sensing their local environment in a distributed fashion. The authors noted that the goal of participatory sensing is not only to collect data, but also to allow common persons and professionals alike to analyze data and consequently, share knowledge. Data collected by the mobile devices not only provide quantitative values—for example, an image indicating the number of vehicles at an intersection of a road at the peak hour—but can also endorse the authenticity with the geotagged location and timestamp of the image.

Campbell et al. [109] presented the architecture of MetroSense, a scalable platform for people-centric *urban sensing*—data about people and their immediate surroundings in urban landscapes. MetroSense leverages *opportunistic sensing*, which arises due to uncontrolled mobility of the human beings in the sensor field. Campbell et al. [107] noted that in people-centric sensing, people act as the primary component of the sensing system so formed. MetroSense has three operational stages:

1. Sense: The sensor devices engage in sensing their immediate surrounding. This can be done by stationary sensors or the mobile sensing devices carried by people.
2. Learn: The sensed raw samples are statistically analyzed—often using machine-learning algorithms—to extract a higher level interpretation. Moreover, social ties among the users are also possibly considered to improve the performance.
3. Share: Enable visualization and/or sharing of the information so gathered.

The authors further distinguish between two types of sensing—participatory and opportunistic—based on the users' role. In participatory sensing, a user is more involved with the sensing process, for example, decision-making on application requests, data sharing, and privacy issues. In contrast, under the opportunistic sensing paradigm, most of the decision-making process is delegated to the system with predefined constraints, for example, privacy settings.

Srivastava et al.'s [108] work followed along the similar direction, where the authors discussed several emerging sensing scenarios. In particular, they observed that in human-centric sensing, people play three distinct roles, although not necessarily mutually exclusive:

1. Sensing targets: Human beings themselves act as the sensing targets. This finds application beyond the security domain, for example, personal health monitoring.
2. Sensor operators: People use the sensors present in their smartphones and other portable devices—together with relevant applications—to sense the environment.
3. Data sources: In many cases (for example, social networking sites), human beings disseminate and collect data about themselves without using any actual sensor.

Moreover, the authors pointed out several challenges in context of the human-centric sensing ranging from energy aspects to participant selection and privacy concerns.

It may be noted that in the above kind of networks, there has been a subtle shift of focus with increased highlight on people. While conventional research on computer networks typically considered human beings as the end users, in the modern scenario, they have increased interaction with the network, and a greater role to play.

4.1.2 Mission-Oriented Sensor Networks

Advances in sensing and communication technologies have made WSNs desirable and useful to a larger audience. WSNs have been deployed to serve a diverse class of interests, both military and civilian, including (but not limited to) surveillance, environmental or habitat monitoring, intrusion detection, and target tracking. WSNs, in general, however, suffer from different problems (for example, energy constraints and coverage), which may affect the goal of mission-oriented networks that use WSNs.

Mission-oriented WSNs [16], unlike other networks, require that the sensors involved in the network coordinate with one another and align themselves altogether with the mission objectives. Such inter-device coordination becomes essential to achieve complex missions (for example, shooting a hostile target without causing any collateral damage). Rao and Kesidis [110] considered a mobile ad hoc sensor network together with the "purposeful mobility" of the nodes to achieve mission objectives. The nodes in the network had multiple roles—tracking, scanning, and data forwarding. The authors considered that the nodes executed a distributed simulated annealing algorithm, which helped them to position themselves in a fashion so as to minimize the networkwide transmission cost. The proposed distributed algorithm was able to achieve global optimization using only locally available knowledge at the nodes.

Eswaran et al. [111] explored the case of marginal utility-based rate adaptation in mission-oriented WSNs and extended the well-known network utility maximization (NUM) model for the purpose. Although several works have addressed different scenarios using the NUM framework, the authors enumerated three key features of mission-oriented WSNs that have been previously unexplored. Often, a single mission's execution is dependent on multiple sensors, which calls for the definition of a joint utility function considering the data received from all such sources. The missions—the receivers eventually consuming the sensed data—can be heterogeneous. While multiple receivers can feed on data provides by a single sensor, they can have different individual utilities defined. Moreover, when multiple missions are considered—many of which may arrive in quick succession—the rate adaptation algorithm must have fast convergence to promptly respond to the changes.

Liu et al. [112] focused on the quality of information (QoI) requirements for mission-oriented WSNs. The authors presented a QoI-centric architecture targeted for military applications with critical decision-making. The proposed architecture can provide an optimal balance among multiple objectives—QoI aspects of the missions, controlling the admission of new missions, and the capacity of the deployed WSN.

Pignaton de Freitas et al. [113] discussed the design of a middleware for mission-driven heterogeneous sensor networks. The authors represented a global mission with a quadruple $M_G = \langle M_N, N_S, F, Q \rangle$, where M is the set of all possible node missions, $M_N \subseteq M$, N_S is a subset of all the nodes in the network. The function F maps a given node mission to a certain node, and the function Q evaluates the said mapping. A user is allowed to specify the requirements (mission) in a high-level Mission Description Language (MDL), which are then translated into a global mission. The Mission Interpreter component reads the MDL statements, and breaks down into node missions, i.e., submissions to be executed by the individual nodes.

On the other hand, the problem of coverage in WSNs has been heavily explored by different researchers. Ammari and Das [114] considered K-coverage in mission-oriented mobile WSNs. The term K-coverage in a WSN implies that any point in the field is covered (sensed) by at least K nodes, while K-connectivity implies that the K nodes covering the entire field form a connected network. The authors in [114] proposed strategies for placement of sensors and movement of nodes toward the region of interest in an energy-efficient manner.

4.2 Disaster Scenarios and Their Aftermath

A disaster of large scale often renders the existing network communication infrastructure unusable due to disruptions, congestion, and blackouts. Thus, post-disaster scenarios provide a fertile ground for the use of OMNs/DTNs. In this chapter, we look at some of the mobility models that have been proposed to simulate post-disaster scenarios. We also look at the delay-tolerant routing protocols proposed for such environments. In this section, we at first review some recent innovations in sensor networks for environmental and disaster monitoring. Subsequently, we look at some of the mobility models proposed in the literature to represent the movements of concerned parties during and after disasters. Finally, we look at the communication aspects—involving traditional social media and opportunistic networks—in such scenarios.

4.2.1 Sensor Networks for Environmental and Disaster Monitoring

Human beings have been active in monitoring their surrounding environment since a long time. For example, devices like barometer and thermometer are used to measure air pressure and temperature, respectively. Many of us still wait to watch the weather report in TV or read the weather forecast in newspapers. Essentially, these all have been pre-sensor network technologies. However, the rapid growth of sensor networks has immensely changed the method and scope of how our environment is monitored.

Figure 4.1 shows a typical WSN with hierarchical structure. In particular, the sensor nodes (for example, seismic and pressure sensors) are divided into multiple clusters. Each such cluster has a cluster head, which could be one of those sensors or some special device. Whenever the individual sensors have some data to send, they do send it to their respective cluster heads, which, in turn, transmits to the base station. Such a paradigm of clustering and multihop transmission can be effective to improve the lifetime of a WSN in contrast to the direct transmission scenarios.

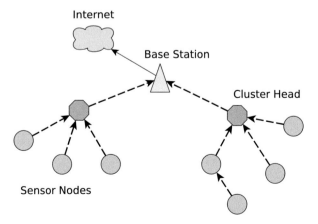

Fig. 4.1 A typical hierarchical WSN consisting of sensor nodes, cluster heads, and a base station

However, whether or not to choose such an architecture should be decided by the experts based on the concerned target application and environment.

The base station in a WSN is typically connected to the Internet with a wired link, whereas communication until the base station is via wireless links. The latter communication is usually via electromagnetic waves when the deployment is terrestrial or airborne. However, in case of underwater sensor networks, the signal strength of electromagnetic waves fade even over a short distance. Therefore, acoustic modems are typically used in such scenarios although the data rate is comparatively low. Moreover, the communication from a sensor node to the base station may not necessarily be via end-to-end paths. These two latter scenarios give rise to delay-tolerant WSNs.

Perhaps one of the pioneering works in the environmental sensing domain was the proposal of using sensor networks for remote and automated monitoring of soil water [115]. More recently, Kim et al. [116] made use of several technologies for developing a distributed WSN to remotely monitor and control an irrigation system. Nevertheless, such usage of sensor networks have not remained limited to small areas, and have expanded to provide a greater coverage. In the following, we review some applications of WSNs for environmental monitoring.

4.2.1.1 Environmental Sensor Networks (ESNs)

ESNs [117]—network of arrays of sensor nodes fitted with radio—are a promising enterprise for environmental monitoring specifically for remote, inaccessible, and dangerous areas. Martinez et al. [118] discussed a generic architecture for an ESN consisting of sensor nodes, base stations, and a sensor network server. The latter can be integrated with the Web and can use various web services so that the users can have easy access to the data collected. The authors observed that while sensing, computing and communication technologies are essential for all kind of sensor networks in general, environmental sensor networks additionally require domain knowledge in order to ensure proper design, development, deployment, and sustainability.

The sensor nodes in such networks can be either static or mobile depending on the scenario [117]. Moreover, the size of the nodes can vary as per the target application. For example, sensor nodes used in a weather station can be up to 0.1 m, whereas the smart dust[1] sensing nodes have dimensions of about 0.001 m. The ESNs can span globally or have coverage limited to certain regions and localities. For example, the Global Seismographic Network[2] uses seismometer accelerometer nodes of dimensions about 50×50 sq. cm [117]. 136 such stations are spread across 59 countries giving a global coverage to the network.

Based on the scale and function, Hart and Martinez [117] classified ESNs into several categories. The first category, *Large Scale Single Function Networks*, span over a large geographic extent and are dedicated for a single purpose. The nodes of

---

[1]http://robotics.eecs.berkeley.edu/~pister/SmartDust/.
[2]http://www.iris.edu/.

this type of networks are usually large and costly, and they measure one or more variables. The Global Seismographic Network mentioned above belongs to this category. This type of networks generally do not suffer from resource constraints—they are connected with wired links and are powered by the main supply. As such, a simple star network topology is often the choice.

The second category of ESNs is the *Localized Multifunction Sensor Networks*. These are generally wireless ad hoc networks formed of relatively smaller sensor nodes whose objectives are to measure some simple physical parameters. The geographical coverage of these networks is often limited to smaller regions say, about 20 km. The volcano monitoring sensor network [119]—discussed later in this subsection—is an example of this category of ESNs. The third category, *Biosensor Networks*, are ESNs formed of biological sensing nodes. These type of sensors can be used to measure biological, chemical, and physiological parameters of the environment. Finally, the authors envision *Heterogeneous Sensor Networks* [117], where essentially the previously described three categories merge toward a larger and comprehensive scale of environment monitoring.

4.2.1.2 SensorWeb

NASA's SensorWeb (http://sensorweb.nasa.gov/) [120] integrates variously deployed sensors (for example, satellite, terrestrial, and airborne) with the Web for remote and automated monitoring of several environmental events, such as volcanic activities, fires, and floods. Such an integration also helps in scientific investigation by choosing the required sensors and accessing the resulting data over the Internet. The project essentially provides a simple, cost-effective, and efficient platform to access the international data related to environmental phenomena and disasters.

At the lowest level of the SensorWeb architecture are the individual sensors, which are connected to the Internet via a set of web service wrappers. In particular, the sensors and data processing algorithms are wrapped with Open Geospatial Consortium (OGC) standards. It also consists of a cross-domain scripting language for orchestrating the web services. The campaign manager component helps to trigger the sensors manually among others. The component also provides an application processing interface (API) for remote control. The system uses Open ID for authenticating users and scripts over the Web. Sample pilot projects of the SensorWeb include the Namibian Early Flood Warning SensorWeb[3]—a flood management decision support system providing flood and disease forecasting tools, and Volcano SensorWeb[4] for surveillance and monitoring the 50 most active volcanoes on the earth [121].

---

[3]http://sensorweb.nasa.gov/Namibia%20Flood%20Early%20Warning%20SensorWeb%20Pilot%20as%20of%20Nov%2010%202010%20v7.pdf.

[4]http://ai.jpl.nasa.gov/public/projects/sensorweb/.

4.2.1.3 Volcano Monitoring

Volcanic eruptions emit multiple signals useful for their detection, such as seismic waves along the ground and acoustic waves into the atmosphere nearby the volcano [119]. The combination of these two signals is found to be efficient in monitoring volcanic activities and interpreting related trends [122]. Werner-Allen et al. [119, 123] discussed their experiences in monitoring volcanic activities of an active volcano in Ecuador using a WSN consisting of acoustic sensors. The authors noted that monitoring a volcano requires that data be collected much more frequently than other environmental monitoring activities. Moreover, when acoustic sensors (such as microphones) are used for monitoring, care must be taken to differentiate actual volcanic activities from noise and other related signals, for example, sound produced due to mining activities. In general, the following typical characteristics can be associated with WSNs used for monitoring volcanic activities [123]:

- **High data rate**: Bandwidth guarantee is a critical issue in these types of WSNs. Werner-Allen et al. [123] observed that while sensor nodes equipped with IEEE 802.15.4 radios offer data rates up to 30 KBps, in practice, achievable data rate tends to less than 10 KBps considering various overheads, such as packet framing and MAC controls. On the other hand, the sensor nodes have a very limited internal storage capacity and the buffer space can get depleted soon enough depending on the type of volcanic activity.
- **High data fidelity**: High fidelity of data collected by the nodes and their accurate time stamping is required. Any missed out or wrong or incorrectly time stamped sample can lead to invalidation of the records or incorrect erroneous analysis of the data.
- **High spatial separation**: Considerable spatial separation among the sensor nodes is required to record infrasonic and seismic signals propagating over different regions. As a consequence, the concerned WSN has a sparse topology in contrast to typically considered dense deployments. Therefore, node failure is a critical issue in such networks since data from a larger region of the network could not be obtained.

The system considered in [119] is comprised of a set of wireless infrasonic monitoring sensor nodes that can record sound waves having frequency up to 50 Hz. An aggregator node receives such sampled signals and, in turn, transmits them to the base station over long-distance wireless links. The base station has various software available for storage, visualization, and real-time analysis of the received data. Time synchronization across the WSN is performed using a GPS receiver node. Analyzing the data obtained from their real-life deployment, the authors observed varying degree of packet loss rate for the sensor nodes, which was accounted to the possible variations in weather and temperature. A good degree of correlation was found between the data obtained through wireless infrasonic sensors and colocated wired seismic sensor nodes.

4.2.1.4 Disaster Response

Ramesh et al. [124] developed and deployed a hierarchical WSN consisting of geo-physical sensors for detecting landslides. Devices such as, pore pressure transducer and dielectric moisture sensor were used to measure pore pressure and moisture content of the soil. The region of interest was divided into three subregions where the concerned sensors were placed. Energy-efficient algorithms for real-time monitoring were used to enhance the network life time. Such systems are suitable for issuing warnings before a potential disaster actually takes place. The usability can be further extended by integrating the system with existing communication systems such as, cellular networks and television broadcasts. One can consider geocasting in this context for efficient dissemination of the warning.

Quaritsch et al. [125] developed *collaborative microdrones* for different scenarios, such as environmental monitoring, surveillance and law enforcement, and disaster management. Collaborative microdrones are a collection of multiple autonomous, small-sized aerial vehicles that fly in a formation and cooperative with one another to accomplish given mission objectives. Such a cooperative network of airborne sensors has multiple advantages over a single microdrone, such as larger scale, coverage and robustness of the collected information. The system may contain either image or video sensors, or perhaps both, if required, and can relay captured data in real time.

Quaritsch et al. [126] considered the use of collaborative microdrones for active monitoring of the region of interest for disaster management applications. The bird's eye view of the landscape provided by unmanned aerial vehicles (UAVs) is helpful in obtaining a holistic view as well as deciding upon the scale of disaster and subsequent planning. Aerial sensor networks, however, face additional challenges in comparison to the traditional sensor networks. In particular, a major fraction of the input power is consumed by the UAVs' propulsion system. Such aspects put constraints upon the UAVs flight time. Moreover, winds cause hindrance to their motion. Therefore, it is essential to plan an energy-efficient trajectory of the UAVs for optimum performance while maintaining necessary quality of different parameters. On the other hand, collaborative microdrones flying in a formation require that the individual UAVs precisely know the location and other relevant vectors of their counterparts. Therefore, the microdrones should be equipped with communication systems that have low latency and communication range of a few hundred meters.

In particular, Quaritsch et al. [126] presented the integer linear programming (ILP)-based problem of obtaining high-quality (for example, in terms of coverage) overview images while optimizing the quality of the resulting images and resource (for example, energy) consumption. The outcome of the optimization problem decided the precise locations where the UAVs should take pictures as well as the size and orientation of the corresponding images. The authors used a genetic algorithm to define an almost optimal trajectory based on the previously computed locations. The result in terms of coverage of the observation and forbidden area was found to be better in this case as compared to the naive approach. However, there were some overheads involved in terms of speed of computation.

Castillo-Effer et al. [127] considered the use of WSNs for alerting flash-flood incidents. The authors noted that such systems have soft real-time constraints and should be operational almost all the time. This is unlike other systems, for example, flood and storm alerts, which typically do not require to be in real time. The requirements of such a system for alerting occurrence of flash floods include a sensor field deployed to the region of interest, communication links to transmit sensed data, a storage system for storing such data acquired, and related software to analyze the data and generate alerts, if necessary. Therefore, on a high level, the system is comprised of several components, such as monitoring subsystem, communications subsystem, data collection and analysis system, and alerting subsystem.

WSNs for alerting flash-flood incidents, unlike many other scenarios, require stationary sensor nodes deployed along the river banks and mountains [127]. The area of node deployment, however, is typically uninhabited by human beings. As such, it is required that deployed nodes have long operational lifetime. On the other hand, the WSN should be scalable to deploy more nodes and/or cover larger region. Finally, the routing protocol should be able to accommodate a scalable network. Therefore, energy efficiency, fault tolerance, and scalability are some of the desired characteristics of such sensor networks.

The proposed flash-flood alerting system in [127] is comprised of multiple types of sensor nodes. Hydrological sensor nodes are placed along the river banks to monitor the flow and level of river water. The meteorological sensor nodes deployed in the near vicinity of the river monitor several atmospheric parameters including, temperature, humidity, pressure, and wind speed. Finally, the landslide nodes—containing geophone, soil moisture, and creep sensors—are deployed in the nearby mountains.

4.2.1.5 Practical Implications

Thus far we looked at different applications and usage of wireless sensor networks in environmental monitoring and disaster response. The underlying network architecture and constraints corresponding to different scenarios were also pointed out. Indeed, there are several other applications that, unfortunately, cannot be accommodated within the scope of this text. Nevertheless, certain patterns emerge from this discussion. In general, based on the above review of the subject matter, the following observations could be made:

- WSNs are not only a domain of research interests, but has a lot of potential applications in real life.
- On one hand, WSNs can be used for efficient monitoring of our environment. On the other hand, they can be used for alerting people before the occurrence of disastrous incidents such as, land slides and flash floods, and, therefore, can save human lives.
- In particular, aerial WSNs can be highly useful to asses the scale and degree of a disaster. Such assessments are helpful for rescue workers in order to plan their operations.

- Integration of WSNs with the Web not only allows collection of sensed data, but also provides access to anyone and anywhere on the earth. This opens a window of opportunity for researches, planners, designers, and other interested people.

4.2.2 Post-disaster Mobility Models

In a post-disaster rescue operation, several groups of human rescue workers (for example, firefighters and paramedics) are involved to locate the victims, move them to with safe locations, and provide medication. For a fast and efficient rescue operation, relevant information should be exchanged among the rescue workers playing different roles. A natural disaster of large scale often reduces the available communication infrastructure to zero, which, in turn, calls for ad hoc communications. Further, such communication should be delay tolerant as well—possible lack of end-to-end paths would require the collected information to be buffered at individual nodes for long durations until a communication opportunity is found.

Multiple post-disaster mobility models have been proposed in the literature [128–132]. These models attempt to capture various movement scenarios, for example, before, during, and after a disastrous event. In the following, we look at some of the relevant mobility models.

Nelson et al. [128] proposed a mobility model for the post-disaster scenarios considering the effects of the environmental events (for example, fire) and the role of the entities (for example, civilians). The unique combinations of roles and events give rise to different movement patterns, where the underlying mobility model is assumed to be Random Walk. The authors identified three different movement patterns that can be observed in such scenarios:

1. Moving away from the site of the disaster event as observed in case of the civilians,
2. Moving toward the disaster site (for example, police and firefighter reaching the spot in order to evacuate or control the situation),
3. Oscillatory movement from the event location to some other location (for example, terrestrial and aerial ambulances).

Based on these, roles of the individual players are defined. Furthermore, these roles can be extended and customized to simulate other types of reactions, for example, while both police and cleanup crew move toward the location of the disaster event, they do so at different times.

To capture how a particular event affect objects located at different locations, the authors used a gravitational force-based model. An event with an intensity I is considered to exert a "force" I/d^2 on an object that is located at a distance d. Moreover, each event has an *event horizon* to reflect the fact that the effect of any event is spatially bounded up to a certain location. The authors also introduced *disaster radius*—the radius of the circular area from the event location inside which all nodes become immobile.

Uddin et al. [129] developed a similar mobility model for DTNs considering a scenario with disaster forewarning consisting of survivors and rescue workers. The model consisted of various centers (for example, coordination and evacuation) and different mobility patterns of the mobile agents. The mobile agents primarily consist of the rescue workers together with different vehicles such as ambulances and police cars. The authors identified four primitive movement patterns in such a context:

1. Oscillatory movement from one center to another usually by the vehicles,
2. Nonoscillatory movement of the agents to the associated center at the triggering of an event,
3. Cyclic movement along certain regions of interest and returning back to the concerned home center, for example, as typically seen with police patrol, and
4. Convergence (divergence) movement of the agents to (from) a center possibly at given time instants.

The model considered that at the announcement of an evacuation procedure, people probabilistically move toward the nearest evacuation center, where they spend a random duration of time before returning to their respective homes. The survivors stayed in the evacuation centers for a long time before returning to their homes. Relief operations are launched after a disaster to help the survivors by providing food and water supplies. To capture the effect of a disaster, the authors considered that a disastrous event (for example, a hurricane) passes along a curved path over a given terrain. Based on this, the damage suffered at any point P at a distance d from such disaster path is given by

$$D = min\{1.0, \frac{d_0^2}{10} \times \frac{I}{(d - d_0)^2}\}, \qquad (4.1)$$

where I is the intensity of the disaster, $1 \leq I \leq 10$, and d_0 is the critical distance such that any location within distance d_0 from the disaster path is entirely destroyed. Now, when the damage ($0 \leq D \leq 1.0$) suffered by any path segment is more than some threshold limit, the corresponding path segment is considered to be unusable. In other words, the damage factor determined the movement paths in the concerned scenario.

Aschenbruck et al. [130] presented a mobility model bearing close resemblance with the real-life rescue operations in post-disaster scenarios. The authors identified several distinct regions typically observed in such scenarios, for example, incident location, transport, and hospital zones as shown in Fig. 4.2. These tactical regions— together with the variously controlled and confined motion of the nodes—give rise to a heterogeneous area-based movement model.

The *incident site* consists of one or more incident locations where the concerned disaster event has actually occurred. Affected people and victims of the disaster are found in this location. Moreover, effort is put forward to minimize the effect of the disaster event, if relevant. For example, in case a fire breaks out, firefighters would try to contain its spreading as soon as possible.

Fig. 4.2 A typical
post-disaster mobility
scenario consisting of several
tactical areas typically
observed in a post-disaster
scenario

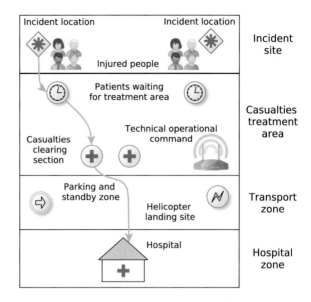

The gray-colored path (depicted with arrowheads) in Fig. 4.2 shows the movement of the affected people. Rescued people are moved from the incident location to the *casualties treatment area*. After some first aid operations, they are moved to the *transportation area* from where they are taken to the hospitals.

Based on the roles, users (or vehicles) are found to move from one of these regions to another. The authors further considered the presence of obstacles in the region. To determine the shortest path from one point to another avoiding the obstacle in the way, Dijkstra's shortest path algorithm is used on the *visibility graph* connecting the two points. An implementation of this mobility model can be found along with the BonnMotion[5] tool.

In general, any movement model exhibits different patterns, such as randomness and spatiotemporal constraints. Conceição and Curado [131], however, noted that, in many cases, post-disaster mobility models involve random movement of nodes, which is often not practical. Instead, the authors in [131] proposed a mobility model mimicking search operations based on realistically observed human behavior for disaster areas (HBDA). The authors noted that a search operation for victims has some underlying characteristics.

- A group of people performing a search operation begin from an initial point and gradually move toward the target point.
- At any point of time, those people are physically separated.
- Any two people maintain a minimum distance of separation between themselves. This ensures that the individual search areas do not have much overlapping, while,

---

[5]http://sys.cs.uos.de/bonnmotion/.

at the same time, it is possible to maintain contact (communication) with one
another.

- On the other hand, the separation between any two people does not exceed a
  maximum threshold distance. Intuitively, at such separation, two people engaged
  in the search operation would not be able to communicate. Moreover, it is likely
  that the potential search area would not be thoroughly searched giving rise to
  "search holes" (unsearched areas).

To achieve such a spatial separation among the nodes, Conceição and Curado
considered a force vectors-based model. In the said model, nodes exhibit force to
one another. The magnitude of the force exerted on a node by another node increases
as their distance of separation decreases; the direction of the force vector is opposite
to that of the neighbor vector. After forces corresponding to the individual nodes have
been determined, the resultant force vector is computed. In the scenario when a node
has multiple neighbors such that the distance to any neighbor from that range lies
within the permissible range of the minimum and maximum values, the concerned
node is said to be *optimally* placed. In such a case, the node keeps on following the
target. Otherwise, a node adjusts its position as described above.

The HBDA mobility scheme proposed in [131] produced interesting results when
compared to the Random Waypoint mobility (RWP) model. The density distribution
of nodes in HBDA was more uniform as compared to RWP; the average node degree
(number of nodes within range) was relatively uniform for HBDA than RWP. On
the other hand, HBDA provided greater area coverage, which is an important metric
in case of search operations. Thus, in comparison to RWP, the HBDA mobility
model is more suitable for capturing a search operation. Conceição and Curado also
considered a variation of HBDA together with obstacles [132], which was again
found to be superior to the RWP mobility model.

> When simulating a post-disaster scenario, it is desirable to avoid the use of
> simple synthetic mobility models, such as Random Waypoint and Random
> Walk. However, at the same time, perhaps no real-life post-disaster mobility
> model is available. In such a situation, a better approach is to use a movement
> model specifically designed for disaster- or crisis-like scenarios.

4.2.3 Communication Aspects

In the recent years, different social media sites, including Twitter[6] and Google Person
Finder,[7] have been found as useful tools [133] in information gathering in the aftermath of large-scale disasters (for example, the tsunami and subsequent earthquake

---

[6]http://www.twitter.com/.

[7]http://google.org/personfinder/.

in Japan, 2011). Multiple characteristics (for example, short length, geotagging,[8] and *hashtags*) of Twitter make it viable as a distributed social sensor network [134, 135]. Sakaki et al. [134] regarded each Twitter user as a *sensor* and used filtering approaches to estimate the locations associated with the tweets. Such data can be, consequently, used to monitor disaster events (for example, trajectory of a typhoon) in almost real time and the users can be warned a priori accordingly. Crooks et al. [135] performed similar analysis and found that users can provide relevant information (for example, impact area of an earthquake) in timely fashion and can complement traditional forms.

Lorincz et al. [136] envisioned that the nodes of a sensor network can play multiple active roles in response to a disaster scenario including triage of the patients—stratifying them based on different factors (such as severity of injury, age, and survival chances), storage of patients' medical information, and monitoring of the environment and location tracking of the patients as well as the first responders, for example, firefighters. In this context, the authors discussed the design and implementation of CodeBlue—a software infrastructure that integrates together various types of sensors (for example, a heart rate monitor) and wireless devices (for example, a PDA) to form a wireless ad hoc network for disaster response. Lorincz et al. [136] observed that there are several desirable characteristics of such networks including:

- Flexible and decentralized device discovery process to establish communication paths between sensor and monitoring devices.
- Efficient and cooperative ad hoc routing mechanisms so that data from a larger geographic region can be obtained.
- Traffic should be prioritized based on their category (for example, an SOS message) since the given bandwidth is limited.

Additionally, there are other aspects, such as authentication of the involved parties and continuous tracking of the location of the devices. CodeBlue is essentially an information plane that sits atop these functional blocks and enables seamless communication among the concerned parties.

George et al. [137] observed that situational awareness is an import aspect for effective response to a disaster. While acknowledging the applicability of wireless ad hoc and sensor networks (WASNs) in multiple domains, such as monitoring, target tracking, and health care, the authors noted that the use of WASNs for disaster response scenarios has been limited due to several factors. Deployment of such networks is an issue since disasters are unpredictable and infrequent. On the other hand, scalability of such a network is also a challenge. Depending on the extent of a disaster, such a network may not only span over a larger geographic terrain, but also over multiple jurisdictions. Collaboration and cooperation are important in such a case. Moreover, there should be commonly agreed upon standards of protocol and architecture to enable interoperability and avoid the adverse effects of heterogeneity. Since existing communication infrastructure is likely to be affected, the deployed WASN should provide robust communication facilities to the concerned region. Such

---

[8]Attaching current geolocation information to a message.

communications should also be possible when the users are mobile. Finally, high volume of data (for example, calls made by the users, sensor readings, and operations update) could soon congest and cripple the network. Therefore, congestion control schemes should be in place together with opportunistic and delay-tolerant communications. To overcome such challenges, the authors proposed DistressNet [137]—an architecture for WASNs for use in post-disaster scenarios.

DistressNet essentially integrates different types of networks formed in a post-disaster scenario. A *BodyNet* is a wearable WSN used for health and status monitoring of the users, such as first responders and support staff. A *TeamNet* is a network formed among the BodyNets of the members of a given team, vehicular nodes, and equipments located in the same area. The concerned team usually has a single focused mission, for example, search-and-rescue operation. *VehicleNets* are formed among the networking devices fitted to the vehicles. VehicleNets form a backbone of the DistressNet and contributes toward mobility-assisted routing. Moreover, VehicleNet helps in bridging heterogeneous protocols, such as IEEE 802.15.4 and IEEE 802.11, to enable interoperability in the DistressNet. *AreaNet*, too, provides similar bridging capacity. An AreaNet is formed among stationary nodes deployed to known positions, and forms the other backbone of the DistressNet. AreaNet has higher available bandwidth, and, therefore, can carry high volume data, such as video. Finally, *SenseNet* is a WSN used for sensing, monitoring, and tracking objects/people of interest in the post-disaster scenario. Typical applications include audio monitoring and chemical level sensing.

A disaster of large scale often renders the existing network communication infrastructure unusable due to disruptions, congestion, and blackouts [138]. In such scenarios, smartphone/featurephone-based communication was found to increase sharply in contrast to the PC-based communication [133]. Since restoring cellular towers would take some days, smartphone-based PSNs can serve as a "quick response" network. Moreover, there is always deployment issues with a new network [137]. Smartphones, on the other hand, are almost always carried by their users, and, therefore, do not suffer from the deployment issue. Perhaps the only prerequisite to engage in communication in this case is to have the relevant application installed in the device.

Noting the wide usage and popularity of Twitter, Hossman et al. [99] developed *Twimight*, a Twitter client for the Android phones, which can engage in opportunistic communications in the immediate aftermath of a disaster. In the normal mode, *Twimight* supports several basic operations (for example, send tweets, reply to and view timeline) observed in Twitter. When the disaster mode is enabled by a user, the tweets created by him/her are marked as *disaster tweets*, and are flooded to the other users using Epidemic [18] routing through opportunistic communications. Subsequently, when a user switches the application from disaster to normal mode, the disaster tweets created are published to the Twitter server. Moreover, various useful sensor data (for example, location of drinking water and evacuation centers) can be shared in this way. The authors raised two interesting questions in the work—what if the disaster prevents mobility of the users and how would users react to the delays.

In Chap. 2, we reviewed the Epidemic routing protocol, which is based on flood-ing. A node containing an undelivered message replicates the message to all the nodes in the network it (node) comes in contact with not having the message already. Therefore, Epidemic routing consumes a lot of network resource and has a high overhead. Large number of transmission and receptions deplete the device energy quickly. While multiple modifications to the Epidemic protocol and other energy-efficient routing protocols have been proposed, here we look at a particular one—n-Epidemic [139]. The rationale behind the design of the n-Epidemic protocol is that a sender can save energy by forwarding a message when it has at least n neigh-boring nodes. This is in contrast to the simple Epidemic protocol ($n = 1$), where message replication takes place even when there is just a single neighbor. When the average number of neighbors per node is 5, i.e., $n = 5$, the delivery performance of n-Epidemic is found to be about 4.34 times better than the simple Epidemic protocol.

Saha et al. [140] advocated a preplanned 4-tier architecture to provide communi-cation facilities in the aftermath of a disaster, and showed that the optimum number of Wi-Fi towers obtained through preplanning provides lower latency compared to their random placement. It may be noted that, in a large-scale disaster-struck region, erecting Wi-Fi towers and using them may not be possible until a few days. More-over, even if such infrastructure had been deployed as preventive measures, their existence could not be guaranteed after a disaster has struck. Finally, placement of those towers should also take into account the prevalent scenario after the disaster.

To summarize this section, we looked at different aspects related to disasters. Human history, on several occasions have been severely affected by natural (or man-made) disasters. However, with the advance of technology and proliferation of WSNs, it is becoming possible to monitor the environment and detect the occurrence of a disaster before it actually happens. Any forewarning based on such technology would be helpful to save precious human lives.

On the other hand, human race is not yet in a stage to completely shield themselves from the impact of disastrous events. Therefore, it is of utmost importance that more and more realistic mobility models pertaining to disaster scenarios are developed and evaluated. A more rational model of such mobility would be helpful in plan-ning search-and-rescue operations. However, a major hindrance in getting close to reality is that, unlike "normal" scenarios, collection of mobility traces in a pre- or post-disaster scenario is rather difficult, if not impossible. Models that are macro-scopically close enough (for example, [130]) can be considered in such cases. Finally, an understanding of the underlying movement of people (and vehicles) would help to evaluate the efficiency of ad hoc communication mechanism as used in OMNs.

4.3 The Notion of Intelligence

One of the fundamental aspects of human beings is their innate intelligence. Intel-ligence can have a very simple manifestation—for example, switching on the fan when feeling hot, or it could be complex—for example, deciding whether or not to make a road trip based on the current weather conditions. Moreover, intelligence often involves learning from the past decisions—for example, if John had bought a product X thrice in the past and found it to unsatisfactory, he might consider buying an alternative product Y the next time.

It is, therefore, a logical next step that intelligence manifested in human beings should be mapped to the computing devices in one form or another. In fact, the concept of agents is central in the field of artificial intelligence [141]. In this section, we briefly look at how intelligent agents are useful in the context of post-disaster scenarios. Subsequently, we would have a glimpse at the notion of *intelligence*.

4.3.1 Agent-Based Systems

Russell and Norvig [141] described an agent as something that can perceive its environment and acts upon the environment based on its perception. An agent has some attributes, behaviors, possibly some memory, and resources.[9] An agent-based model additionally has another component [142]—a set of relationships among the (autonomous) agents. Thus, in agent-based modeling, agents can interact with their environment, as well as, one another. When the agents interact with one another, they often learn from their interactions, and accordingly adapt their own behavior [142].

Agent-based systems have been widely used to model and evaluate various scenarios related to disasters (natural or otherwise), for example [143–151]. In the following, we take a look at some research projects that use agent-based systems in the context of disaster preparedness and response and recovery operations.

Simulating disaster scenarios using a multiagent system is often helpful for planning rescue operations in emergency situations. The RoboCup Rescue[10] project can be mentioned in this context. However, a major concern is how faithfully reality can be replicated in such simulated environments. Skinner and Ramchurn [143] discussed about the features of the RoboCup Rescue simulation platform. The authors listed a series of interesting features (such with as dynamically changing environment and limited and uncertain information) that bring the simulation platform close to reality. Such a platform is quite useful for research in multiagent-based systems domain.

Fiedrich and Burghardt [144] noted that agent-based systems can find their application to support various processes across different stages of a disaster. They identified two primary areas in this context—agent-based simulation systems and agent-based decision support systems. Agent-based systems can be used to model human and systems behavior during a disaster or after it has occurred. Such a process is highly scalable since thousands of agents can be considered together for study and analysis. For example, one can verify how a given rescue strategy would perform in the aftermath of a large-scale disaster in a busy region. Such studies can be useful for improved preparedness and training. The authors noted that agent-based systems can also be used to create a supersystem by combining different information systems, communication networks, and human decision makers.

---

[9]http://www.mcs.anl.gov/~leyffer/listn/slides-06/MacalNorth.pdf.

[10]http://www.robocup.org/robocup-rescue/.

Fiedrich and Burghardt observed that in the contemporary time, agent-based systems are hardly used in any real-life disaster scenario. The authors, further, acknowledged that whether at all agent-based systems can be used in such scenarios is an ongoing debate. Finally, the authors argued that advances in both technology, as well as, human organizational structure during disaster events can help bridge the gap.

The role of Information Technology (IT) in our work and life is ever increasing. It is, therefore, desirable that continued advances in IT can be adopted in disaster recovery operations for saving precious human lives and critical infrastructure. In this context, Massaguer et al. [145] proposed DrillSim, a multiagent-based simulation platform for evaluating the effectiveness of IT solutions and models in disaster response and recovery operations (for example, in terms of casualties).

DrillSim [145] aims to absorb the best practices from both simulation-based approaches and disaster response drills to create emergency scenarios. To account for higher realism, DrillSim uses virtual/augmented reality mechanisms. In particular, DrillSim takes input from, and can be integrated with, real-life drill scenarios. The physical space, where such a drill takes place, is equipped with various sensors, such as audio and video sensors, RFID tags, and visualization and communication interfaces. Thus, as a drill operation progresses, the corresponding spatiotemporal information are continuously fed into the simulator. This achieves the goal of augmenting the simulation space with real-life people and events. On the other hand, the reverse—augmenting real world with the simulated virtual space—is also accomplished by allowing real-life users to interact with the simulated world projected in their devices.

Real persons (for example, rescue operators) are simulated in DrillSim using agents. However, since the simulated world represents a dynamic environment, the $state^{11}$ of an agent changes with time. DrillSim uses various information (present geolocation, different devices carried with, health status, available information about the world, decisions made, and corresponding plans) to represent the state of an agent at any given point of time. The information about the observed world, however, itself is a complex parameter and represented with related finer details. An agent has a role, which determines its actions/decisions. The decision-making module of an agent has been modeled using a recurrent neural network. Moreover, like the real world, agents interact and have social ties among themselves. DrillSim allows editing the roles of the agents so that agents with new roles can be created within the simulation environment. Results of simulation-based experiments indicated that having a few (1–3) agents with the role of "floor warden", together with several agents with the role of "evacuee", helped in achieving a better degree of evacuation against the backdrop of a fire alarm.

One may recall from Fig. 4.2 that a post-disaster rescue operation involves a command-and-control center from where the entire operation is managed. Multiagent-based systems are also being developed with the aim of training incident commanders for such operations. For example, Schurr and Tambe [146] developed Demonstrating Effective Flexible Agent Coordination of Teams via Omnipresence

---

[11] In general, in the domain of artificial intelligence, representing the state space is a common task.

(DEFACTO), a multiagent-based training system, and discussed their evaluation experiences while working with the Los Angeles fire department. The authors noted that training incident managers for rescue operations against the backdrop of large-scale disasters is of utmost importance. Such training facilitates the preparedness of prompt and correct decision-making in real life in case of an actual disaster. Schurr and Tambe further noted that DEFACTO, unlike a traditional training process, does not require a large number of firefighters to be involved in the training program. Therefore, available firefighters can rush to service whenever required. Moreover, a simulation can enact a more realistic scenario (for example, a fire continually spreading with time), unlike the ones in traditional training programs, against which the incident commander can train.

One of the major components of DEFACTO is Omni-Viewer, a visualizer, as well as, interface for interacting with the agents. Omni-Viewer allows the incident commander to have local and global perspectives of the disaster site in 2D and 3D modes. Such local and global views help the incident commander in making better decisions. For example, the firefighters at the disaster site might not notice the smoke arising from the back side of the building, which the incident commander can inform.

Teamwork is managed using the Machinetta[12] proxy. Each member of the virtual firefighting team has its own proxy and coordination activities within a team are achieved via the proxies. Such proxies also allow for "adjustable autonomy" where the onus of decision-making can transfer from humans to agents, and vice versa. Simulations using DEFACTO provided various key insights (for example, a human commanding more agents is not a guarantee of better performance) [146], and the system was further improved based on inputs from the fire department.

A disaster of large-scale often brings down available infrastructure, which, in turn, results in loss of human lives. Cheng and Wu [147] noted that in modern towns and cities, bridges play a crucial role in connecting different geographic regions without which life would be critically affected. However, large-scale disasters—especially in earthquake-prone areas—pose potential threat toward damage of such bridges, and thus, disruption of normal life. Consequently, the authors developed an intelligent agents-driven platform [147] for data exchange among different related systems with the aim of preventing bridge disaster events. Although the case study presented was specifically for Taiwan, the general framework can be potentially adapted for other environments too.

Cheng and Wu noted that it is not the acute crisis of data that prevents containing the impact of large-scale disasters involving bridges. In fact, in the context of the case study in [147], data are available from various sources, such as weather monitoring system, bridge monitoring system, earthquake loss analysis system, and so on. The problem is elsewhere—aggregating data from various systems and deriving actionable and spatiotemporal intelligence, which would help in controlling the traffic across potentially dangerous bridges. However, performing such aggregation manually is a tedious task and often error prone. Therefore, the authors focused on developing intelligent agents to address the problem. In particular, Cheng and Wu

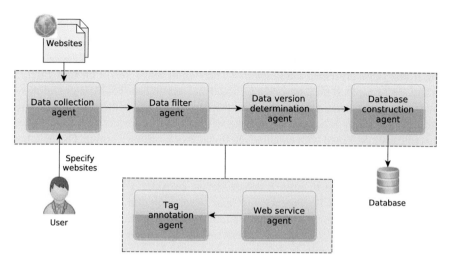

Fig. 4.3 A multiagent intelligent system for handling bridge related disasters in Taiwan [147]

developed six agents to achieve their goal (such as those for collecting data, filtering, and annotating—see Fig. 4.3) exhibiting different attributes. The "data version determination agent", for example, is able to identify whether or not the available data is recent.

Additionally, the authors also developed a data exchange platform with the aim of achieving efficient flow of data during an ongoing disaster event. Thus, when relevant data is timely available with appropriate authorities, alerts can be generated about potentially dangerous bridges and mishaps can be prevented.

In summary, we observe that agent-based systems play several different roles in the context of disasters and crisis managements. On one hand, such systems are useful to provide more realistic trainings to incident commanders to handle a crisis scenario. Efficient allocation of rescue resources and making a quick but wise decision in a real crisis is a difficult task. Such trainings can help in imparting better preparedness to the incident commanders. Realistic simulations can also help in evaluating the efficiency of evacuation strategies. Moreover, modern technologies allow both way integration of real and virtual worlds. Such strides are hoped to have critical impact on the contemporary disaster response and rescue operations.

On the other hand, we noted that agent-based systems can be leveraged for automated, efficient data collection and aggregation. It is not often the case that data are not available. However, such data are often spread across several systems, exhibit nonuniformity and lack of structure, and often difficult to determine the age of data. Cheng and Wu [147] observed that assimilation of data from diverse sources and taking timely actions based on that would be useful. In fact, with such variety, volume, and velocity of data, it is expected that Big Data [152–155] would play a pivotal role in this context in the coming years.

4.3.2 Situation Awareness

In our daily lives, we often use the term "intelligence." But what *is* intelligence? Legg and Hutter [156] presented a collection of 71 different definitions of the term offered by different people. It may, however, be noted that intelligence has different implications in the domain of communication networks as compared to the field of cognitive and neurosciences. When we are dealing with intelligent devices or protocols, we are typically concerned with machine intelligence or machine learning. On the other hand, users of such devices in a network may also leverage their intelligence, as we shall in the next section.

Without attempting to raise a debate, let us consider a simple, nontechnical, general description of intelligence—to learn something and apply those learning. Whether we are talking about human intelligence or machine intelligence, under any condition, an object (human or otherwise) needs to understand its environment and act/decide accordingly. In this section, we briefly discuss about a closely related topic—situation awareness.

Endsley [157–159] presented an overview of situation awareness and the cognitive theory behind it. Endsley described situation awareness as an "internalized mental model" that an operator[13] forms of his/her current environment. Such a model is formed by taking into consideration different inputs gathered from the environment, systems, team members, and others. A situation model formed in this way captures the big picture of the current environment based on which an operator takes decisions and actions. For example, in an examination hall a student might observe that his/her fellow examinees are switching off their mobile phones and thus, he/she might do the same. In this case, the student considered the data obtained from his/her team members. On the other hand, the examiner might write clear instructions on keeping mobile phones switched off on the white board. The students might notice this and act accordingly. In either case, a student is aware of the current situation, and the decisions/actions of the student are based on the situation model developed by him/her.

Based on what discussed until now, a reader now might be tempted to think that situation awareness is a rather simple phenomena—form a largely working model and keep taking decisions based on it. Unfortunately, the case is not so [159]. The problem is that in real life, more often than not we find ourselves in dynamic and frequently changing environments. With such changes occurring around, one has to accordingly update his/her situation model, too. In other words, situation awareness can be treated as a spatiotemporal process. Endsley [159] noted that such updates to the situation model may require a large number of factors to be taken into consideration, and human history has witnessed several failures due to faulty situation awareness.

Endsley [159] described three distinct levels of situation awareness.

---

[13]The term "operator" has been used by Endsley; we follow the same terminology here. Perhaps the nomenclature could be understood by realizing the fact that the concept of situation awareness first originated in the aviation domain [157].

1. **(Level 1) Perception**: The first level is about perception (*sensing*) of the critical factors (*signals*) relevant to the given environment. Endsley observed that in order to support the level 1 of situation awareness, related information should be conveyed in an efficient way to the operator.
2. **(Level 2) Comprehension**: In the second level, one attempts to interpret/understand what has been perceived. In other words, different perceptions are like different pieces of a puzzle, which must be arranged together to get a big picture and understand their implications with respect to a given goal of an operator.
3. **(Level 3) Projection**: In the final level, based on the understanding of the current situation, an operator attempts to project/forecast the likely status of the system or environment in the near future. This is the highest level of situation awareness an operator can have.

Endsley noted that while domain experts can have fast and accurate situation awareness, those for the novices can be often slow and incomplete. Subsequently, Endsley considered the interrelation [159] among the different levels of situation awareness, and observed that at times, situation awareness levels 2 and 3 can be leveraged to complement missing data from level 1 by using their default values.

> Situation awareness can be performed by human beings or computers. A general difference between the two is what we do with the situation awareness process. With human beings involved, it needs to be measured. However, with computers, definition and implementation of the process are necessary [160].

Acquiring situation awareness is a fundamental necessity for an efficient response and rescue operation in the aftermath of a large crisis. In the recent time, a lot of effort has been focused toward social media, and several methods and tools have been proposed (for example, [161–166]) to improve situation awareness in the context of natural disasters.

Vieweg et al. [161] analyzed tweets pertaining to two large crisis scenarios in USA. The authors described their methodology in detail starting with tweet filtering (based on related search terms) and collection. Subsequently, the samples were further refined to include "on-topic" tweets by users "local" to the events. It was found that about 18–40 % tweets contained geolocation information [161]. However, a high percentage of users were found to post at least a single tweet that had geolocation information. Interestingly, it was observed that the tweets contained additional information that the authors labeled as *situation updates*, and categorized them into several high-level groups (such as warning, preparatory activity, and so on). It was found that about 49–56 % of the on-topic tweets contained situation updates.

On a similar note, Yin et al. [164] observed that signals emanating from social media during crisis scenarios can be harnessed to improve an operator's situation awareness. For example, updates from survivors can help in gaining first-hand information about the ground zero, whereas posts from people in nearby areas can provide pseudo real-time updates as the crisis scenario unfolds. However, it is a mammoth task if one has to keep monitoring Twitter 24×7 for possible relevant updates!

Yin et al. [164] designed an intelligent system to capture the tweets and automatically process them in order to acquire situation awareness in the face of a crisis. At a high level, the system consists of three components.

1. **Capture**: Capture the ever-generating data from Twitter using the streaming and location-based search APIs provided by it.
2. **Process**: As tweets are captured, they pass through several stages of processing.

   a. The first stage is bursty event detection—whether a word appears as a bursty feature in a given time frame. Such bursts can indicate whether or not any not-quite-normal incident(s) took place.
   b. In the next stage, text classification is performed to evaluate the impact of incidents so discovered.
   c. In the subsequent stage, tweets are clustered based on likely topics to which they belong. Similarity metrics combined with temporal distance between two tweets are used for this purpose.
   d. In the final stage, a tweet is displayed at a relevant location based on its geotag.

3. **Visualization**: The final component allows crisis managers to have an interactive visualization of the tweets and information collected from them.

Various social media platforms, such as Twitter and others, are often projected as social sensors since they make available crisis information (for example, those related to earthquake and cyclone) in quasi-real time. Dong et al. [167] compared Twitter to the underlying sensor network, and the front-end application to a base station.

To summarize, we observe that having situation awareness is of utmost importance while dealing with natural disasters or crisis scenarios, in general. Such tasks becomes relatively easy for domain experts, but novices often face a lot of challenges. Endsley [159] have provided suggestions in this respect. Moreover, in situations where one has acquired levels 2 and 3 of situation awareness, it is feasible to substitute the missing data from level 1 with default values.

On the other hand, there is a growing interest of leveraging signals available from social media to extract situation awareness. Several methodologies have been proposed for this purpose especially, with respect to Twitter. It was found that the peaks of tweets volume can be correlated with contemporary usually-not-normal events of importance. Interested readers might also look at [168] for a detailed methodology on capturing tweets generated by Twitter users. A practitioner, however, should ensure availability of high storage and computing capacities if he/she plans to capture tweets for a considerable time period.

The two closely related terms—situation awareness and context awareness—are sometimes differentiated [169–171]. Originally, context awareness aimed at enabling a software to identify and adapt to the surroundings of a mobile user. For example, location of a user can act as a context. Situation, however, usually deals with knowledge state information. In other words, the focus of situation awareness is to understand the contemporary situation, whereas in context awareness, the objective is to exploit relevant contemporary information to provide better interaction with a system.

4.4 Intelligence-Induced Movement in MOONs

So far in this chapter, we have looked at monitoring disasters, and mobility and communication aspects in their aftermath. We have also briefly explored the notion of intelligence specifically in the form of situation awareness. We have also observed that agent-based intelligent systems can be quite helpful to deal with such scenarios.

Traditionally, communication networks—whether DTNs, MANETs, cellular networks, or plain old wired networks—have been viewed as a network of connected devices. In other words, such devices had been the primary actors in a network considered. However, as we noted in Chap. 1, MOONs have a different view toward networking. To recollect, a MOON is a PSN with a mission, and where human-network interplay has consequences upon the network performance. Thus, with MOONs, we make transition toward a paradigm where innate intelligence of human users is also of significance, unlike conventional networks, where we had limited ourselves to intelligence of the devices.

In this section, we take a look at how human movement induced by their intuitive form of intelligence can help in increasing communication opportunities in MOONs and, in turn, help in enhancing the accomplishment of the mission objectives. The study would be limited to typical post-disaster scenarios encompassing a large terrain.

The aftermath of a disaster, in general, can witness movement of two broad categories of people:

- Rescue workers and volunteers engaged in rescue operations (together with the victims being transported), and
- Unaffected citizens or victims with minor injuries.

The latter category, through their movement, provides scope for opportunistic communications throughout the disaster area. In other words, they can act as *message ferries* [172]. The scope of the first category is limited in this context, since the rescue efforts are often localized in certain geographic areas. While several mobility models have been proposed, they do not focus on how the movement of the human users, and therefore, the impact on opportunistic communications, would be when their innate intelligence is considered. This is particularly relevant in the case of MOONs, where human beings are considered to be a part of the network. This motivates us to explore a post-disaster movement pattern that considers the effects of human intelligence on their motion planning.

Let us consider a post-disaster scenario, where human rescue operators are involved in either locating trapped people or say, taking measures of radiations at different areas. It is essential that not only the entire geographic region be covered by the rescue team, but also the acquired information be passed through. As noted earlier, a natural disaster of large scale often reduces the available communication infrastructure to zero, which calls for ad hoc communications. Further, such communication should be delay tolerant as well—possible lack of end-to-end paths would require the collected information to be buffered at individual nodes for long durations until a communication opportunity is found.

It may be noted that while *coverage* guarantees are critical in WSNs and its applications (for example, hostile target tracking), in a post-disaster rescue scenario, the primary requirement would perhaps be an uniform network coverage (availability) over the entire region to have the victims identified and necessary information communicated. In this section, we would explore the issues related to such communication opportunities in the context of a post-disaster rescue scenario, where human rescue operators are involved.

4.4.1   Representation of MOONs

We define a MOON as follows. Let, $\mathcal{M} = (\mathcal{G}, \mathcal{T}, \mathcal{S}, \mathcal{P})$ be a mission, where $\mathcal{G}$ is the set of mission goals, $\mathcal{T}$ is the timeline of the mission, $\mathcal{S}$ denotes the set of strategies applied to execute a mission, and $\mathcal{P}$ is the set of performance parameters related to $\mathcal{G}$. The underlying MOON is a network $\mathcal{N} = (N, \mathcal{M})$ with a mission $\mathcal{M}$ and a set of nodes N. Herein, we consider the MOONs with a single mission only.

To illustrate, let us consider a MOON formed in the aftermath of a large-scale disaster. A key objective here is to increase the communication opportunities among the mobile and stationary nodes in such a scenario. In this context, increasing the communication opportunities represents the mission goal. Whereas, intuitive intelligence-based movement of the mobile nodes represent the strategy of the mission.

The mobile nodes in this scenario represent the rescue workers, paramedics, fire-fighters, and unaffected or minimally affected people. On the other hand, the stationary nodes indicate the trapped victims, rescued persons requiring immediate medical care, various camps, and so on. Therefore, the increase in communication opportunity between a mobile and a stationary node ensures that the likelihood of availing the relevant services is enhanced. For example, a paramedic, after coming to know about the location of a recently rescued victim, would move to attend the person.

In the remainder of this chapter, we consider a square-shaped terrain in the context of a post-disaster recovery scenario, which is divided into multiple square-shaped zones of equal size. Each such zone is identified with (x_id, y_id), where x_id and

Fig. 4.4 Illustration of the zones

1	2	3
8	9	4
7	6	5

y_id, respectively, denote the zone index along the x- and y-axes from a chosen origin. In particular, the left-bottom zone is identified with (0, 0). Further, we consider that the human rescue operators (mobile nodes) form a MOON in such a scenario, and all the nodes are aware of their current locations.

The following definitions are provided in regard to the geography of the terrain.

Definition 4.1 *K*-**left zones**: The *K*-left zones of any zone in the network represent the *K* neighboring zones horizontally toward the left of the concerned zone. Thus, if the identifier of a square zone in the network is (*xid*, *yid*), its 2-left zones would be (*xid* − 1, *yid*) and (*xid* − 2, *yid*), provided the zones exist in the terrain.

Definition 4.2 *1*-**near zones**: The *1*-near zones of any zone in the network represent the adjoining neighboring zones, which are one block away. Thus, if the identifier of a square zone in the network is (*xid*, *yid*), then it could have up to eight possible *1*-near zones: (*xid* − 1, *yid* − 1), (*xid* − 1, *yid*), (*xid* − 1, *yid* + 1), (*xid*, *yid* + 1), (*xid* + 1, *yid* + 1), (*xid* + 1, *yid*), (*xid* + 1, *yid* − 1), and (*xid*, *yid* − 1).

Figure 4.4 illustrates these terminologies considering nine adjacent zones in a terrain, which are identified with a single number for ease. The 2-left zones of the zone numbered 3, for example, are the ones numbered 1 and 2, whereas the *1*-near zones of the zone numbered 9 are the ones numbered from 1 to 8.

The definition of *1*-near zones could be generalized to *K*-near zones, where any zone could have up to $(2K+1) \times (2K+1) - 1$ neighboring zones. Other terminologies (for example, *K*-top zones) can be similarly defined.

4.4.2 Opportunistic Communications with Intelligence

A mission-oriented opportunistic network reflects characteristics, such as involvement of **human beings**, and thus, human intelligence, which are not usually seen in a general WSN or any other wireless network. Unlike machines, it is hypothesized that the presence of human beings in such a network would have a positive effect on accomplishing the mission. For instance, in the context of our post-disaster recovery operation example, when a rescue worker finds a trapped victim, he/she would possibly look in the adjoining places for the availability of any more victim(s) requiring assistance.

To depict such scenarios, we define a set of *levels* of human intelligence. Once again, we are not looking for a technically accurate definition of intelligence. Rather, the *levels* of intelligence considered here manifest simple intuition-based schemes that a human being could adopt when it comes in contact with a victim. Here, coming to know a victim's location indicates the learning phase. The application of learning is determined by the different levels,[14] as described below.

- **L1**: When a mobile node learns about the location of a stationary node, it moves toward that node. This could represent the case when a trapped person has been recovered, and requires medical attention. In terms of network communication, this could imply tracking of updates of one's medical reports.
- **L2**: A mobile node with intelligence level L2, on learning the location of a new stationary node, visits the node as well as a randomly chosen zone from the stationary node's *1*-near zones.
- **L3**: A mobile node with intelligence level L3, on learning the location of a new stationary node, visits the node as well as its K-left neighboring zones
- **L4**: A mobile node with intelligence level L4, on learning the location of a new stationary node, visits the node as well as all of its *1*-near zones.

We note here that the L2 scheme described above is loosely based on the idea of "Neighborhood Search Based Exploration" discussed in [173]. To summarize, in the neighborhood search-based exploration, an action in the current state is probabilistically chosen in random from the neighborhood of the previous action state.

Figure 4.5 presents an overview of how a node behaves in a MOON. We note that in a simulation framework we cannot distinguish between humans and devices. However, the architecture presented here closely relates to reality. To illustrate this, in real life, a human being would be working with the application. Therefore, the interfaces among application, router, and database remain valid. The only difference arises in the case of the mobility module, because a real human being moves by himself/herself. Nevertheless, the proposed application could suggest to its owner the locations it should travel next.

When the application (Fig. 4.5) running in a device receives the coordinates of a new stationary node, the application stores the location in its database. It, then, passes the information to the router module by creating a message (containing the newly found coordinates) destined to all other mobile nodes. Finally, it instructs its mobility module to go to the location. The intelligence level determines whether the node should only visit that particular location or any neighboring zone(s) as well.

When the above node comes in contact with a mobile node, the latter's router receives a message containing the new coordinates, and passes upward to the application. The application running in this node then checks whether it has already visited that particular stationary node (identified by location). If not, it takes steps as discussed above.

---

[14]The levels considered here are not related to the different levels of situation awareness as described earlier. Use of similar terminologies in these two cases are merely coincidental.

Fig. 4.5 A node's behavior
in a MOON

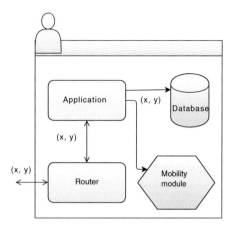

Algorithm 4.1 presents the actions taken by the application module of a node under intelligence level L2. For instance, line numbers 7–10 show that the application executing in the mobile node randomly selects a zone from the concerned zone's immediate neighbors and subsequently visits that selected zone.

Algorithm 4.1: Purposeful mobility under the L2 scheme

Input:
- s: A stationary node's location
- *database*: Database of locations of stationary nodes stored by the node

Output:

- Movement as per the L2 level

1 // Identify the zone where the stationary node is
  located
2 $ix, iy = $ getZoneID($s.X, s.Y$)
3 // Compute the list of the neighboring zones
4 *neighbors* $= $ getNeighbors(ix, iy)
5 **if** s **not in** *database* **then**
6   *database*.addLocation(s)
7   moveTo(s)
8   // Randomly select a neighbor
9     $r = $ randomIntegerInRange(*neighbors*.size())
10   *zone* $= $ *neighbors*[r]
11    $c = $ *zone*.getCenter()
12   // Visit the neighboring zone
13   moveTo(c)

4.4.3   Comparative Study

Let us now briefly look at some quantitative performance results concerning the proposed intelligence levels. For this purpose, we simulated different movement scenarios using the ONE simulator [51], as discussed in the following. We considered a 5×5 sq. km. simulation terrain, which was divided into multiple square-shaped zones, each of size 10×10 sq. m. The stationary nodes were placed randomly within the terrain. We considered two cases in regard to the density of nodes—(1) 45 stationary and 20 mobile nodes and (2) 60 stationary and 30 mobile nodes. We considered the nodes to be moving with three different speed ranges: 0.5–1.0 m/s, 1.0–2.0 m/s, and 2.0–2.5 m/s. It may be noted that human walking speed is about 1.5 m/s, on an average. The subaverage speed would be obtained in scenarios with obstacles, whereas higher speed indicated movements like rushing toward a spot. The scenarios were simulated over two simulation time durations—12 and 24 h. The Epidemic [18] routing protocol was used for message dissemination by the mobile nodes.

To closely reflect the real-life scenarios, we also considered the stationary nodes to be located in clusters around seven different points in the terrain. For example, the case of multiple victims trapped in vicinity reflects such scenario. Moreover, we also took into account some heterogeneity factors that more often than not are manifested in our daily lives. Specifically, we considered the case when certain fraction of the mobile nodes used their intelligence level L1, while the remaining used L4. Thereafter, we studied the scenario where not all the mobile nodes had the relevant application available as shown in Fig. 4.5. In this case, we considered homogeneous intelligence levels, i.e., all the nodes using a single level of intelligence in order to isolate the effects of any other factor.

We contrasted our human intelligence-based schemes with the Random Waypoint (RWP) mobility model. By doing this, we, however, do not contradict our earlier suggestion that RWP should better be avoided for simulating post-disaster scenarios. Note that our primary objective in this problem is to evaluate the efficiency, if any, of aforementioned exploratory movement. Thus, RWP fits in this case. In fact, it has been shown that limited intentional exploratory movement in realistic scenarios is likely helpful [174].

In case of intelligence level L2, we considered 3-left neighboring zones. The results for all simulation scenarios were obtained by taking an average over 30 random scenarios. We look at the performance of different scenarios by considering the following metrics:

- Average number of encounters between a mobile and a stationary node.
- Number of unique encounters between a mobile and a stationary node. For example, if a mobile node m encounters a stationary node s for say, n times, then the total number of encounters for this pair is n, whereas the number of unique encounters is 1.

Fig. 4.6 Encounters of mobile nodes with stationary nodes with **a** 45 stationary and 20 mobile nodes and **b** 60 stationary and 30 mobile nodes for 12 simulation hours

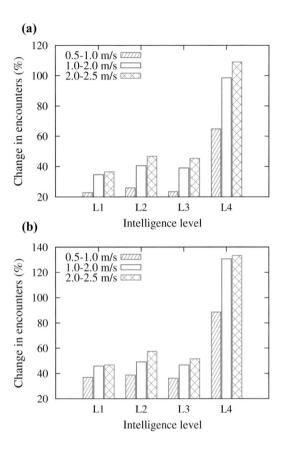

Both these metrics help in evaluating the extent to which intelligence-induced movement helps, if at all, in accomplishing the previously discussed mission objectives. In the following, we take a look at some of the representative results.

Figure 4.6 shows the relative increment in encounters—with respect to the RWP mobility model—between the mobile and the randomly located stationary nodes, on an average, when the four different intelligence levels were used. The proposed intelligence levels are shown along the horizontal x-axis. Reduced performance is observed using the L1 scheme, in which a mobile node only visits a new stationary node. On the other hand, the improvement is maximum for the L4 scheme, where a mobile node searches the *l*-near zones of a zone where a stationary node is located. Figure 4.6a and b correspond to two different node densities, as indicated. Figure 4.7 plots the same performance results when the simulation was executed for 24 h. Similar trends are observed in both figures.

Figure 4.8 plots the result of performance evaluation when 60 stationary nodes were clustered around different locations in the terrain; 30 mobile nodes were considered. From figure, we can observe that the number of unique encounters between

Fig. 4.7 Encounters of
mobile nodes with stationary
nodes with **a** 45 stationary
and 20 mobile nodes and **b**
60 stationary and 30 mobile
nodes for 24 simulation
hours

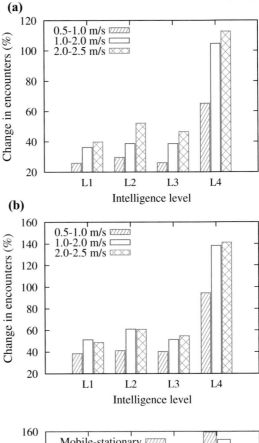

Fig. 4.8 Change (%) in
encounters with clustered
static nodes

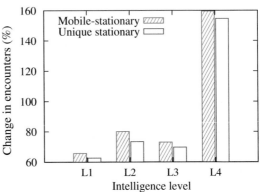

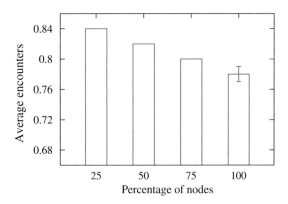

Fig. 4.9 Performance when different fraction of nodes had L1 and L4 levels

a pair of mobile and stationary node, on an average, improved to different extents for the different intelligence levels. Indeed, when a mobile node visits a stationary node and one or more of its neighboring zones, the chances of encountering new stationary nodes also increases.

As we shall see later in a subsequent chapter, it is a difficult job to guarantee network homogeneity in real life. Therefore, we also investigated some scenarios pertaining to the MOON together with heterogeneity. Figure 4.9 shows the case when certain fraction of the mobile nodes—shown along the x-axis—moved according to the intelligence level L1, while the remaining mobile nodes used the level L4. It can be observed that when 25 % of the nodes used the level L1, and the remaining used L4, the number of encounters between a mobile and a stationary node, on an average, was highest. This is because, with more nodes visiting the 1-near zones of the discovered stationary nodes, the chances of mobile-stationary node encounters increased. On the other hand, reduced performance was obtained when all the nodes used the intelligence level L1.

Finally, in Fig. 4.10, we studied the performance obtained when certain percentage of the mobile nodes—shown along the x-axis—in the MOON had the relevant application available (Fig. 4.5). The average encounters achieved for each such fraction of nodes for the different intelligence levels are also shown. As can be intuitively explained, the mission performance increased when the increasing number of nodes in the MOON had the application available. The maximum performance was obtained for the 100 % case, i.e., when all the mobile nodes had the application available with them.

In this context, it can be observed that the performance under the L1, L2, and L3 schemes were quite close, although the same under the L4 level was statistically different from the others. As such, a question regarding the necessity to have the three levels L1, L2, and L3 may arise. We note that the primary objective here has not been to put forward a "best" exploration scheme. Rather, it has been to underscore the fact that movement decisions made by intuitive forms of human intelligence can help in accomplishing the mission objectives by enhancing the communication opportunities.

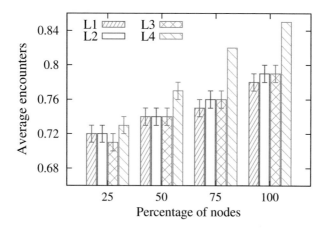

Fig. 4.10 Performance when different fraction of nodes had the relevant application available

To summarize, this study found that when users are expected to act (move, in this study) based on their simple forms of intelligence, there could be significant improvement in the performance of MOONs in terms of the number of contact opportunities. This study assumed realistic human walking speeds and clustered locations of nodes, as typically observed in post-disaster scenarios. However, situations can be different in a real post-disaster aftermath. In particular, movement to the neighboring zones might be infeasible due to obstacles or otherwise. Therefore, the implications of the results of performance evaluation should be interpreted accordingly. Nevertheless, the study indicates that some form of intentional movement by the human users in a MOON (or OMN, as a matter of fact), in general, could be useful.

4.5 Summary

This chapter presented a tour of some of the recent advances in WSNs. These emerging with paradigms, such as human-centric sensing, urban sending, and opportunistic sensing, are finding applications in our daily life—be it our personal health monitoring, citywide pollution monitoring or large-scale environmental monitoring. Well planned and tested WSNs can be helpful in continuous monitoring and possible disaster warning. On the other hand, signals emanated from social media, such as Twitter, which act as social sensors, can be effectively used to identify any crisis/disaster scenario, and keep track of it in near-real time.

However, there is no substitute for preparedness. Thus, a lot of efforts have been put forward in modeling realistic mobility scenarios in the aftermath of disasters. Ad hoc communications, especially OMNs, can play a greater role in such scenarios in the absence of any kind of network infrastructure. Simulation-based studies are performed to evaluate the effectiveness of such schemes. With similar objectives, agent-based systems are being developed to train for better preparedness, response to disasters, and efficient rescue operations.

In this chapter, we also touched upon the topic on intelligence. In particular, situation awareness is an essential criteria for handling disaster scenarios. To be situation aware essentially means to notice what is happening around, understand them, and project the impact on future scenarios. Finally, we presented a formal definition of MOON. We considered a MOON formed in the context of a post-disaster scenario with the mission of increasing communication opportunities with stationary nodes. We considered different levels, L1 to L4, as manifestation of intuitive intelligence based on which users make their movement decisions. It was observed that such "intelligent" decision-making by human users helped in achieving a better performance measure.

4.6 Review Terms

- ■ Human-centric sensing
- ■ Opportunistic sensing
- ■ Environmental Sensor Networks
- ■ Agent
- ■ Situation awareness
- ■ Comprehension
- ■ Situation updates
- ■ K-left zones
- ■ Urban sensing
- ■ Mission-oriented WSNs
- ■ SensorWeb
- ■ Agent-based systems
- ■ Perception
- ■ Projection
- ■ MOONs
- ■ 1-near zones

4.7 Exercises

4.1 What is the difference between participatory and opportunistic sensing? Nike+ Sportwatch GPS[15] measures, among other things, the distance that you have ran. Discuss whether such sensing is participatory or opportunistic.

4.2 What is a proximity sensor? Why do smartphones typically have such a sensor?

4.3 Discuss how Twitter can act as a "social sensor".

4.4 What is an agent? How are agent-based systems useful?

4.5 What is situation awareness? Discuss the three levels of situation awareness relevant to an aircraft pilot.

4.6 Suppose that you have a smartphone that has an app, which advises you the optimum time to withdraw cash from a nearby ATM center. Now, you are not always

---

[15]https://secure-nikeplus.nike.com/plus/products/sport_watch/.

available to access the ATM, for example, you might be in the office or university. Again, if it is too hot or cold outside or, if it is raining heavily, you are unlikely to go. Moreover, the app should disable all notification when you are sleeping at night. In this scenario, what are the different inputs—from sensors or otherwise—the smartphone can provide to that app to make it context aware?

4.7 How are MOONs different from OMNs?

4.8 In Fig. 4.4, which are the 2-right zones of zones 8 and 9?

4.8 Programming Exercises

4.9 *Coordinate to zone conversion*: Consider a terrain of dimensions $L \times L$. The terrain is divided into multiple zones each of dimension $l \times l$, where $L \mod l = 0$. Each zone is identified with an integer starting from 1. Formulate a scheme such that, given a 2D coordinate (x, y), it returns the identifier of the zone to which the coordinate belongs.

4.10 *Restricted random waypoint mobility model*: Once again, consider the scenario from the above mentioned scenario. Suppose that zone i is an *obstacle*. Thus, no node can go to any location inside that particular zone. In other words, the mobility is *restricted*. Create a new mobility model in the ONE simulator (based on the `RandomWaypoint` class) so that movement to a particular zone is disallowed.

Hints: When a new destination coordinate is generated, check whether it is within the boundary of the obstacle zone. If so, discard that destination and generate a new one.

Advanced level: You might realize that the above formulation has a loophole. In particular, it is possible for a node to *cross* the obstacle and reach a destination location that is not within the bounds of the obstacle zone. How would you fix it?

Chapter 5
Aspects of Human Emotions and Networks

One of the fundamental aspects that characterizes human beings is emotion. Human emotions and their surrounding environment have a two-way interaction. It is not only a common knowledge that our emotions often affect our actions, but several scientific experiments have established the same. On the other hand, different events taking place around us, in turn, affect our emotions. For example, one may feel sad on hearing about the news of some accident.

Research on emotions have crossed the domain of psychology and has witnessed increasing cross-disciplinary approaches. Apart from psychology, the domains of artificial intelligence (AI), physiology, software engineering, behavioral science, online social networks, and others have been intertwined by this very topic. Thus, it deserves importance to look at certain aspects of emotions to the extent permitted by the scope of this book.[1]

We begin this chapter by looking at some of the popular models of human emotions proposed in the field of psychology. In particular, we discuss about the works of Ekman, one of the pioneers in this domain. We also look at the circumplex model of emotions proposed by Plutchik. Subsequently, we look at a few computational models of emotions and their usage. Computational models of emotions are very popular in the AI domain among others, where the aim is to create "believable" agents that can reflect emotions similar to real human beings. Apart from the Markovian models, we also look at one of the computational models proposed based on Plutchik's theory of emotions.

A significant portion of this chapter discuss about one of the contemporary and popular research topics—emotion detection. In general, merely knowing about what emotion is perhaps of limited use. It is desirable that one (or adaptive systems) should adapt themselves (or their activities) in response to the prevailing emotions. In this chapter, we take a look at different modern approaches of detecting peoples' emotions. Existing research reveals that not only our conversations/interactions (verbal,

---

[1] We keep our discussion at a high level so that the reader does not require a prerequisite in biological sciences or similar subjects. In particular, we do not reflect upon the neurological and related theories of emotion.

© Springer International Publishing Switzerland 2016
S. Misra et al., *Opportunistic Mobile Networks*,
Computer Communications and Networks, DOI 10.1007/978-3-319-29031-7_5

written, or otherwise) are emotion laden, but even the way we use our smartphones can hint at our contemporary emotions or moods.

In the latter part of this chapter, we study an aspect of human–network interplay in MOONs, by considering the effects of human emotions on the network performance. The study uses Meftah's computational model of the emotions, which itself is based on Plutchik's circumplex model of emotions proposed in psychology. In particular, we consider how the variation in individual emotions affects the traffic generation rate in the network. Further, we investigate the scenarios in which users switch off their devices' radios, and effects of such actions on the message delivery ratio. Moreover, we also look at how information received in the MOON affects emotions of the individuals.

5.1 Models of Human Emotions

Psychologists view emotion either as a mental state or process [175]. Multiple theories have been put forward to model and understand emotions, for example, evolutionary, social, and cognitive theories. In this section, we look at some of the well-known models of human emotions proposed in the domain of Psychology till date. More specifically, these models can be classified under the "dimensional" model of emotions. A detailed characterization or analysis of these and other models of emotions is, however, scoped out of this work. Moreover, our focus is on the conceptual representation—we do not look into the underlying neurobiological process of emotions.

5.1.1 Emotions and Facial Expressions

Paul Ekman has been a pioneer in detecting emotions from facial expressions of human beings [176–178, 180]. Based on decades of research, Ekman and his colleagues formulated methods to identify emotions currently experienced by someone by merely looking at his/her face.[2] Moreover, contrary to the contemporary belief, Ekman's research showed certain emotions are not culture specific, but span across diverse cultures.

Ekman and Friesen [177, 178] studied the cross-cultural relationship between human facial expressions and their emotions involving members of the Fore tribe, New Guinea, and some college students. Their study revealed that the association between certain emotions and the corresponding facial expression is typically observed across different cultures giving it a universal notion. Similar experimental results were already obtained in different urban settings. However, the authors thought that in the modern age, with people having exposure to various media, could

---

[2]See [181], Chap. 4, or [182] for detailed discussions.

possibly introduce some form of experimental bias. Hence, the isolated location of New Guinea, where the tribe members were detached from the rest of the world, was chosen. Further experiments [179] revealed that facial expressions related to certain emotions are universal irrespective of the cultures. For example, among other characteristics, happiness is often expressed with a smiling face and the corners of the lips being pulled upwards. Such identified emotions were happiness, sadness, anger, fear, disgust, and surprise.

The above list of emotions constitute the basic emotions as considered by Ekman and his colleagues. Ekman and Cordaro [180] noted that basic emotions are discrete in nature, distinguishable from one another and have evolved typically by adapting to our environment. Ekman ([183], pp. 45–60) discussed in details on how to differentiate between different basic emotions. The author noted that "each emotion is not a single affective state but a family of related states." The emotions belonging to a family share common characteristics. Ekman identified 11 such characteristics—for example, distinct universal signals—(see Table 3.1, [183]; subsequently, two more characteristics were added in 2011 [180]) to distinguish among the basic emotions. However, the author observed that any such single characteristics by itself may not be sufficient to distinguish between two basic emotions.

Facial action coding system (FACS) is a comprehensive tool devised by Ekman and Friesen for distinguishing among, and encoding, different types of facial expressions. Any such expression is composed of movement several facial muscles (for example, a voluntary smile involves contraction at the corner of eyes) that are indexed by Action Units (AUs). Moreover, different emotions can be identified by the predefined combination of the AUs. FACS is widely used for emotion real-time recognition, and also has a related certification program available.

5.1.2 Plutchik's Circumplex Model

Plutchik [184] proposed a color combination-based "circumplex model" to represent eight basic emotions—joy, sadness, trust, disgust, fear, anger, surprise, and anticipation—and their combinations. Plutchik observed that each basic emotion can exist with different levels of intensities. For example, as shown in Fig. 5.1, a higher intensity of joy represents the emotion ecstasy, whereas a lower intensity of it indicates serenity. It may be noted that in the circumplex model, the basic emotions of opposing polarity are placed opposite to one another. In particular, there are four such opposing pairs—joy and sadness, trust and disgust, fear and anger, and surprise and anticipation. This aspect is roughly shown in Fig. 5.1. The colors in the figure, however, are representative, and varies from what Plutchik had originally used.

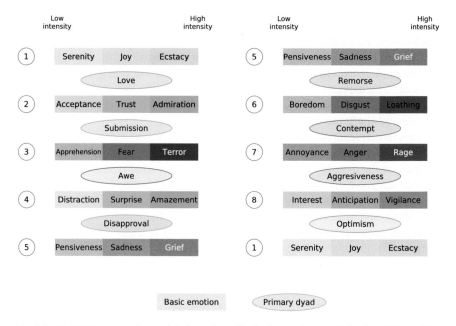

Fig. 5.1 Plutchik's circumplex model of emotions. The basic emotions, variation in their intensities, and primary dyads generated by their combination are shown here

Moreover, advanced emotions can be derived from combination of these eight basic emotions. The primary dyads of emotions are obtained by combining two adjacent basic emotions. For example, a mix of the emotions joy and trust gives rise to the emotion love. Similarly, optimism can be represented as a combination of anticipation and joy. The primary dyads are often felt by us. In a similar way, secondary and tertiary dyads can be obtained by combining two basic emotions 2- or 3-place away along the wheel.

Plutchik's circumplex model of emotions, often represented as wheel or flower, can be folded to give rise to a spinning top-like structure.[3] In particular, the pointed bottom of the so obtained top-like structure, as well as the center of the upper flat surface, represent the points of "emotional zero."

5.1.3   Pleasure-Arousal-Dominance Model

Mehrabian [185] proposed the three-dimensional Pleasure–Arousal–Dominance (PAD) model to represent the emotional states of individuals. In the PAD model, three almost orthogonal axes in the 3D space represent the dimensions pleasure,

---

[3]http://dragonscanbebeaten.wordpress.com/2010/06/04/plutchiks-eight-primary-emotions-and-how-to-use-them-part-1/.

arousal and dominance in relation with any emotion. The Pleasure–Displeasure dimension indicates how pleasant a particular emotion is. For example, while joy is a pleasant emotion, disgust indicates dislike or displeasure. The second dimension, Arousal–Nonarousal, gives an measure of the intensity of the emotion felt. For example, although fear and rage indicates almost similar emotional states, rage has a higher intensity than fear. However, both these emotions are unpleasant [185, 186]. Finally, the Dominance–Submissiveness dimension represents the controlling (or lack thereof) tendency related to an emotion. As an example, anger represents a dominant emotion, but fear typically has a submissive nature associated with it.

Based on the dimensions of the PAD model, Mehrabian [187] further proposed the PAD temperament model. Temperament typically represents the long lasting emotional characteristics (traits) [188]. The model proposes three dimensions— Trait Pleasure–Displeasure, Trait Arousal–Nonarousal, and Trait Dominance– Submissiveness—to represent an individual's emotional trait or temperament.

5.2   Computational Models of Emotions

In the previous section, we looked at some of the models of human emotions. Such models help us to represent or understand emotions in a better way. In relation to this, several computational models of emotions [189–196] have also been proposed by the researchers. Unlike "conceptual models", a "computational model" is helpful for purposes of computations. As an example, let us consider that we have a robot that talks with us. However, talking with a real human being and a robot are largely different experiences since conversations with the latter does not reflect any emotion—a human touch is being missed there. Thus, it would be nice to have the robot emulate certain forms of emotions as well. In other words, based on the current context and situation, the robot's voice, expressions or actions should be emotionally coherent. Computations models of emotions are useful in such scenarios. This domain has been a very popular topic in AI and agent-based research.

5.2.1   Computational Model Based on Plutchik's Theory

Meftah et al. [194] used Plutchik's model in Psychology to develop an algebraic representation of emotions. In [194], the basic emotions are represented with a basis vector $\mathbf{B} = \left(Joy\ Sadness\ Trust\ Disgust\ Fear\ Anger\ Surprise\ Anticipation \right)$, and an emotion vector $\mathbf{e} = \left(e_1\ e_2\ \ldots\ e_8 \right)$, where $e_i \in [0, 1]$ indicates the intensity of the ith emotion in $\mathbf{B}$, $\forall i \in \{1, 2, \ldots, 8\}$. Therefore, a basic emotion, for example, joy, can be represented by the intensity vector $\left(0.6\ 0\ 0\ 0\ 0\ 0\ 0\ 0 \right)$. In general, an intensity vector represents a basic emotion if it contains a single non-zero coefficient [194]. Moreover, any emotion can be expressed as a linear combination of the basic emotions.

Meftah et al. also presented multiple operations that can be performed upon the emotion vectors including vector addition, scalar multiplication, and Euclidean distance. The authors noted that Plutchik's model is simple and intuitive. Moreover, the basic emotions from Plutchik's model can be used to generate all possible combinations of emotions. Their proposed algebraic model of human emotion provides ease of computation as well. Further, the different operators proposed by them help in combining the effects of network dynamics together with the emotional intensities.

Example 5.1 Let us consider the emotion intensity vector $\mathbf{e} = \begin{pmatrix} 0.6\ 0\ 0\ 0.5\ 0\ 0\ 0\ 0 \end{pmatrix}$. Here, the basic emotions joy and trust have non-zero intensities. From Fig. 5.1, we find that combination of these two basic emotions give rise to the secondary emotion love. Other secondary emotions can be represented in a similar way.   □

5.2.2   Markovian Model of Emotions

Researchers have proposed stochastic models of emotions based on Markov models [196–198]. Chandra [196] represented different emotions as discrete states of a Markov chain. Whereas, Kühnlenz and Buss [197] proposed a hidden Markov model (HMM)-based emotion core consisting of four states representing the emotions joy, anger, fear, and relief. Banik et al. considered similar approaches in the context of multi-robot team work.

Chandra [196] proposed a computational model of emotions based on Markov models. As shown in Fig. 5.2, the individual states of the proposed Markov model represent a different state of human emotion. The edges among the states represent transition from one such state to another, and the edge labels indicate the corresponding transition probabilities. Chandra noted that the memoryless property of Markov models is useful since our actions are largely dependent on our current state of emotion rather than the previous history of emotions.

Fig. 5.2 A typical Markovian model of emotions

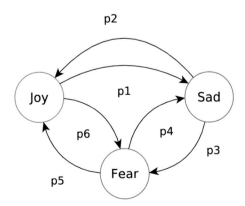

In Chandra's model, each edge is assigned an initial transition probability, which changes with time based on the positive and negative stimuli or experiences. For example, in a two-state model consisting of the emotion states joy and sad, let p_1 and p_2, respectively, be the initial transition probabilities from joy to sad, and vice versa. Further, let α and β, respectively, capture the effects of negative and positive stimuli. Then, the updated transition probability from joy to sadness becomes $p_2 + \alpha p_2 - \beta p2$. Similarly, the transition probability from the state sad to joy becomes $p_1 - \alpha p_1 + \beta p_1$.

Chandra noted that the concerned emotion model can be used in two modes of operation—emotion update mode and behavior query mode. In the emotion update mode, values of the coefficients α and β change based on the external stimuli received by the person/system. As a consequence, the transition probabilities from one state of emotion to other also changes. Such dynamism in the emotional states affect the subsequent actions of the person/system. As an example, Chandra noted that when a person receives a promotion, the value of α would decrease, whereas that of β would increase.

In the behavior query mode of operation, a query stimuli and time period (τ) are given as input based on which the emotion model is solved. If the value of τ is small, probabilistic short-term solution is obtained; otherwise, long-term responses are computed. For example, if $\mathbf{P}$ be the given state transition matrix, then based on Markovian dynamics, the states after time τ can be obtained. Moreover, the steady-state solution can also be derived as τ approaches infinity. Finally, based on the state probabilities, the response is generated as an integer between 0 and 1.

Chandra outlined the architecture of the emotion engine that can be used for realizing this model. The emotion engine takes as input an input stimuli or behavior query, which are processed by the stimuli processor. As an output of the processing, the α and β values are generated and stored in respective registers. Similarly, when operated in the behavior query mode, the solution time, τ, is also stored in a register for further use. The core component of the emotion engine is a matrix buffer to store the emotional states. If there are n underlying emotions, the state transition matrix would have the dimension $n \times n$. The matrix processor unit is a hardware component that reads in the transition matrix and the values of α, β, and τ, and generates the solution. The solution is stored in a set of registers consisting of the states and their respective probabilities. Chandra evaluated the proposed emotion engine using a software implementation, but recommended the use of hardware when the number of emotion states is large.

5.2.3 Emotion and Adaptation

One of the popular models of emotions is based on appraisal theory, which looks at emotions as *processes* in contrast to other schools of thoughts that treat them as states [192, 199]. To give an example [200], one may notice that a person is running toward him/her. As long as the former person does not recognize the latter, he/she may feel fear or anxiety caused by the latter. However, when the person realizes that the latter person is a friend, he/she would become joyful. Thus, three stages can be noticed here [200]. The event that someone was running toward the observer was *perceived* by the observer. Based on the perception, the observer undergoes *appraisal* of the events. Finally, in the *mediation* stage, the person's appraisal outcomes affect his/her emotions. In general, it is the intertwined dynamics of two processes, appraisal and coping, that lead to generation of emotions and subsequent response [192].

Gratch [201] developed Emile, a plan-based agent using emotion appraisal. Few years later, Gratch and Marsella [202] proposed Emotion and Adaptation (EMA), a computational model of emotions based on appraisal theory. The authors noted that EMA is a suitable (and extensible) model for representing human behavior in interactive environments. The proposed EMA algorithm covers five stages starting with construction of causal interpretation of ongoing events to aggregation of emotions and adopting a suitable coping strategy. Gratch and Marsella presented a mapping of appraisal frames into six emotions, and their intensities, based on the desirability (utility) and likelihood (probability) of the perspective (viewpoint) upon which appraisal variables are conditioned. Several coping strategies, such as action, procrastination, and denial, were considered as responses to the appraised events and their induced emotions. The authors noted that the proposed model has the capacity of not only getting affected by the cognitive processes, but also affecting them in turn as well. The EMA model was found to perform well when its behavior was compared against typical expected behavior [192].

5.3   Emotion Detection

So far in this chapter, we have looked at different conceptual and computational models for representing emotions. The reader might be wondering at this point that what purpose does understanding the aspects of emotions serve. Undoubtedly, it is essential that a practitioner knows about the basic concepts. Nevertheless, there are several dimensions to research involving emotions than merely modeling them. For example, a quick survey of the literature reveals that a huge volume of research efforts have been directed toward (automatic) detection of emotions under various contexts [203–213]. Moreover, such research spans across various diverse disciplines, such as AI, psychology, and physiology.

5.3.1   Overview and Applications

One of the commonly used methods of detecting emotions of a person involve physiological sensors [203, 208–210]. Such sensors help in measuring different physiological signals,[4] which, in turn, can be correlated to the emotional states. Li and Chen [203] conducted an experiment where four physiological signals of the subjects were measured—skin temperature, skin conductance, heart rate, and electro-cardiogram (ECG). The authors noted that skin temperature can indicate emotional states quite effectively. Interestingly, it was observed that the waveform for skin conductance almost remained constant when the neutral emotion was considered, whereas heart rate became high and remained so when fear was considered.

The reader might recollect about Ekman's experiments of classifying emotions based on facial expressions. With the advance of technology, research methodology has evolved and later on multimodal approaches have been involved for emotion detection. Gunes and Piccardi [204] concluded that bimodal approach of fusing facial expressions and body movements together can provide better emotion recognition rather than using a single modality. Sebe et al. [205] discussed about the challenges of multimodal approaches (for example, involving users' facial expressions and voice) for detection of emotions. Baig et al. [211], on the other hand, discussed about techniques for emotion detection of crowds. Such innovations are useful in apprehending and possibly controlling or preventing difficult situations.

Chanel et al. [208] presented an interesting study of emotions in the context of games. The authors considered three emotion states in such contexts—boredom, engagement, and anxiety. The authors observed that transition from one emotion state to another can occur due to various reasons. For example, when a player become well versed and competent with a game, but the difficulty level of the game remains the same, then the user apparently gets bored. On the other hand, when the game's difficult level rises up, but the player is not yet competent enough to play it, the player is likely to get anxious. It is when the competence of a player matches with the difficulty level of a game, the player is said to be engaged with the game. The authors verified their hypotheses using a Tetris game that enabled collection of physiological signals, as well as self-reported data. The difficulty level was varied by controlling how fast a block slides downwards. Results of experiments performed on 20 subjects revealed that players have varying skills with the game and different players engaged with the game at different levels. Measurements of physiological signals indicated that different levels of the game (easy, medium, and hard) resulted in different levels of pleasure, pressure, arousal, and motivation. Moreover, the authors were able to classify the emotion states based on the physiological signals with up to 63 % accuracy.[5]

Kołakowska et al. [212] presented a survey of diverse methods of emotion recognition and related applications. Apparently, one may think that there could be no possible connection between the domains of software engineering and emotions. However, Kołakowska et al. noted that, indeed, there is such a connection. For example, the first impression that a website creates on a visitor often determines

---

[4]For example, cardiac activity, respiration, body temperature, skin conductance, and electrical activities of the brain and muscles [210].

[5]Interested readers may also look at [209] where Koelstra et al. discuss in details about collection of physiological signals and their analysis. The concerned database is also publicly available.

its (website's) subsequent use by the visitor [214], and different color themes and text styles have different impacts upon the visitors [215]. Kołakowska et al. reviewed similar impacts of emotion in other different contexts, such as e-learning and video games.

5.3.2 Smartphone-Based Emotion Detection

Measurement of physiological signals can indicate states of emotions with high accuracy. Moreover, such measurements can be done in a more spontaneous environment, for example, a person experiencing fear while watching a horror movie. However, use of physiological techniques has some drawbacks. The devices (sensors) are intrusive—subjects have to fit different electrodes at different parts of their body to record the signals. This requires that the subjects must be present at the experimental zone. Thus, for example, it may not be possible to recognize the emotion of a man walking down the street. In some cases, cost of such sensor devices can also be a factor. Moreover, such techniques may not be scalable or suitable to gather a large data set. For example, depending the experimental details, it may take weeks to collect data from a sizable user base.

On the contrary, in today's world, smartphones are becoming our ubiquitous companions. We carry it with us all the time and remain engaged with it quite frequently throughout the day. To say with a pinch of humor, we are growing a unique type of *human-device relationship*!

Motivated by this, LiKamWa et al. [216] studied whether or not statistics on smartphone usage can be leveraged to infer the mood of its owner. A focus group study involving 25 users revealed interesting characteristics. For example, most participants were found to be interested in sharing (most of) their moods with their friends, as well as in knowing others' moods. On the other hand, depending on their mood, they were found likely to communicate with different people.

For the actual study, LiKamWa et al. deployed two software into the users' smartphones. The first software was a logger that runs in the background and records different usage of the smartphone. The other software was a GUI-based application wherein users can input their contemporary moods. Four different mood states were considered based on the degree of pleasure and activeness reported by the users. When a generalized model was used to infer moods of all the users, an accuracy of 61 % was reported. However, the authors noted that the mood inference models should be personalized—the model should be trained with data from the same user whose mood is to be inferred. For example, a particular user may be more likely to make phone calls, while another user might prefer sending SMS. Such personalized considerations resulted in an average accuracy of about 91 %. In other words, our smartphone usage pattern, in general, tells a lot about our moods!

Lee et al. [217] took a machine-learning approach for unobtrusive emotion recognition of users from sensors' data of smartphones. As a proof-of-concept, the authors developed an affective Twitter client named, AT Client, wherein users were able to post a tweet, as well as report their contemporary emotion. Seven emotions were considered for this purpose—the six basic emotions as suggested by Ekman and a neutral emotion. The objective was to classify a user's emotion into one of those seven categories based on the data collected from various sensors.

From the available sensor inputs, Lee et al. extracted 14 features (for example, typing speed, frequency of special symbol usage, level of device shaking and time) from the behavior and context of smartphone users that are potentially related with users' emotions. Subsequently, 10 features were shortlisted that were found to exhibit stronger correlation with users' emotions. It was found, based on the available data, that Bayesian network resulted in better classification in comparison to others, such as Naïve Bayes classifier. Experimental results indicated that, on an average, the classification worked with about 67.5 % accuracy. In particular, the two emotions neutral and happiness were identified with high accuracy—about 82 and 78 %, respectively. However, the emotions fear and sadness were classified with a very low accuracy. The authors observed that their corresponding accuracies may further be improved by considering additional data. Nevertheless, the consideration of sensor data for emotion recognition opens a new dimension for research in the related field.

5.3.3 Emotion Detection in Online Social Networks

Tang et al. [218] developed MoodCast, a system to predict users' moods based on their activities in online social networks. The authors collected two sets of data—one from an actual mobile social network (MSN) consisting of 30 users at a university, while the other from LiveJournal,[6] a social networking site, where data from about 469K users were analyzed. The MSN data set recorded various attributes, such as communication logs, geolocation, calendar events, and information about the moods of the users. On the other hand, in LiveJournal, users can tag their posts with a "mood label," which the authors classified into three broad categories—positive, negative, and neutral.

In subsequent analysis of the MSN data set, Tang et al. considered three types of correlations—attribute, temporal, and social. In attribute correlation, it is hypothesized that environmental factors and activities would affect a user's emotional states. It was observed that certain activities were highly correlated with some emotional states. For example, it was highly likely to feel "terrible" during a "meeting," whereas the chances of feeling "wonderful" was greater while "shopping" or "playing." Another interesting trend emerged from the data set—the likelihood of making a phone call or sending an SMS drastically increased when the contemporary emotion of the users was "wonderful" as compared to the "normal" emotion. The emotional states were found also to be temporally correlated—past emotions affected the current emotions. Moreover, various degree of social correlation was also observed. In general, it was found that negative emotion was more contagious than positive

---

[6]http://www.livejournal.com/.

emotion. If a user feels bad, then his/her friend is highly likely to have a similar feeling as well. The authors presented a learning mechanism, and a technique based on Metropolis–Hastings algorithm to forecast users' moods. It was found that Mood-Cast resulted in a higher degree of precision with the MSN data set as compared to similar other contemporary methodologies.

Ortigosa et al. [219] developed SentBulk, a Facebook application that supports a wide range of sentiment analysis tasks (for example, classification of sentiments into positive, negative, and neutral categories) based on the messages retrieved from users. The authors noted that such sentiment analysis techniques can be very helpful in the context of e-learning. For example, if the e-learning module identifies that a student is experiencing negative emotions, it can launch some activities (for example, games and simulations) that can motivate the student. Such user engagement could be beneficial to alleviate the negative feelings. Ortigosa et al. observed that, in fact, this can be further generalized in the case of collaborative learning. Often students form study groups to work on their studies together. Having an idea of the group's sentiment would be useful in such collaborative learning scenarios. For example, similar to the previous case, the e-learning system can engage the group in activities to motivate them. Moreover, course instructors can also be appraised of the sentiments of the students. Ortigosa et al. discussed one of the methods for collecting students' sentiments—among the messages retrieved from SentBulk, filter those that belong to the students enrolled to a course. Subsequently, among those messages, filter them further based on the topics. Alternatively, if the concerned course already has a Facebook page, messages posted in that page can be retrieved and analyzed for sentiments.

Thelwall et al. [220] investigated a more general question—how likely it is for social networking interactions to reflect emotions? The authors studied several randomly selected public comments from MySpace[7] for this purpose. Additionally, the authors looked at whether at all there is any difference based on age and gender upon the emotions so expressed. The positive and negative emotions were classified based on a five-point Likert scale,[8] for example, a rating 1 indicates nothing positive, whereas a rating 5 indicates to extremely positive. Several interesting results were unearthed in this context. For example, about 66 % of the comments reflected positive emotions in the rating range 2–4. Interestingly, becoming extremely happy (or sad) was found a rare, if not nonexisting, occurrence. On the other hand, positive emotions, on an average, are mostly reflected when females interact with males. Whereas, male-to-male interactions indicated relatively more negative emotions, on an average. However, Thelwall et al. noted that negative emotions cannot conclusively be associated with gender. Thus, in general, social networks—and MySpace in particular—interactions tend to be emotion-rich.

---

[7]https://myspace.com/.

[8]A Likert scale is a bipolar rating system commonly used in questionnaires. Typically, there are five or seven rating points. The middle point represents a neutral option. Whereas, those to the left and right indicate points with varying degrees of negative and positive biases.

Physiological sensors can provide quite accurate measurements of contemporary emotion of a person. On the other hand, smartphone- and social network analysis-based techniques, among others, are unobtrusive and easily scalable.

5.3.4 Emotional Response of Human Beings

As described earlier, emotions constitute a dynamic process. Not only our emotions do get affected by external events, but often our emotions dictate how we respond to external entities. In the following, we look at a few works highlighting human emotion–action interrelation.

Gao and Liu [221] presented an analysis of emotional reactions of human beings in response to disastrous events. The authors collected series of publicly available materials from BBC and ReliefWeb[9] during the period of Japan earthquake in 2011. Everyday during the period of experimentation, a word matrix was generated from the articles containing the keyword "Japan." Potential stopwords were removed before creating the matrix. Subsequently, the asymmetric similarity of the words were measured and the similarity matrix was clustered. The clustering followed a greedy approach where clusters of words indicating related meanings (for example, earthquake and tsunami) were merged together to form a bigger cluster. Finally, a measure of the frequency (or relative frequency) of emotional words for a given day provided a notion of peoples' emotional response on that particular episode of the disastrous event.

As naturally expected, Gao and Liu found that the words reflecting negative emotions dominated those reflecting positive emotions. In particular, sadness was found to dominate other emotions in both the data sets from BBC and ReliefWeb. It was found that during the earthquake, people mostly came to know about the events by hearing from others. Moreover, propelled by stress and in order to reduce the prevailing uncertainty, people communicated more with others to acquire a clear picture of the ongoing scenario. However, when no clustering-based approach was used, it was rather difficult to identify which specific event episode induced which emotions. This was circumvented by the proposed clustering-based approach by Gao and Liu. Analysis revealed that sadness prevailed more during earthquake, whereas anxiety due to the nuclear crisis was clearly visible.

Beaudry and Pinsonneault [222] investigated the effects that human emotions have on the use of information technology (IT). In particular, the authors conducted a survey-based study upon 249 account managers of two banks. The concerned managers were already working at those two banks before their respective banks decided to introduce a new information system. The authors classified different emotions into

---

[9]http://reliefweb.int/.

four broad categories—challenge, achievement, loss, and deterrence. Four specific emotions namely, excitement, happiness, anger, and anxiety, belonging to each of the four broad categories, respectively, were considered.

Beaudry and Pinsonneault considered different hypotheses and (in)validated them against the experimental data. The authors observed several interesting results after analyzing the responses to the questionnaires. For instance, while anger was not found to be directly related with IT use, anxiety was found to be directly negatively related. Moreover, although happiness had a direct strong association with the IT usage, the association between excitement and IT usage, however, was not found to be significant. Savolainen [223] studied works of other contemporary researchers and arrived at similar conclusions. Savolainen observed that when negative emotions prevail, users may avoid seeking information, whereas emotions, such as fear and anxiety, may lead to seek for information. In other words, our emotions, in most cases, often dictate how we do we act—whether it is using IT, or seeking information, or otherwise.

To summarize this section, we find that it is not only psychologists who should be concerned about emotions. Rather, an understanding of emotions, and its interplay with different external stimuli, would be helpful in different aspects of life and work as well. Intelligent and adaptive systems should be able to tune themselves according to the emotions experienced by the users. For example, it might not be the best idea to present a hard exercise to a student who is feeling a lack in confidence, or a very difficult level of game to a bored player. Moreover, it would be good idea to research the visual elements of a website (such as color scheme and font style) so that they can stimulate the positive emotions in users. Finally, our emotions affect our actions more often than not.

5.4   Effects of Emotion in MOONs

Until now in this chapter, we have looked at various aspects and models of emotions and their applications. Although there are several works that deal with emotion detection in social networks or using smartphones, a different aspect largely remain unaddressed in the literature—how emotions affect ongoing communications in a network. And its reverse—how ongoing communications affect the emotions of the users in the network? This is more relevant in MOONs, where the presence of human beings is inherent to the network. In the remainder of this chapter, we study these aspects in details.[10]

---

[10]Sections 5.4, 5.5, and 5.6 have been reproduced by permission of the Institution of Engineering and Technology.

S. Misra, B. K. Saha: "On emotional aspects in mission-oriented opportunistic networks," *IET Networks*, 2013, vol. 3, no. 3, pp. 228–234.

5.4.1 Relevance in MOONs

It may be noted that this problem is of particular interest in MOONs (or OMNs, in general) unlike traditional networks, such as the Internet and cellular networks. The reason behind this is largely their architecture; let us try to understand why. Let us assume that Alice is making a phone call to Bob. Now, suppose that Bob is sad, and does not respond to Alice's call. What are the implications of such action? Of course, Alice would not be able to pass her message to Bob. In other words, we can say that Alice generated (or had) a message, which, unfortunately, could not be delivered to Bob. Therefore, this is a loss for the Alice-Bob pair considered. However, it may be observed that Bob's refusal to accept the call in no way impact another ongoing call between say, Carol and Dave. Other users can engage in phone calls or send SMS irrespective of Bob's actions or similar actions by any other people. To generalize, one or more user(s) refusal to participate in communication does not affect communication between any other pair of users. This is only possible because cellular networks have a centralized infrastructure, which ensures end-to-end communication paths. Similar arguments go for any message transferred over the Internet.

Now, the reader might recall from Chap. 1 that unlike conventional networks, OMNs/DTNs have vastly different characteristics. A major feature is that OMNs (and thus, MOONs) lack in end-to-end communication paths more often that not. Such constraints in communication mechanisms require that every node in OMNs cooperate with one another. In particular, if a node refuses to receive and forward messages, not only the it fails to receive messages that were destined for itself, but also negatively influences the message delivery to other nodes. This impact may not be large if only a single node behaves so, but could be overwhelming if several nodes act similarly. Thus, it is an interesting problem to evaluate the effects of emotions on the communications in an OMN, and vice versa. Such a study is presented in the remainder of this chapter.

5.4.2 Terminologies

We consider Plutchik's circumplex model of emotions and the computational model [194] of emotions by Meftah et al. as our basis. In particular, we represent the emotion states of individual users of a MOON using intensity vector as in [194]. In this section, we present a few definitions (Definitions 5.1 and 5.2 are adopted from [194]) and other theoretical characterizations that help in modeling the effects of emotion.

Definition 5.1 (*Addition of Emotion Vectors [194]*) Let, $\mathbf{e}_1 = (e_{1,1}\ e_{1,2}\ \ldots\ e_{1,n})$ and $\mathbf{e}_2 = (e_{2,1}\ e_{2,2}\ \ldots\ e_{2,n})$ be two emotion vectors. The addition of the vectors, $\mathbf{e}_1$ and $\mathbf{e}_2$, denoted by $\oplus$, is defined as:

$$\mathbf{e}_1 \oplus \mathbf{e}_2 = (max(e_{1,1}, e_{2,1})\ max(e_{1,2}, e_{2,2})\ \ldots\ max(e_{1,n}, e_{2,n})).$$

Definition 5.2 (*Scalar Multiplication with Emotion Vector [194]*) The scalar multiplication of an emotion vector $\mathbf{e} = \begin{pmatrix} e_1 \ e_2 \ \dots \ e_n \end{pmatrix}$ with a scalar c, denoted by $\odot$, is defined as:

$$c \odot \mathbf{e} = \begin{pmatrix} c \times e_1 \ c \times e_2 \ \dots \ c \times e_n \end{pmatrix}.$$

Definition 5.3 (*Dominant Emotion*) Given the intensity vector $\mathbf{e} = \begin{pmatrix} e_1 \ e_2 \ \dots \ e_n \end{pmatrix}$ of an individual's emotions at any time t, the dominant emotion (DE) of the person at that time instant is defined as:

$$DE(\mathbf{e}_i(t)) = \begin{cases} \underset{k}{\mathrm{argmax}}(e_k), & e_k > e_j, j \neq k, \forall j \\ \varnothing_N, & e_k = e_j, \forall j, k \\ DE(\mathbf{e}_i(t-1)), & \#(max(\mathbf{e}_i(t))) \in (1, n), \end{cases} \tag{5.1}$$

$\forall j, k \in \{1, 2, \dots, n\}$. Here, e_k indicates an element of $\mathbf{e}_i(t)$, and $\#(A)$ denotes the count of any event A. In other words, when all the emotions have unique intensities, the DE is the one with the maximum intensity. Here, $\varnothing_N$ denotes the *neutral* emotion, and $DE(\mathbf{e}_i(t-1))$ indicates the dominant emotion of the person at a previously counted time instant. In case, at any time instant, there exists multiple emotions with equally highest intensities, we consider the previously dominant emotion to sustain.

Definition 5.4 (*Scalar Exponent of Emotion Vector*) The scalar exponent of an emotion vector $\mathbf{e} = \begin{pmatrix} e_1 \ e_2 \ \dots \ e_n \end{pmatrix}$ with a scalar c is defined as:

$$c \circledast \mathbf{e} = \begin{pmatrix} c^{e_1} \ c^{e_2} \ \dots \ c^{e_n} \end{pmatrix}.$$

Definition 5.5 (*Diagonal Form of an Emotion Vector*) Given an emotion vector $\mathbf{e} = \begin{pmatrix} e_1 \ e_2 \ \dots \ e_n \end{pmatrix}$, its representation as a diagonal matrix is denoted by:

$$\mathrm{diag}(\mathbf{e}) = \begin{pmatrix} e_1 & 0 & \dots & 0 \\ 0 & e_2 & \dots & 0 \\ \vdots & \vdots & \ddots & 0 \\ 0 & 0 & \dots & e_n \end{pmatrix}.$$

A summary of the different notations used in the remainder of this chapter is presented in Table 5.1.

5.4.3   Influence on Network Dynamics

Let us now look at the interplay between users' emotions and network dynamics in details. To elaborate, on one hand, the contemporary emotional states affect aspects of users' participation in the MOON. On the other hand, the dynamics of MOON, in turn,

Table 5.1 Emotion and network interaction—notations summary

Symbol	Description
$\mathbf{e} = \begin{pmatrix} e_1 & e_2 & \dots & e_n \end{pmatrix}$	Emotion intensity vector
$\beta = \begin{pmatrix} \beta_1 & \beta_2 & \dots & \beta_n \end{pmatrix}$	Amplification in the traffic load
$\gamma = \begin{pmatrix} \gamma_1 & \gamma_2 & \dots & \gamma_n \end{pmatrix}$	Degree of cooperation
$\mathbf{a} = \begin{pmatrix} a_1 & a_2 & \dots & a_n \end{pmatrix}$	Emotion decay coefficient vector
$\mathbf{r} = \begin{pmatrix} r_1 & r_2 & \dots & r_n \end{pmatrix}$	Effect of the received messages upon the basic emotions
M	Set of messages generated
$R(t)$	Traffic generation rate at time t
$k(t)$	Degree of cooperation at time t
τ	Temporal decay function

affect the emotions of users. Such considerations are motivate by the appraisal theory of emotion, which states that emotions affect the cognitive process and behavior, and itself (emotions) get affected by them.

Here, specifically, we consider two network-specific aspects—traffic generation rate and degree of users' cooperation. We assume that the intensity of traffic (i.e., message generation rate) varies with different emotions. Sometimes the rate is above normal, while sometimes it goes below normal. By "cooperation",[11] we indicate to what extent a user keeps the radio of his/her device ON. Note that, message exchanges are possible only when two devices are in proximity and both have their radios ON.[12] Once again, we assume that such behavior is affected by the contemporary state of emotion

A careful reader might observe here that we did not consider any state-based emotion model (for example, the Markovian model of emotion). However, note that we have already defined the term dominant emotion (Definition 5.3). The dominant emotion of a user corresponds to a particular state of emotion, or it is neutral otherwise.

We elaborate on these aspects in the following.

Let, $\mathbf{e}_i(t) = \begin{pmatrix} e_1^i & e_2^i & \dots & e_n^i \end{pmatrix}$ denote the emotion vector of the ith person at time t. Let, $R_i(t)$ and $k_i(t)$, respectively, be the message generation rate and level (degree) of cooperation of the ith person at time t. Let us consider that $R(0)$ and $k(0)$, respectively, represent the message generation rate and level of cooperation under "normal conditions." Specifically, let us assume that $k(0) = 1$. In other words, under normal circumstances, all users are assumed to be fully cooperative.

---

[11] We shall revisit this again in Sect. 5.5.3.

[12] Additionally, certain guarantees on homogeneity may also be required, as we shall see in a later chapter.

Let, $\beta = \begin{pmatrix} \beta_1 & \beta_2 & \dots & \beta_n \end{pmatrix}$ be the coefficients of amplification in (individual) traffic load, so that

$$R_i(t) = \beta_j \times R(0), \tag{5.2}$$

where $j = DE(\mathbf{e}_i(t))$, is the dominant emotion for any user i at time t, $\beta_j \geq 0$, $\forall j \in \{1, 2, \dots, n\}$. Thus, the aggregate traffic injected at any time t in the network is given by $R(t) = \sum_{i=1}^{N} R_i(t)$, where N denotes the total number of users.

Further, let $\gamma = \begin{pmatrix} \gamma_1 & \gamma_2 & \dots & \gamma_n \end{pmatrix}$ be the degrees of cooperation corresponding to different basic emotions, $0 \leq \gamma_j \leq 1$, $\forall j \in \{1, 2, \dots, n\}$, such that

$$k_i(t) = \gamma_j \times k(0). \tag{5.3}$$

Thus, for any individual i at time t, (5.2) and (5.3), respectively, give the variation in traffic generated and cooperation extended as an effect of his/her the then *dominant emotion* j.

Further, the network dynamics—in the form of information (messages) received— can, in turn, alter an individual's state of emotion. Let, M denote the set of messages generated in the MOON. Let, the vector $\mathbf{M}_i(t) = \begin{pmatrix} M_{i,1} & M_{i,2} & \dots & M_{i,n} \end{pmatrix}$ be a collection of the set of messages received by any node i till the time instant t, such that the $M_{i,j} = \{m : m \in M\}$ messages affect the jth emotion, $\forall j \in \{1, 2, \dots, n\}$.

Further, let $\mathscr{I}$ be a function that captures the effects of a given set of messages (information) on a particular emotion so that $\mathscr{I}(M_{i,j}) \in [0, 1]$, $\forall j$. Then, the updated emotion vector at time t:

$$\mathbf{e}_i(t) = \mathbf{e}_i(t') \otimes \tau(t - t') \oplus \mathbf{r}_i(t), \tag{5.4}$$

where $\otimes$ denotes the matrix multiplication operator and $\mathbf{r}_i(t) = \begin{pmatrix} r_{i,1} & r_{i,2} & \dots & r_{i,n} \end{pmatrix} = \begin{pmatrix} \mathscr{I}(M_{i,1}) & \mathscr{I}(M_{i,2}) & \dots & \mathscr{I}(M_{i,n}) \end{pmatrix}$; $t > t'$.

The function τ represents the temporal decay in the intensities of the emotions and is considered to be exponential, as indicated in [224]. Therefore, with $\Delta t = t - t'$, we have

$$\tau(\Delta t) = \begin{pmatrix} e^{-a_1 \Delta t} & 0 & \dots & 0 \\ 0 & e^{-a_2 \Delta t} & \dots & 0 \\ \vdots & \vdots & \ddots & \vdots \\ 0 & 0 & 0 & e^{-a_n \Delta t} \end{pmatrix}$$

$$= \text{diag}(e \circledast (-\mathbf{a} \odot \Delta t)).$$

Here, the constant vector $\mathbf{a} = \begin{pmatrix} a_1 & a_2 & \dots & a_n \end{pmatrix}$ denotes the coefficients of decay of the basic emotions, $a_j \in (0, 1)$, $\forall j \in \{1, 2, \dots, n\}$, and e is the base of natural logarithms. Thus, the greater the value of any a_j, the quicker the decay of that basic emotion. In other words, given an initial emotion vector $\mathbf{e}$, the vector $\mathbf{a}$ determines which basic emotion, if any, would be dominant in the long run when not affected by any other external factors.

The vector addition operator is used with $\mathbf{r}$ in (5.4) because the effects of the received messages can act as stimuli that could altogether alter the emotion of an individual.

If e'_j represents the intensity of the jth emotion at time t', $\forall j$, then, on simplification, we get:

$$\mathbf{e}(t) = \left(e'_1 \times e^{-a_1 t}\ e'_2 \times e^{-a_2 t}\ \dots\ e'_n \times e^{-a_n t}\right) \oplus \left(r_1\ r_2\ \dots\ r_n\right). \quad (5.5)$$

Thus, using the computational model of Meftah et al. [194], we can determine the emotional state using a few algebraic operations.

Example 5.2 Consider the following emotion intensity vector, $\mathbf{e} = \left(0.3\,0\,0\,0\,0\,0\,0\,0\right)$. Also, let the coefficients of decay of basic emotions be given by the vector, $\mathbf{a} = \left(0.001\ 0.001\ 0.001\ 0.001\ 0.001\ 0.001\ 0.001\ 0.001\right)$. For simplicity, we assume all the decay coefficient values to be same. Therefore, after a single unit of time, the updated emotion vector would be, $\mathbf{e} = \left(0.3 \times e^{-0.001 \times 1}\ 0\ 0\ 0\ 0\ 0\ 0\ 0\right)$, which approximately is $\left(0.29\,0\,0\,0\,0\,0\,0\,0\right)$. $\qquad\square$

5.5 Application Scenario

The above-discussed model is applicable in the case of a MOON formed in a post-disaster rescue operation. In such a scenario, the survivors and the rescue workers, equipped with smartphones, engage in opportunistic communications. The messages exchanged among the users include, but not limited to, information about the rescued victims and queries for missing persons.

In the remaining sections, we consider a subset of the three emotions namely, joy, sadness, and fear, i.e., $n = 3$, typically occurring in the post-disaster scenario. In such a scenario, a user feels happy if he/she is unhurt and his/her family is safe. However, if a person is unable to get someone's information, he/she may panic. News of any injured family member would sadden a person. We note that these assumptions could be easily generalized for all the emotions—and could be used in other scenarios as well—by considering the effects of the received messages upon them, if any.

5.5.1 Variation in Emotion

Let, the messages exchanged in the MOON be classified into three types—*Good*, *Bad*, and *Unknown*. The *Good* messages represent the case when someone intends to inform that he/she is safe, or perhaps a missing person has been located. Likewise, the *Bad* and *Unknown* represent the cases of dissemination of related information. Let, m_G, m_B, and m_U, respectively, be the counts of such messages received by any individual until time t. To determine $\mathbf{r}_i$, we consider that the effect of any kind of message on a basic emotion is proportional to the number of such messages received. Thus,

$$\mathbf{r}(t) = \left(\frac{m_G}{m_N+1} \quad \frac{m_B}{m_N+1} \quad \frac{m_U}{m_N+1} \right), \tag{5.6}$$

where $m_N = m_G + m_B + m_U$, is the total number of messages received, $m_N \geq 0$.

While the intensity of emotions decay with time, the rate of decay may not be the same for all. For example, one may quickly shift from a joyful state to a neutral state, but the sadness may sustain comparatively longer. Here, we consider that a_1— the decay rate of joy—to be greater than the decay rates of sadness and fear, i.e., $a_1 > a_2, a_3$. It may be noted here that this is in contrast to the findings in [218], where it was reported that normal and good emotions sustained longer than the negative emotions. The apparent deviation can be explained as follows. The study in [218] was performed in a university setting, and the distribution of the emotions exhibited there are largely different from that of the aftermath of a large-scale disaster. Thus, the observations would also vary as per the scenario considered. Indeed, it can be recalled from an earlier section that negative emotions dominate in typical scenarios considered here [221].

5.5.2   Variation in Traffic Load

We consider the time distribution of message generation in the network to be Poisson, with the rate λ messages/second. For simplicity, we further consider that, on an average, all the users create same number of messages. Thus, if there are N users in the network, we have $X = \sum_{i=1}^{N} X_i$, where $X_i \sim \text{Poisson}(\lambda_i)$, and $\lambda_i = \frac{\lambda}{N}, \forall i \in \{1, 2, \ldots, N\}$. The inter-arrival times of X are exponentially distributed with mean time as $\frac{1}{\lambda}$. Thus, under normal circumstances, the average duration between two consecutive generation of messages for each individual is given by

$$\bar{T}_i = \frac{1}{\lambda/N}. \tag{5.7}$$

We consider that a user generates messages in the "normal rate" with the average intermessage duration as $\bar{T}_i$ when he/she is happy. The amplification in traffic generation rate in this case is unity, i.e., $\beta_1 = 1$. A user, however, is considered to stop generating any message when sad so that $\bar{T}_i = \infty$. In this case, $\beta_2 = 0$. Finally, any user is assumed to double the message generation rate when the dominant emotion is fear, which could be explained by the rationale that an anxious person sends out more queries to the different people in the MOON. This case is equivalent to halving $\bar{T}_i$, and is represented with $\beta_3 = 2$. Thus, $\beta = \begin{pmatrix} 1 & 0 & 2 \end{pmatrix}$. Once again, we apply similar reasoning as earlier to justify its divergence from [218]. Moreover, this assumption is further supported by the observation by Savolainen [223] that anxious behavior leads more toward information seeking, while negative feelings discourage that.

Finally, let $\mathbf{p_m} = \begin{pmatrix} p_{m_1} & p_{m_2} & p_{m_3} \end{pmatrix}$ denote the probabilities of different categories of messages generated in the MOON. In other words, on an average, p_{m_1} gives the proportion of *Good* messages generated in the MOON. Similarly, p_{m_2} and p_{m_3}, respectively, denote the proportion of the *Bad* and *Unknown* messages generated.

In this context, the following lemma and theorem are stated.

Lemma 5.1 *In a MOON with N nodes, where the network-wide rate of traffic generation is Poisson(λ), and the source and destination nodes are chosen uniformly at random, the traffic arrival rate at any destination node is less than Poisson(λ_e), where $\lambda_e = \frac{(2N-1)}{6N} \times \lambda$.*

Proof Let us consider that the N nodes in the MOON form a complete graph, where the messages could be delivered *instantaneously*. Also, let us assume that a particular node, X, is selected as destination by the other $(N-1)$ nodes. Let, P_i denote the probability that i nodes in the MOON select the node X as the destination. Such selection by the source nodes is independent of each other. Then, the expected rate of traffic destined for the node X is $\sum_{i=1}^{N-1} i \times P_i \times Poisson(\lambda/N) = \sum_{i=1}^{N-1} i \times \frac{i}{N-1} \times Poisson(\lambda/N) = \frac{2N-1}{6} \times Poisson(\lambda)$. Now, since each of the N nodes is equally likely to be chosen as the destination, on an average, the traffic arrival rate at any destination node is $\frac{1}{N} \times \frac{2N-1}{6} \times Poisson(\lambda) = Poisson(\frac{2N-1}{6N} \times \lambda) = Poisson(\lambda_e)$.

In real-life MOONs, however, messages cannot be delivered *instantaneously*, since the contacts among the nodes are intermittent and the messages are transferred over multiple hops. Moreover, contact opportunities are lost due to the effects of the emotions sadness and fear. Assuming that the traffic arrival at the destination is *still* a Poisson process, let us denote the arrival rate as $\lambda_e - \mu$, where $\mu > 0$ is a parameter that accounts for the characteristics of the MOON considered. Thus, the traffic arrival rate at the destination nodes is less than $Poisson(\lambda_e)$. □

The above reflects the maximum traffic in flow to a node, which can be useful for the buffer management schemes.

Theorem 5.1 *The rate of change in intensity of the ith emotion, over a time interval Δt, is either $\frac{e_i(t) \times (e^{-a_i t} - 1)}{\Delta t}$ or $\frac{p_{m_i} - e_i(t)}{\Delta t}$, where p_{m_i} is the proportion of messages generated in the MOON that affects the ith emotion. It is assumed that the proportion of the corresponding messages received is also the same.*

Proof Let us consider equally spaced time instants t_j, so that $t_j - t_{j-1} = \Delta t$. Then, from (5.4) and (5.6), the intensity of the ith emotion at time $t_{j+1} = t_j + \Delta t$ is $e_i(t + \Delta t) = max\{e_i(t) \times e^{-a_i t}, p_{m_i}\}$. Therefore, the time rate of change in the intensity of an emotion is $\frac{max\{e_i(t) \times e^{-a_i t}, p_{m_i}\} - e_i(t)}{\Delta t}$. Thus, if $e_i(t) \times e^{-a_i t} \geq p_{m_i}$, the rate of change is $\frac{e_i(t) \times (e^{-a_i t} - 1)}{\Delta t}$; otherwise, it is $\frac{p_{m_i} - e_i(t)}{\Delta t}$.

5.5.3 Changes in User Cooperation

We now elaborate on the notion of "cooperation" used in this work. As discussed previously, Beaudry and Pinsonneault [222] established that while happiness is positively associated with the IT usage, anxiety of human beings is negatively associated with it. Inspired by this, we consider the case in which the users deactivate networking capabilities of their devices depending on their respective dominant emotions, and thus, loses potential opportunities of both sending and receiving messages. We consider the scenario where a user deactivates networking capabilities of his/her device—by turning his/her device, or just the radio, off—depending on his/her dominant emotion. This, however, is not selfish behavior—a user does not look for his/her own gains. Rather, by switching off the radio, he/she loses potential opportunities of both sending and receiving messages.

We consider that a user keeps the radio of his/her device active when the DE is joy or neutral. In this case, the user fully "cooperates" with the network operations, i.e., $\gamma_1 = 1$. Any user, whose DE is sadness is assumed to switch off the device or its radio, i.e., $\gamma_2 = 0$. Finally, when the DE of any user is fear, he/she switches off the device radio with a probability 0.5, i.e., $\gamma_3 = 0.5$. In this case, the device remains inactive for a random duration of time in between $\{0, \ T_{off}\}$ seconds. However, a user is not expected to turn off the radio every second. Indeed, if she does so, then no, or perhaps immensely less number of, messages could be transferred. Therefore, we assume that any user opts to switch off the radio periodically after time T_{Fear}.

Thus, in terms of user cooperation, we have $\gamma = \begin{pmatrix} 1 & 0 & 0.5 \end{pmatrix}$. We note that, when the DE is neutral, the individual traffic generation rate is λ, and the device radio remains on. Algorithm 5.1 shows the different actions taken by any user when the DE is fear.

It may be noted that the choice of β and γ is context-sensitive. For example, if one considers winning a gold medal by a particular country during the Olympics, people in the audience from that country would become happy and may result in traffic explosion while trying to communicate with others. In this work, we choose the values of β and γ such that they can represent typical post-disaster scenarios.

It may be further noted that certain effects of emotion are permanent, or at least linger for long without being affected by other factors. For example, in the context of the post-disaster scenario considered in this study, if a person gets news about losing someone very close to his/her, the individual would become sad. In that case, no other external messages or emotions could surpass his/her sadness, at least for a long duration of time. As an effect, the user's device would remain inactive for a long period of time. Proposition 5.1 formalizes this effect for any emotion, in general. However, we do not consider this phenomenon further in our work.

Proposition 5.1 (Permanent Effect on Emotions) *The permanent effect of an event, occurring at time t_c, on the intensity of the jth emotion, $j \in \{1, 2, \ldots, n\}$, of any individual is $max\{e_j(t), u_{t_c}(t)\}$ or $min\{e_j(t), 1 - u_{t_c}(t)\}$, where $u_{t_c}(t)$ is the unit step function, i.e., $u_{t_c}(t) = 0$, if $t_c < t$; otherwise, $u_{t_c}(t)$ is 1. The choice of min or max depends on the event and emotion considered.*

Algorithm 5.1: Actions taken when dominant emotion is fear

Input:
- **e**: Emotion vector at time t
- **a**: Decay coefficients
- λ: Normal message generation rate (Poisson)
- t': Current time

Output:

- Changes in traffic load and cooperation

1 $\Delta t = t' - t$
2 Evaluate **r** using (5.6)
3 Update **e** according to (5.5)
4 Determine dominant emotion DE
5 **if** DE **is** *fear* **then**
6 $\quad$ $\lambda' = 2 \times \lambda$
7 $\quad$ Generate messages with Poisson(λ') process
8 $\quad$ $t_L =$ Last time when the radio was switched off
9 $\quad$ **if** $t' \geq t_L + T_{Fear}$ **then**
10 $\quad\quad$ $t_L = t'$
11 $\quad\quad$ $p =$ Random number in the range $[0, 1)$
12 $\quad\quad$ **if** $p < 0.5$ **then**
13 $\quad\quad\quad$ $t_{off} =$ Random number in range $[0, T_{off})$
14 $\quad\quad\quad$ Switch off the radio of the device for time t_{off}

5.6 Practical Implications

In this section, we take a look at the effects of emotion on the dynamics of MOONs. Out objective is to consider an initial traffic pattern, an initial set of emotion intensity vectors, and determine how the system evolves over time. We used the ONE simulator [51] with real and synthetic mobility scenarios for this purpose. We considered the NCSU's [225] KAIST [85] real-life mobility traces for 24 h collected in a terrain of about 33×23 sq. km area. We, however, treated the traces bit differently. Since there is no way to identify which trace belonged to which device, and all these traces were generated independently of one another, we considered the 92 traces to belong to 92 different individuals. In other words, 92 nodes moved according to the pre-specified coordinates.

It may be noted that the use of mobility traces over connection traces was advantageous in this scenario. This is because, with mobility traces, one has the ability to control what the transmission range of a node should be. However, when connection traces are used, the connection patterns are well defined, and one does not have any

control over the nodes. Therefore, in general, one should also think upon which kind
of real trace is suitable under which scenario.

We assigned an initial emotion intensity vector to each node in the OMN. Their
emotions decayed naturally in an exponential fashion in the absence of any external
event. We also assigned the initial traffic generation rate and the degree of cooperation
to each node. The proportion of different types of messages created in the MOON—
Good, *Bad*, and *Unknown*—were indicated by the vector $\mathbf{p_m}$. Based on the messages
received, the individual emotions were affected. This, in turn, influenced the traffic
rate and cooperation, as discussed in Sect. 5.5.

We also studied at the performance of the MOON considering heterogeneity in the
network. Specifically, we investigated the scenario when all the all human owners of
the devices were not affected by their respective emotions, and, therefore, reflecting
heterogeneous actions. Moreover, we also took the heterogeneity of the devices into
consideration. We considered two types of network interfaces—*if*1 and *if*2—that are
incompatible with one another. The transmission range of the former was taken to be
10 m, while the latter's range was 20 m. We considered a group of devices equipped
with the interface *if*1, and the remaining having *if*2. We evaluated the performance
of such a heterogeneous MOON under different scenarios.

Additionally, we used the post-disaster mobility model developed by Nelson et al.
[128]. This scenario involved 60 civilian nodes and 2 ambulances moving in a terrain
of size 1000×1000 sq. m. with 2 hospitals for 7 h. A single disaster event of intensity
10000–20000 unit was considered. It may be noted here that this model induces severe
partition of the nodes in the network.

Individual emotions of the nodes were updated every 300 s. Further, the nodes
checked every 900 s whether their DE is fear, with $T_{off} = 900$ s. The changes in
the traffic generation rate and switching the device radio on/off were implemented
accordingly. All users were considered to have same initial intensities of the basic
emotions, and the rate of decay of a given basic emotion was considered to be same
for all. The message generation rate was Poisson with mean intermessage creation
time as 25 s in the normal scenario. The transmission speed and range, respectively,
were 2 Mbps and 10 m.

The Spray and Wait [42] routing protocol (in binary mode with 16 copies) was
chosen, since it has a upper limit on the message replication count. However, the
emotion-induced actions would affect the message delivery using other routing protocols as well.

In all scenarios, wherever appropriate, average of 10 samples were considered
and the 95 % confidence interval (CI) was computed. It may be noted here that
although the real-life movement traces considered here would always generate a
deterministic topology, in certain cases, random numbers have been used here—for
example, while implementing the effects of user degree of cooperation—which, to
some extent, results in randomization of the simulation scenarios considered.

At first, we present the results obtained using the KAIST traces. Figure 5.3 shows
the variation in the average count of messages generated every hour in the network
using Poisson traffic. The effects of emotion on cooperation were not considered in
this case. The case where $\mathbf{e} = \begin{pmatrix} 0.5 & 0.5 & 0.8 \end{pmatrix}$, i.e., fear has the highest initial intensity,

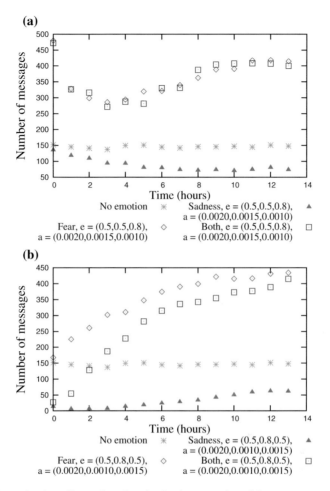

Fig. 5.3 Fluctuation in traffic rate based on the dominant emotion of the users

reflects the situation that the users exhibited fear/anxiety after a disaster has struck. The figure clearly shows that there is a steep increase in the message generation rate when only the effects of fear are considered.

In Fig. 5.3a, the effects of the emotions fear, and fear and sadness both, appear to be similar. This is due to the reason that the intensity of fear in this case is higher ($e_3 = 0.8$), and its corresponding decay coefficient is the minimum ($a_3 = 0.001$). Thus, fear tends to be the dominant emotion in the network in this case. However, in Fig. 5.3b, sadness is the dominant emotion ($e_2 = 0.8$, $a_2 = 0.001$), and as a result of which, traffic generation almost stops when only sadness is considered.

We also studied at how the messages created in the MOON varied when different traffic amplifications (β)—as a consequence of fear—were considered. The effects of sadness on the traffic fluctuation was not considered here. Figure 5.4 plots the

Fig. 5.4 Variation in traffic rate based on amplification factor when **a** **e** = (0.5 0.5 0.8) and **b** **e** = (0.5 0.8 0.5). Here, **a** = (0.002 0.0015 0.001)

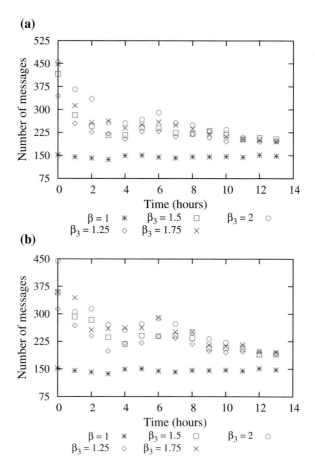

temporal rate of messages generated in the MOON for different amplification factors. The initial intensity of the emotions (**e**) and their decay coefficients (**a**) are as indicated in the figure.

In Fig. 5.4, the result with $\beta = 1$ represents the scenario when no effects of emotions upon the users were considered and traffic was generated at the normal rate following the Poisson process. The other results depict the fluctuation for different amplification factors (β_3), as shown in the figure. It can be observed that, with increasing simulation time, the amplified traffic rates tend to converge to the normal rate. This can be explained by recollecting that the messages in the network were generated for a given time interval. Therefore, once all such messages are delivered—or lack thereof—the decay of individual emotions cease to be affected by the external stimuli. As a consequence of such mostly-natural decay, after one point of time, intensities of all the basic emotions tend to zero and thereby, making neutral emotion ($\varnothing_N$) the dominant one in most cases. As described earlier, the traffic generation rate was considered to be invariant in such a scenario.

Fig. 5.5 Emotion-induced
variation in cooperation and
its effects on the message
delivery ratio when (**a**)
a = (0.002 0.001 0.0015)
and (**b**)
a = (0.002 0.0015 0.001)

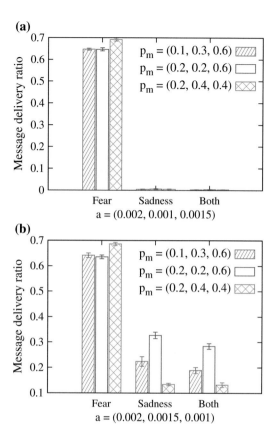

We studied the delivery ratio of the messages obtained in the MOON considered
here. As noted in Chap. 3, this is considered as a typical mission performance parameter. Figure 5.5 shows the delivery ratio of the messages, when the users switched
their devices' radios on/off as per their contemporary dominant emotion. Emotion-
induced traffic fluctuations were not considered here. We note that the delivery ratio
of the messages when no effects of emotion were considered, was 0.7249 ± 0.00423
with 95 % CI.

The plots in Fig. 5.5a, b, respectively, correspond to the decay rates, $\mathbf{a} =$
$(0.002\ 0.001\ 0.0015)$ and $\mathbf{a} = (0.002\ 0.0015\ 0.001)$. In other words, sadness sustained for a long time in the former case. As a consequence, when only the effects
of sadness were considered, as shown in Fig. 5.5a, the device radios were turned off
for longer time, and delivery of the messages became impossible.

On the other hand, in the second case (Fig. 5.5b), the effect of fear was prominent.
So, when only the effects of sadness were considered, moderate delivery ratio of
the messages could be observed. Further, in case of sadness, the delivery ratio was
maximum when the number of *Bad* messages generated—as indicated by $\mathbf{p_m}$—was
minimum.

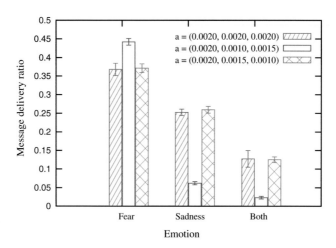

Fig. 5.6 Effects of emotions on the message delivery ratio considering the variation in both traffic and cooperation

Figure 5.6 presents the case when the effects of emotion upon both traffic generation and cooperation were considered, with $\mathbf{e} = (0.5\ 0.5\ 0.8)$. The case when the decay coefficients were $\mathbf{a} = (0.002\ 0.001\ 0.0015)$ presents some interesting results. When only fear was considered, although the initial intensity was high, slow decay of sadness reduced the traffic generation rate, and thus, the delivery ratio of the messages was effectively increased. However, when only the effects of sadness were considered for the same decay coefficients, the impact of switching off the device radios prevailed over the reduced traffic generation rate. As a result, the delivery ratio of the messages was substantially reduced. Finally, when the effects of both sadness and fear were considered for the same $\mathbf{a}$, traffic spike together with device radio switching off for longer durations resulted in the lowest delivery ratio.

Furthermore, if we look at these three cases for the other two decay coefficients, we find that the performance has been almost similar for both the decay coefficients. This is due to the reason that given the initial intensities of the basic emotions and the decay coefficients, sadness did not become the dominant emotion for most nodes in the network.

Figure 5.7 captures the effect of turning off the device radio on the performance of the MOON. Here, we considered $\mathbf{p_m} = (0.2\ 0.4\ 0.4)$, $\mathbf{e} = (0.5\ 0.5\ 0.8)$, and $\mathbf{a} = (0.002\ 0.0015\ 0.001)$. In other words, we let fear to be the dominating emotion among most users in the MOON. Moreover, the effect of fear on the traffic amplification was not considered—all the nodes generated messages with the normal Poisson rate. The probability of turning off the radio is shown along the x-axis. The maximum delivery ratio of the messages was obtained when the probability was 0.0, i.e., the case when the users did not turn off their device radios. It can be observed that the performance degraded steadily as the probability kept on increasing. This

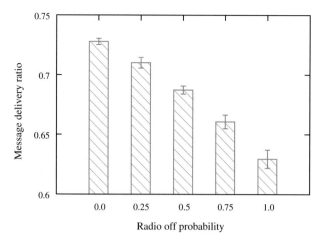

Fig. 5.7 Effect of radio off probability on message delivery ratio

result bears significance in relating the performance with the action of turning off the device radios only, since we have deviated from several other considerations for this specific case.

Next, we studied how the individuals' dominant emotions varied with time as a result of the network dynamics. The heat maps in Fig. 5.8 show the temporal fluctuations in the dominant emotions of the users, sampled at every hour, for two random cases. The indices along the color scale indicates the different emotions, starting with neutral ($=-1$), joy, sadness and fear ($=2$). We note that this variation is not due to the *natural decay* of the emotions, but rather due to the impact of the messages received. This could be clearly understood from Fig. 5.8a, where, although fear has the highest initial intensity and decays at the slowest rate with $a_3 = 0.001$, the greater number of *Bad* messages exchanged in the network ($p_{m_2} = 0.6$) result in the overall dominant emotion as sadness. Reverse is the case shown in Fig. 5.8b.

Figure 5.9 presents the empirical CDF of the contacts among the devices in the network when $\mathbf{p_m} = \begin{pmatrix} 0.1 & 0.3 & 0.6 \end{pmatrix}$ and $\mathbf{e} = \begin{pmatrix} 0.5 & 0.5 & 0.8 \end{pmatrix}$. It could be observed from the figure that, while about 50 % of the total contacts occurred within 5 h, when no emotion was considered, the corresponding fraction increased to about 75 %, when only the effects of sadness were considered. This is due to the reason that sadness dominated in this case. Thus, the total number of contacts in the network were reduced due to switching off the radios and most of the contacts occurred during the initial time period.

Figure 5.10a shows the effect of traffic explosion (due to fear) and changes in cooperation (due to sadness) on the delivery ratio of the messages when the actions of the different percentage of nodes were considered to be affected by their emotions. The results indicate that the presence of few non-emotional agents (for example, throwboxes [226]) in such scenarios may aid the communication process.

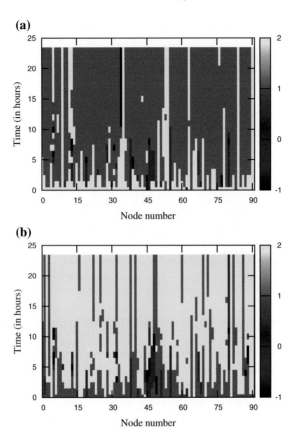

Fig. 5.8 Variation in the dominant emotions of the users when **a** p_m = (0.1 0.6 0.3), **e** = (0.5 0.5 0.8), **a** = (0.002 0.0015 0.001), and **b** p_m = (0.1 0.3 0.6), **e** = (0.5 0.8 0.5), **a** = (0.002 0.001 0.0015)

Throwboxes are battery-powered stationary devices that come with storage, processing capability, and usually large transmission range. Throwboxes not only allow storage of larger number of messages in an OMN with resource-constrained devices, but also create contact opportunities among them by virtue of larger transmission ranges. It was found that placement of such devices can increase the capacity of DTNs [226].

Figure 5.10b presents the delivery ratio of the messages in MOON when the devices had heterogeneous network interfaces. The first group of the nodes—shown along the x-axis—used the interface *if*1, while the remaining used *if*2. All the nodes were assumed to be affected by their emotions with **e** = (0.5 0.5 0.8) and

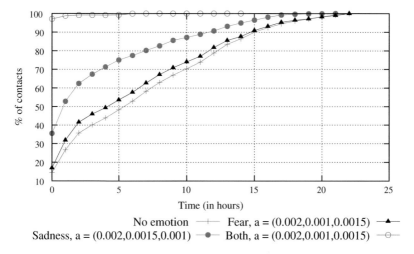

Fig. 5.9 CDF of the contacts among the devices

Fig. 5.10 Performance
degradation considering the
effects of traffic and
cooperation when **a** different
percentages of the nodes
were affected by emotions,
and **b** different network
interfaces were used by the
devices

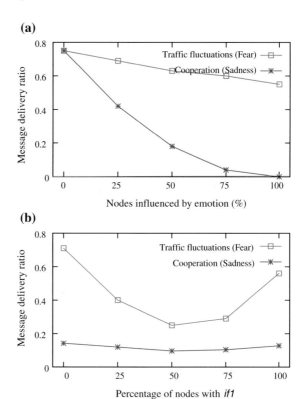

$\mathbf{a} = (0.002\ 0.0015\ 0.001)$. We, individually, considered the effect of fear on traffic generation rate and the effect of sadness on user cooperation. It is observed that the delivery ratio reached the maximum when 0 % of the devices had *if* 1 implying that all the devices in the network had a higher transmission range of 20 m. The minimum is, however, reached when half of all the devices used either network interface resulting in the least number of message transfer opportunities. A similar "U" shaped curve was also obtained considering only the effects of sadness. We note here that different values of $\mathbf{a}$ were taken while simulating the scenarios in Fig. 5.10a, b. This was done in order to highlight the impact of sadness on cooperation.

Finally, we discuss a few observations obtained using the synthetic mobility model. Since the model results in very low contacts due to network partitioning, we only considered the effects of emotion on cooperation; traffic explosion due to emotions was not considered. In the base case—when the nodes did not had any emotion—the delivery ratio of the messages were found to be about 13.89 % with $\mathbf{p_m} = (0.2\ 0.4\ 0.4)$. When only the effects of sadness were considered, the delivery ratio was nearly 0 with $\mathbf{a} = (0.002\ 0.001\ 0.0015)$; with $\mathbf{a} = (0.002\ 0.0015\ 0.001)$ it was about 8.31 %. Sadness being dominant in the former case, device radios were switched off for longer duration resulting in almost no delivery of the messages. Considering only the effects of fear, the message delivery ratio were about 12.43 % and 10.24 %, respectively, for $\mathbf{a} = (0.002\ 0.001\ 0.0015)$ and $(0.002\ 0.0015\ 0.001)$.

5.7 Summary

This chapter presented an odyssey of human emotions, their detection, and applications. On one hand, our emotions get affected by events taking place around us. On the other hand, our emotions often dictate how we behave. For example, our information seeking tendencies increase when we are anxious, while it decreases when negative emotion prevails.

Psychologists view emotions in different ways, such as states and process. With advance in technology, it has been possible to measure physiological signals detect and/or establish different aspects of emotion theories. In fact, it has been shown that emotions can quite accurately detected based on our smartphone usage pattern. And of course, our interactions in online social networks and else do often hint at our contemporary feelings. Emotion detection has wide uses, for example, suggesting activities in an e-learning program or controlling the difficult level of a game based on the player's engagement.

However, recent progresses in research and development go beyond emotion detection. It has been possible to consider synthetic emotions and embed them to the so called "believable" agents. Thus, a robot would not only be able to speak with us, but also adapt its behavior based on our emotions. Computational models of emotions are particularly useful in this context. We looked at few such models in detail. Subsequently, we looked at some of the scenarios of emotional responses of human beings. For example, we looked at how the emotions of people varied during the great earthquake in Japan in 2011.

In the latter portion of this chapter, we looked at an interesting study on the effects of human emotions in MOONs. The study used the computational model of emotions proposed by Meftah et al. together with relevant network parameters. Specifically, we considered the effects on the offered (Poisson) traffic load and cooperation in the network. Additionally, we also looked at how information received by the users changed their contemporary states of emotions. This study implicates that traffic load management schemes should be in place, especially in the typical post-disaster scenarios that we have considered. Moreover, if the entire population is prone to be influenced by their emotions, deployment of emotion-neutral entities may help the communication scenarios.

5.8 Review Terms

- Basic emotions
- PAD model
- Emotion intensity vector
- Emotion and adaptation
- Physiological signals and sensors
- Dominant emotion

- Plutchik's circumplex model
- Computational models of emotion
- Markovian model of emotion
- Emotion detections
- Decay of emotions
- Degree of cooperation

5.9 Exercises

5.1 What are the basic emotions considered in Ekman's and Plutchik's models?

5.2 Draw a Markov chain depicting the basic emotions from Ekman's model. Show all the state transitions.

5.3 List all the combinations of basic emotions that give rise to secondary emotions in Plutchik's model.

5.4 How are emotion and mood distinguished? You may refer to [202] and [216].

5.5 How the physiological signals can be classified based on the nervous systems?

5.6 List a few physiological signals and discuss what they represent. Also, list some sensors that can measure such signals.

5.7 What is emotion contagion?

5.8 Consider the following emotion intensity vector, $\mathbf{e} = (0.3 \ 0.2 \ 0.3)$, and the coefficients of decay of basic emotions, $\mathbf{a} = (0.0025 \ 0.0010 \ 0.0015)$. What would be the dominant emotion after 10 units of time?

5.10   Programming Exercises

5.9 Write an application in the ONE simulator that periodically toggles the network interface radio, i.e., if the radio is on, it is turned off, and vice versa. Simulate a scenario by attaching the created application to the nodes in the simulation settings file.

5.10 The `MessageEventGenerator` in the ONE simulator generates messages based on a uniform distribution. In particular, the message generation interval is uniformly distributed between $[a, b)$, where $a < b$. Extend the `MessageEventGenerator` class to create a new class, `PoissonMessageEventGenerator`, so that the messages generated follows the Poisson distribution with parameter λ. Here, λ specifies the number of messages created per unit time.

   Hint: A Poisson arrival process results in exponential inter-arrival times.

Part III
Cooperation in Opportunistic Mobile Networks

Chapter 6
Evolutionary Game in Wireless Networks

In 1944, John von Neumann and Oskar Morgenstern published a book titled "Theory of Games and Economic Behavior," which is widely regarded as the "classic work upon which modern-day game theory is based."[1] Ever since the systematic and mathematical foundation was laid by Neumann and Morgenstern, game theory has found its application in diverse domains, such as economics, biology, and communication networks, with several researchers being awarded the Nobel Prize for their contributions in this field. In this chapter, we take a look at two different branches of game theory—classical game theory (CGT) and evolutionary game theory (EGT). CGT refers to what we commonly call as game theory. On the other hand, EGT corporates principles from biology and evolutionary processes into game theory. EGT differs from CGT in many ways. For example, in CGT, individual players behave rationally in order to maximize their respective payoffs, whereas the objective in EGT is to adopt the most successful strategy.

This chapter begins with a brief introduction to game theory. We present the common terminologies used in CGT and illustrate them with examples. We also discuss two different areas of CGT—cooperative and non-cooperative games. Subsequently, we delve into EGT and look at some of the core concepts of EGT, for example, evolutionary stable strategy (ESS). Equilibrium analysis of a game is a fundamental requirement. In non-cooperative games, we look for Nash Equilibrium, whereas in EGT, we aim to achieve an ESS. We also look into replicator dynamics, which is at the core of EGT. Followed by this, we discuss about one of the popular games—the Rock-Scissors-Paper (RSP) game. Not surprisingly, EGT has found its application in several diverse research areas, such as biology, economics, and wireless networks. We devote a significant portion of this chapter to briefly look into some among many such notable applications.

The latter portion of this chapter presents an interesting topic—the use of RSP games to represent aspects of internode cooperation in OMNs. Our objective is to study the network performance when different nodes behave in different ways with

---

[1] http://press.princeton.edu/titles/7802.html.

© Springer International Publishing Switzerland 2016
S. Misra et al., *Opportunistic Mobile Networks*,
Computer Communications and Networks, DOI 10.1007/978-3-319-29031-7_6

respect to message replication. In particular, we consider three different behaviors (or strategies) of the nodes namely, cooperate, exploit, and isolate. We formally represent such a network, and discuss in details how the aforementioned three strategies are expected to affect the performance of OMNs in terms of message delivery under diverse scenarios. A mathematical outlook of the resulting network dynamics is presented next. Finally, this chapter concludes with some insights about the performance of RSP-based cooperation game in OMNs.

6.1 Overview of Game Theory

A "game" can be regarded as a joint activity between two or more players, where the individual players engage in some actions following some pre-defined rules. For example, tic-tac-toe is a popular game played by two players in a 3×3 grid. In this game, each cell in the grid can be marked either with a cross or circle, and each player plans their course of action to win the game. Game theory is the study of a series of plan of actions in a game to achieve long-term gains. Specifically, it can be expounded as the mathematical study of strategic decisions of conflict and cooperation among the population. In this section, we look at two major flavors of this subject namely, CGT and EGT.

6.1.1 Classical Game Theory

In game theory—whether CGT or EGT—a game is composed of a set of *players*, a set of *strategies*, and the *payoff matrix*. A game with a single player is known as an *one-person game* and it can be solved by *decision theory*. *Players* are the decision makers of a game. They can choose the best plan of action within the context of the game. A *strategy* is a set of rules for specifying the individual actions/decisions, whereas the *payoff matrix* defines the player's utility functions. For example, in the tic-tac-toe game, if there is a row, column, or diagonal with all the same symbols (cross or circle), the corresponding player wins and gains a score. In real-life, such payoffs or utilities can be in terms of money, energy, latency, efficiency, and so on. Formally, a game is represented with a tuple $G = (N, S_i, \pi_i)$, where N is the set of players, S_i is the strategy set of player $i \in N$, and π_i is the payoff for player $i \in N$.

Cooperative game theory and non-cooperative game theory are two popular branches of game theory. The extensive form games[2] and strategic games (see Footnote 2) are classified under the non-cooperative games, whereas coalitional games (see Footnote 2) belong to the cooperative games category. We briefly discuss the two branches of game theory in the following.

---

[2]Refer to [227] for more details.

6.1.1.1 Non-cooperative Game Theory

Non-cooperative game theory involves a set of players, where each player take their individual decision independently knowing the possible choices of other players and their effect on the player's utilities. Non-cooperative game theory focuses on the study and analysis of competitive decision-makers. These type of games are used in various fields, for example, economics, political science, biology, sociology, and wireless networks.

In non-cooperative game theory, individual players want to maximize their own profit without caring about other players in the system. Each player's success depends on the decision of the other players in the game. For example, in a *zero-sum game*, one player's loss is equal to the other player's gain. The previously mentioned tic-tac-toe is an example of two-player zero-sum game.

Example 6.1 Consider a two-player strategic game, where each player has two actions, as shown in Table 6.1. The numeric values inside the cells indicate the payoffs obtained by the first and second player, respectively.

This is an example of two-player nonzero-sum game. The game results in the following four possible outcomes.

1. If both the players choose strategy S_1, then the outcome of the game is (8, 8).
2. If both the players choose strategy S_2, then the outcome of the game is (5, 5).
3. If Player 1 plays S_1, and Player 2 plays S_2, then the payoffs received by Player 1 and 2, respectively, are 1 and 5.
4. If Player 1 plays S_2, and Player 2 plays S_1, then the payoffs received by Player 1 and 2, respectively, are 5 and 1.   □

Nash Equilibrium

Equilibrium is a state of a game where the players are in the balanced/stable state. Nash equilibrium (NE) is a solution concept for non-cooperative games. Informally, when NE is achieved, a player does not get more profit (benefit or incentive) by switching from its chosen strategy to any other strategy while keeping the strategy chosen by the other player fixed. This holds true for all the players and their strategies in the concerned game.

Table 6.1 Payoff matrix for two-player game

		Player 2	
		S_1	S_2
Player 1	S_1	8, 8	1, 5
	S_2	5, 1	5, 5

Table 6.2 The Prisoner's Dilemma Game

		B	
		Cooperate	Defect
A	Cooperate	3, 3	0, 4
	Defect	4, 0	1, 1

Formally, in a two-player game, the NE is a pair of strategy (s_i, s_j) such that

$$\left.\begin{array}{l} \pi_1(s_i, s_j) \geq \pi_1(s_i', s_j), \forall s_i \neq s_i' \\ \pi_2(s_i, s_j) \geq \pi_2(s_i, s_j'), \forall s_j \neq s_j' \end{array}\right\} \tag{6.1}$$

Here, s_i and s_j, respectively, represent the strategies chosen by players 1 and 2; $\pi_k(s_i, s_j)$ denotes the payoff of player k for playing strategy s_i against strategy s_j of the other player. Note that a game can have zero or more NE. For instance, in Example 6.1, the pair of strategies (S_1, S_1) and (S_2, S_2) constitute the NE. The definition of NE can be extended for multi-player games in a similar way.

Prisoner's Dilemma

Prisoner's dilemma is one of the most well known games in non-cooperative game theory. In this game, two persons say, A and B, are arrested for a serious crime, and kept in separate cells. Both of them are questioned simultaneously, and offered a reduced punishment in exchange of betraying the other. If both the prisoners betray each other (i.e., *defect*), they serve higher jail sentences. However, if neither of them says anything against the other (i.e., *cooperate* with the investigation), both of them will get higher payoffs, i.e., lower term of sentence.

Table 6.2 shows the payoff matrix for the Prisoner's dilemma game. Higher payoff means prisoners spend less time in jail. If A defects B, but B cooperates, then A will get higher payoff, i.e., lower jail sentence, and B will get lower payoff.

A player in this game gets maximum reward when it itself chooses to defect while the other chooses to cooperate. For example, when player A chooses defect and B chooses cooperate, A receives a payoff of 4. So, each prisoner is motivated to switch its strategy from "cooperate" to "defect" for maximizing its payoff. Therefore, the NE of the game is $(1, 1)$ when both the prisoners A and B choose to defect.

6.1.1.2   Cooperative Game Theory

In non-cooperative game theory,[3] each player's individual actions are taken as primitives, whereas in cooperative game theory, the joint actions of group of players are

---

[3] Portions of this chapter are reproduced with kind permission from Springer Science+Business Media: Next-Generation Wireless Technologies, Cooperation in Delay Tolerant Networks, 2013, 15–35, Sudip Misra, Sujata Pal, Barun Kumar Saha.

taken as primitives. In a cooperative game, a group of players (coalition) compete with other groups of players. Hence, the cooperative game theory is known as *combinatorial* game theory [228], whereas non-cooperative game theory is known as *procedural* game theory.

Coalitional Games

Coalitional games (see Footnote 3) are cooperative games, which focus on the behavior of the group of players rather than on individual players as in non-cooperative games. A coalitional game can be represented by the pair (N, v), where N is the set of players and v is a characteristic function specifying the value generated by different coalitions in the game. In this context, any nonempty subset of N is termed as a coalition. For any coalition $S \subseteq N$, $v(S)$ is said to be the *worth* of the coalition S; $v(\varnothing) = 0$. When different players come together to form a coalition, the coalition as a whole gains some value. For example, when multiple researchers work collaboratively (form a coalition), they can produce a very quality of work in less time in contrast to what each researcher could have done individually. The coalitional value, which is determined based on v, is mapped into the amount of utility that individual members obtain from their coalition. A coalitional value can be in three different forms—characteristic, partition, and graph. In the *characteristic* form, the value $v(S)$ of a coalition S depends exclusively on the members of the coalition. In the *partition* form, the coalitional value depends both on the members of the coalition as well as on rest of the players in other coalitions. Finally, in the *graph* form, the value v depends on the connected graph structure among the players in the coalition.

A relevant question that arises here is whether or not it is possible to redistribute the coalitional value among the members of a coalition. Two scenarios are possible in this case. In *transferable utility* (TU) coalitional game, the function v is defined as $v : 2^N \rightarrow \mathbb{R}$. The real-valued payoff, $v(S)$, of a coalition S can be distributed among the members of the coalition. For example, if a coalition gains in terms of money, the gain so made can be divided among the members of the coalition. However, there may exist such scenarios where payoff obtained may not be transferred to individual players of a coalition. As an example,[4] let us consider that several researchers have worked collaboratively on a research paper. As a consequence of the work, every author receives some kind of payoff, for example, global recognition and promotion at his/her institute. Such payoffs, however, are non-transferable. In general, a *nontransferable utility* (NTU) coalitional game is defined as a pair (N, V), where $V(S) \subseteq \mathbb{R}^S$ is the value of a coalition S. Here, $\mathbb{R}^S$ denotes the $|S|$-dimensional real-coordinate space [229].

A TU coalitional game is said to be *superadditive* if for any two coalitions $S, S' \subset N$, we have $v(S \cup S') \geq v(S) + v(S')$, where $S \cap S' = \varnothing$. This implies the value of the larger coalition formed out of the smaller disjoint coalitions must be greater than or equal to the sum of the individual values of smaller coalitions. Due to the superadditive property, the members of the game always try to form the *grand coalition*. The grand coalition is the coalition formed of all members of the game.

---

[4]This example is taken from http://www.math.wsu.edu/math/faculty/lih/464-16pp.pdf.

In other words, the grand coalition of a game is the set N itself. Grand coalition[5] is always beneficial to the members because the payoff received from $v(N)$ is at least as large as the sum of individual payoffs received from the disjoint coalitions form from the set N.

From the above, we can observe that the main objective of the coalitional game is the formation of grand coalition. However, a challenging issue is the division the payoff of a coalition among its member while ensuring it (the division) to be *fair* and *stable*. Here, *fair* means that the payoff must be proportionally distributed among the members of the game based on their contributions. On the other hand, *stable* distribution implies that the grand coalition must be stable so that no coalition member leaves the grand coalition in order to obtain a better payoff. *Core* and *Shapley value* are two main solution concepts proposed for coalitional games. Core guarantees stability, whereas Shapley value guarantees fairness in coalition formation.

When different players decide to form a coalition, the notions of individual and group rationality come into the picture. One defines a payoff vector $\mathbf{x} = (x_1, x_2, \ldots, x_{|N|})$, where x_i indicates the amount to be received by player i. Such a payoff vector $\mathbf{x}$ is said to *group rational* if the total amount received by the individual members sums up to the worth of the coalition, i.e., $\sum_{i \in N} x_i = v(N)$. On the other hand, a payoff vector is *individually rational* if $x_i \geq v(\{i\})$, $\forall i$. In other words, it pays off more to be a member of a coalition rather than acting alone; $\{i\}$ denotes a coalition with just a single player. Accordingly, a payoff vector $\mathbf{x}$ is said to be an *imputation* if it is both group rational and individually rational. Moreover, when an imputation $\mathbf{x}$ is proposed, a coalition S can upset such a division of payoffs if $v(S) > \sum_{i \in S} x_i$. In other words, players are motivated to form such a coalition deviating from the proposed payoff. In this case, the imputation $\mathbf{x}$ is said to be *unstable* and the coalition S, if it exists, *blocks* the imputation $\mathbf{x}$. Interested readers are suggested to look at [231, 232] for further discussion on these topics.

Core

The core of a coalitional game is the set of its stable imputations [231]. Formally, the core, $\mathscr{C}$, of a transferable utility [233, 234] coalitional game is defined as:

$$C = \{\mathbf{x} \in \mathbb{R}^N | \sum_{i \in N} x_i = v(N) \text{ and } \sum_{i \in S} x_i \geq v(S), \forall S \subseteq N\} \qquad (6.2)$$

The core guarantees the stability of the grand coalition and no player has better incentive to deviate from the grand coalition. It may be noted that a core of a coalitional game can have zero or more solutions. A game is said to be stable if its core

---

[5]Please refer [230] for more details.

is nonempty. Similarly, the core of the nontransferable utility coalitional game, C', is defined as:

$$C' = \{\mathbf{x} \in V(N) | \nexists S \subseteq N, \, y \in V(S), \text{ such that } y_i > x_i, \forall i \in S\}. \quad (6.3)$$

Shapley Value

The Shapley value[6] *fairly* divides the payoff of the grand coalition among its members. However, it does not guarantees stability. Shapley value allows the members of a coalition to receive their share based on their marginal contribution toward forming the coalition. In particular, let us consider a value function, $\phi(v) = (\phi_1(v), \phi_2(v), \ldots, \phi_{|N|}(v))$. Shapley intensively characterized this value— the value ϕ is unique and satisfies *additivity, dummy, efficiency,* and *symmetric* axioms. We briefly discuss these axioms below.

- **Additive**: Let (N, u), (N, v), and $(N, u + v)$ are coalitional games, such that $(u + v)(S) = u(S) + v(S)$, $S \subseteq N$, for every coalition S. Then $\phi_i(u + v) = \phi_i(u) + \phi_i(v)$.
- **Dummy**: Player i is a dummy player if its contribution to any coalition is zero, i.e., $\forall S, v(S \cup \{i\}) = v(S)$. Let i be a dummy player. Then for any v, $\phi_i(v) = 0$.
- **Efficiency**: $\sum_{i \in N} \phi_i(v) = v(N)$. The total value is distributed to each player i, $i \in N$.
- **Symmetric**: Players i and j are symmetric for every $S \subseteq N$, $i, j \notin S$, if $v(S \cup \{i\}) = v(S \cup \{j\})$. If i and j are symmetric, then $\phi_i(v) = \phi_j(v)$.

The Shapley value uniquely divides the payoff among the players of a coalition based on the following equation:

$$\phi_i(v) = \sum_{S \subseteq N - \{i\}} \frac{|S|!(|N| - |S| - 1)!}{|N|!} (v(S \cup \{i\}) - v(S)). \quad (6.4)$$

6.1.2 Evolutionary Game Theory

Evolutionary Game Theory (EGT) is a branch of game theory that is widely used in biology, evolutionary psychology, and behavioral ecology. In an evolutionary game, the game is repeatedly played among the players[7] over successive generations. Inspired by biology, the evolutionary process in EGT passes through two phases— mutation and selection. In the mutation phase, the players in EGT may switch their

---

[6]Named in the honor of Lloyd S. Shapley, who has immense contributions to game theory, and was awarded the Nobel Prize in Economic Sciences in 2012 jointly with Alvin E. Roth.

[7]The terms players and nodes are uses interchangeably in the rest of this chapter.

strategies. In EGT, the selection is done by using *replicator dynamics*. In this phase, the players choose the most successful strategy having higher payoff over the other players.

6.1.2.1 Evolutionary Stable Strategy

Evolutionary stable strategy (ESS) is a solution approach of EGT. Here, each player empirically checks different strategies and their corresponding payoff values, and chooses the most successful strategy. After a period of strategic interactions, the group of players converge to a stable equilibrium by choosing a particular strategy, which is called an ESS. In ESS, the players are neither aware of the game, nor have any control over their strategies. ESS is a refinement over NE.

According to Smith and Price [235], a strategy s_i is an ESS, $\forall s_i' \neq s_i$, in a pairwise game if it satisfies either

1. $\pi(s_i, s_i) > \pi(s_i', s_i)$, or
2. $\pi(s_i, s_i) = \pi(s_i', s_i)$ and $\pi(s_i, s_i') > \pi(s_i', s_i')$.

Here, $\pi(s_i, s_i')$ represents the payoff obtained on playing strategy s_i against another strategy s_i'. The first condition specifies that when both the players employ strategy s_i, the payoff received is strictly greater than that in the scenario where one of the players use a different strategy, s_i'. The second condition specifies that if a strategy s_i' is well against the strategy s_i, then playing s_i against s_i' results in a greater payoff in contrast to the scenario where both the players employ strategy s_i'. Here, the first condition indicates equilibrium, whereas the second condition guarantees stability. If, however, only the first condition is satisfied for both the players, and $\forall s_i \neq s_i'$, then the pair of strategies (s_i, s_i) leads to strict NE where the inequalities of (6.1) hold strictly.

Neutrally stable strategy (NSS) [235] is less demanding than ESS. In NSS, the incumbent having strategy s_i performs equally well against the mutant. Mutants are the players who play the alternative mixed strategy other than the stable strategy. A strategy s_i is an NSS, $\forall s_i \neq s_i'$, if it satisfies either

1. $\pi(s_i, s_i) > \pi(s_i', s_i)$, or
2. $\pi(s_i, s_i j) = \pi(s_i', s_i)$ and $\pi(s_i, s_i') \geq \pi(s_i', s_i')$.

6.1.2.2 Replicator Dynamics

Replicator dynamics [236] can be used as a model for selection of population evolution over time. Let S be the set of strategies $S = \{s_1, s_2, \ldots, s_n\}$ available to a player. Further, let k_i be the number of individuals using strategy $s_i \in S$. Then, the total population size is $|N| = \Sigma_{i=1}^{n} k_i$, and the proportion of individuals using s_i is given by $p_i = k_i/|N|$.

Table 6.3 Payoff matrix

	R	S	P
R	0	v	-u
S	-u	0	v
P	v	-u	0

In EGT, the proportion of individuals changing over time is governed by the replicator dynamics, which is expressed by the following differential equation:

$$\dot{p}_i = p_i(\pi(s_i, \mathbf{P}) - \bar{\pi}(\mathbf{P})). \tag{6.5}$$

Here, the vector $\mathbf{P} = (p_1, p_2, \ldots, p_n)$ represents the population state. The average payoff of the population, $\bar{\pi}(\mathbf{P})$, is calculated as:

$$\bar{\pi}(\mathbf{P}) = \Sigma_{i=1}^{n} p_i \times \pi(s_i, \mathbf{P}). \tag{6.6}$$

The evolutionary equilibrium [230] of the replicator dynamics can be determined by equating (6.5) to zero, i.e., $\dot{p}_i = 0$. At evolutionary equilibrium, the proportion of players changing over time comes to an end.

6.1.2.3 Rock-Scissors-Paper Game

Rock-Scissors-Paper (RSP) [237] is a game played by two players. Three strategies are available in this game—rock, scissors, and paper. Each player independently chooses one of these strategies. The objective of a player is to select a strategy that defeats the strategy chosen by the opponent. The strategies are resolved as follows:

- Rock defeats a pair of scissors,
- Scissors defeat paper, and
- Paper defeats rock.

In other words, if, for example, player A chooses rock and player B chooses scissors, then player A defeats player B. If, however, both the players choose the same strategy, there is a draw and the players play again.

Example 6.2 In this example,[8] let us consider a RSP game, as shown in Table 6.3.

Here, $v > 0$ represents the benefit of winning the game, $u > 0$ represents the cost of losing the game, and ties are worth 0 to both players. The Nash equilibrium of this game is $x^* = (1/3, 1/3, 1/3)$, which implies uniform randomization over the three strategies. This game is a *good* RSP game if $v > u$ and x^* is an ESS. In a good RSP

---

[8]The wording of this example is borrowed from [238], p. 3179.

game, the benefit of winning is greater than the cost of losing. If $v = u$, the game is a *standard* RSP game and x^* is a NSS, but not an ESS. If $v < u$, the game is *bad* RSP game and x^* is neither an ESS, nor an NSS.   □

In CGT, a player attempts to maximize its payoff when faced with a set of strategies that itself and other players can adopt. In contrast, the goal of a player in EGT is to adapt to the most successful strategy. Based on the observations made by Gintis [239], we summarize below some of the key differences between CGT and EGT. The quoted phrases are taken from [239].

- The focus of CGT is to describe "socially and temporally isolated encounters," whereas EGT tries to explain "macro-social behavioral regularities."
- Players in a CGT-based game always remain constant. However, players in EGT are continuously changing.
- In CGT, one is concerned with the analysis of Nash equilibrium of "isolated encounters." In EGT, however, one considers the "evolutionary equilibria of the dynamic system."

Moreover, in EGT, the size of the population is very large, which may not be necessarily true for CGT.

6.2 Applications of EGT

Much like CGT, EGT also has found its application in diverse fields, such as biology, economics, and wireless networks. In this section, we discuss some of those applications. In the context of wireless networks, EGT has been used to solve various problems, such as for cooperative communication by forwarding messages of other nodes in Vehicular ad hoc networks, selecting reliable path during routing in multi-hop wireless networks, and multiple access control in slotted ALOHA-based wireless networks. We look at these in details in the following.

6.2.1 Biology and Economics

In biological applications [240], the members of the populations are not involved in pairwise games, but are in a global competition among the whole population. Growth of plants in a population [241] is such an example. Here, plants are the players of the game. Plants grow from the seeds dropped nearby or from the other seeds blown away by the wind. The transience of seeds blown by the wind is greater than that of the seeds dropped nearby. However, the growth rate of the nearby seeds may be dropped due to too many seeds in the immediate neighborhood. So, every plant compete with the other plant for growth in a population.

In biological games [240], strategies are the individuals' behaviors, and payoffs are represented by *fitness*. Fitness is measured by the success rate in reproduction. *Selection* choose the most successful strategy in terms of higher reproductive success. The process of evolution does not always reach the stable equilibrium. However, the focus of EGT is to reach at stable equilibrium. John Maynard Smith and George R. Price described the notion of ESS, as explained in Sect. 6.1.2.1. ESS is the central equilibrium concept of EGT. A strategy is called *evolutionarily stable* if it is stable against mutant's strategy.

Walrasian and Bowles' paradigms [242] are two well-known approaches in economics. Walrasian [243] used an approach similar to CGT. Walrasian paradigms assumes that

1. Individuals make decisions based on self-interested preferences and exogenous determination,
2. Repeated interactions are ignored and social interactions are only in the form of transactions,
3. In most of the applications, returns to scale are ignored and institutions do not evolve, and
4. Individuals are fully informed about their environment.

Bowles' paradigm, on the other hand, supports EGT. Bowles highlights that endogenous preferences involve learning or genetic changes, i.e., individuals evaluate outcomes through learning and change their behavior over time accordingly.

Antoci et al. [244] proposed a mechanism that the public administrator (PA) of a tourist place may follow to attract visitors while maintaining the environment. The authors studied the financial mechanism of public administration and economics agent using an EGT-based approach. The PA sells the tourists an environmental call option. The tourists get reimbursed if the environment quality is unsatisfactory. Similarly, the PA offers the firms to adopt a new nonpolluting technology. PA allows the firms to get a reimbursement for the new technology option. The evolution of visitors' and firms' behavior is modeled using replicator dynamics. The dynamics lead to a Nash equilibrium in which the firms adopt the environmental-friendly technology, or to a Pareto-dominated equilibrium with no technological innovation and no tourism.

6.2.2 Vehicular Ad Hoc Networks

Vehicular Ad hoc NETworks (VANETs) [245–247] extend the concept of MANETs to vehicles moving along roads. These networks are voluntarily created for exchanging data in Vehicle-to-Vehicle (V2V), Vehicle-to-Roadside (VRC), or Vehicle-to-Infrastructure (V2I) communication scenarios. The network structure of VANETs keep constantly changing over time due to continuous addition and removal of nodes (vehicles). Similar to OMNs and MANETs, each node in VANETs helps in forwarding messages of the other nodes. So, cooperation plays a vital role in VANETs.

Shivshankar and Jamalipour [248–250] proposed an EGT-based approach for cooperation in VANETs, and analyzed the effects of nodes' behavior on different aspects of the network. The authors considered two types of nodes—cooperators and defectors. In each round of the game, a node compares its payoff with a randomly chosen neighbor. The node adopts its neighbor's strategy if the latter has a higher payoff than the former. The authors noted that cooperation can be enforced with higher network connectivity, which is achieved with higher density of nodes. In urban scenarios, behavior of the nodes is highly influenced by their neighboring nodes due to frequent changes in the network topology. In case of high clustering among the nodes, for example, in city networks, internode cooperation sustains and reaches to a steady state even if there are minor changes in the network topology. The authors noted that mobility of the nodes in VANETs induce cooperation. However, at higher velocity, the degree cooperation levels among the nodes reduce due to recurrent changes in the network topology.

To augment the capacity of VANETs, computing devices known as Road Side Units (RSUs) are often placed along the sides of the roads. RSUs help in establishing communication between vehicles in VANETs, and passes various information, for example, traffic safety warnings, from one node to another. Maintaining sustained connections between vehicles and RSUs is challenging due to high mobility of the vehicles in VANETs [251]. Moreover, RSUs help in dealing with the load imbalance problem arising due to uneven distribution of vehicle in VANETs. For these reasons, access to RSUs is one of the important issues in VANETs. The underlying challenge is when and which RSU should be accessed for data transfer in VANETs. Wu et al. [251] took an EGT-based approach to address this issue. The authors used the concept of EGT, in which the players (vehicles in the coverage area of RSU) choose the best possible strategy by learning over time. The payoff of a player is the difference between the throughput obtained using a strategy and the cost of accessing an RSU. The latter cost is obtained by considering together the bandwidth occupation cost and handoff cost while requesting service from RSU. For example, the authors in [251] considered a scenario with two RSUs, where a set of vehicles request services from the two RSUs. This scenario belongs to a single community evolutionary game. In a single community evolutionary game, all vehicles' strategy set are same. The road is divided into four different areas based on the coverage of RSUs. A vehicle accesses RSU based on its payoff. Here, a replicator is an individual vehicle, which replicates through the process of mutation and selection. The vehicle with the highest payoff replicates and reproduce faster. In each round, a vehicle observes the payoff of other vehicles, and in the following round, the vehicle adopts the strategy that gives the higher payoff. The ESS is reached when all vehicles receive identical payoffs from different RSUs.

Information dissemination in VANETs heavily depends on the intermediate relay nodes due to its scalable and dynamic nature. Generally, in these networks, information is disseminated via multi-hop paths. So, cooperation among the nodes is crucial in such networks. Banerjee et al. [252] and Zhang et al. [253] proposed information dissemination in vehicular networks using EGT. Similar to [248], Banerjee et al. [253] studied the evolution of cooperation in VANETs using EGT. The authors showed that cooperation can sustain for higher synergy factor[9] ($\alpha > 0.6$). Synergy factor is used to measure the benefits obtained due to cooperation. Clearly, in order to motivate the nodes into cooperation, the benefits of cooperation must be greater than the sum of the contributed costs. Cooperative behavior increases and survives for moderately high values of α. However, as noted earlier, the topology of VANETs changes rapidly due to high velocity of the nodes, which results in disintegration of cooperators. As a consequence, some nodes switch their strategies from cooperate to defect. However, at higher values of α, cooperators are less likely to make such switch. Hence, the evolution of cooperation is slightly affected by velocity of nodes in VANETs.

Zhang et al. [253] showed that EGT-based information dissemination helps in reserving more bandwidth, while maintaining similar packet delivery ratio as compared to the flooding-based schemes. The authors used EGT for guiding the behavior of the nodes. In each round, the nodes revise their strategies. The authors considered a star topology where the central node begins with the cooperation strategy. The central node connects with the other external nodes in the topology. The cooperation ratio increases with an increase in the synergy factor when the central node begins with the cooperation strategy. The cooperation ratio decreases when the central node begins with the defector. When the central node is a cooperator, its payoff is greater than its neighboring defector. Therefore, the neighboring nodes switches their strategies to cooperators and the cooperator maintains its own strategy. However, when the central node is a defector, its payoff is greater than its neighboring cooperator. Therefore, the probability of switching from defector (central node) to cooperator is low and the probability of switching from cooperator (neighboring node) to defector is high. This implies that in a network, higher degree of cooperators or defectors greatly influences the evolution of strategies.

6.2.3   Other Wireless Networks

Having looked at some of the applications of EGT to VANETs, we now conclude this section with a review of different other areas of wireless networks where EGT has been used.

Backhaul Networks

Backhaul networks link the backbone networks and the access points through the gateways stations. Multi-hop wireless backhaul networks (MHWBN) provide global Internet services in a cost-effective manner. During unfavorable channel conditions, cross-layer approach is used between the routing and the physical layers. Anastasopoulos et al. [254] proposed an adaptive routing strategy for MHWBN using

---

[9]Synergy factor (α) is the estimation of profit gained due to cooperation.

EGT. The proposed strategy minimizes the end-to-end packet error ratio (PER) of the agents. Initially, an agent randomly selects any possible route. Subsequently, the agent computes the PER of the current path. The agent periodically samples different paths and compares their PER with that of the current path. If any other path is found to have a lower PER, the agent switches its routing strategy to consider the new path; otherwise, it retains the current path. In rain fading conditions, the signal strength of microwave radio frequency (RF) is reduced and absorbed by atmospheric rain, snow, and ice. The authors proved that under such conditions, uni-path routing is the stable evolutionary strategy.

Slotted ALOHA-Based Wireless Networks

ALOHA[10] is a simple communication protocol where the source node sends data in the form of a frame to a destination node. If a collision occurs while transmitting, then the data frame is retransmitted later. In slotted ALOHA, an improvement over the "pure ALOHA" mechanism, time is divided into discrete slots, and a node can transmit data only at the beginning of a time slot. The size of a time slot is set equal to the transmission time of a packet. In ALOHA-based systems, each mobile node makes its own transmission decisions as well as tries to maximize its own utility. A mobile node can either transmit or remain silent. A packet transmission is successful if other nodes in the vicinity are not transmitting at the same time. Otherwise, collision occurs and the packet is not received by the destination node. For successful packet transmission, a reward is given to the transmitting user (mobile node). If all the users stay silent at same time, they pay regret cost. Tembine et al. [255] studied multiple access control in slotted ALOHA-based wireless networks using evolutionary games. When one node is transmitting packets, others remain silent. The transmitting node receives a reward, whereas other nodes receive nothing in terms of cost. This structure is in equilibrium. So, exactly n-equilibria exists and all these are Pareto optimal,[11] where n is the number of players in the game.

Distributed Adaptive Filter Networks

Distributed adaptive filter networks have been used for data fusion, diffusion, and processing of data over distributed networks, such as environment monitoring using WSNs, event localization in wireless ad hoc networks, and so on. Jiang et al. [256] proposed a game theoretic view for adaptive networks and formulated a graphical evolutionary game for filtering problems in distributed adaptive networks. Here, the players are the nodes in the networks and strategies are the estimation information obtained from different neighbors. In each round, the nodes update their fitness locally after interacting with their neighbors. The nodes use three different rules for updating their strategies: birth-death (BD), death-birth (DB), and imitation (IM). In the BD update rule, a player is chosen for reproduction based on its fitness (Birth process). Then, the player's strategy is replaced by any neighbor's strategy. The DB

---

[10]ALOHA stands for Additive Links On-line Hawaii Area.

[11]In Pareto optimal resource allocation, the payoff of no nodes is better off without at least one node in a less advantageous position.

update rule is just reverse of the BD update rule. In DB, the current node's strategy is the death strategy and this node adopts the neighbor's strategy based on the neighbor fitness (Birth process). In IM update rule, the players adopt strategies of neighbors or continue with their current strategies according to their fitness values. The authors considered two types of nodes—common nodes and good nodes. Common nodes have common noise variance and good nodes have lower noise variance. Based on the behavior of the nodes, two types of strategies are possible. A node can collect information from common or good nodes. The authors proved that the strategy of using information from good nodes is the favorable and evolutionary stable strategy.

Cooperative Spectrum Sensing in Cognitive Radio Networks

In Cognitive radio (CR), a transceiver can intelligently detect and configure the best wireless communication channel in its proximity. Spectrum sensing finds unused spectrum in CR networks. Spectrum sensing in cognitive radio networks [257] uses only the radio-frequency spectrum. The basic task of CR is to detect the PUs.[12] CR also identify the available spectrum if PUs are absent. Cooperative spectrum sensing improves spectrum utilization of CR. Wang et al. [258] proposed cooperative spectrum sensing for CR users using EGT and studied how secondary users (SUs)[13] collaborate in cooperative spectrum sensing. The goal of this work is that SUs protect the PUs from interference. The SUs detect and sense the unused spectrum. The payoff of the game is the throughput achieved by the SUs. In each round, an SU try different strategies and select the best strategy that gives higher payoff. This way the secondary users solely approaches to ESS and achieve higher system throughput.

6.3 RSP Game in OMNs

Until now, in this chapter, we have looked at different aspects and applications of game theory. In the context of wireless networks, one of the major areas where game theory finds its application is cooperative communication among the nodes. Arguably, in real life, not all may be willing to cooperate. This could be due to inherent selfishness of a node, or due to its resource constraints. Nevertheless, in the long run, it is desirable that they do reflect cooperative behavior so that the underlying communication network performs efficiently.

In the remainder of this chapter, we consider the case of cooperation in OMNs using an EGT-based approach. We model the problem of cooperation among the nodes in an OMN as a Rock-scissors-paper (RSP) game [237] from EGT. In particular, we consider that all the nodes in an OMN can be broadly classified into three groups based on their behavior (strategy) in terms of message replication. The strategies of these three groups of nodes are termed as *cooperate*, *exploit*, and *isolate*. These strategies are elaborated below.

---

[12]Licensed users are known as Primary Users (PUs).

[13]Secondary users are the cognitive and unlicensed users.

- *Cooperators* help delivering messages created by other nodes in the OMN by replicating those messages. Of course, cooperators also replicate messages created by themselves to others.
- *Exploiters* use other nodes as "free-riders." An exploiter replicates messages created by itself to other nodes irrespective of their strategies. However, when it receives a message from any other node, the exploiter does not further replicate those messages.
- *Isolators* neither take help, nor provide so to the other nodes in replicating their messages. In other words, isolators directly deliver their messages to their corresponding destinations if and when an opportunity becomes available.

As the reader might have guessed, the ideal scenario would be when all the nodes in an OMN become cooperators. Having all nodes as exploiters is definitely bad. In fact, our goal would be to reduce the number of exploiters in the network if not make it completely zero. On the other hand, isolators seem interesting and fair. However, direct delivery of messages greatly reduces the network performance. Therefore, having all nodes to behave as isolators is undesirable. Note that irrespective of their strategies, all nodes receive any message that is destined for themselves.

Before proceeding further, let us try to understand how the aforementioned scenario is related to an RSP game. Let us consider a set of nodes consisting of different percentages of cooperators, exploiters, and isolators. Now, cooperators replicate messages generated by both cooperators and exploiters. However, an exploiter replicates messages created only by itself.[14] On the other hand, messages created by isolators are replicated by none. This has the following consequences.

- When the proportion of cooperators in an OMN is large, exploiters gain at their expense. This is due to the reason that given limited contact duration and buffer constraints, an exploiter has a chance of replicating—and therefore, delivering—more messages created by itself. Similar chances for cooperators are diminished in their generous attempt to deliver their messages of their own as well as exploiters. In other words, exploiters outperform cooperators with respect to message delivery ratio.
- When the proportion of exploiters is large in the network, isolators make the most profit in terms of percentage of the (isolators') messages delivered. Note that since the number of cooperators is less in this case, they do not get enough chances to replicate messages created by themselves. Moreover, available contact durations are distributed between their messages and those created by exploiters. On the other hand, although an exploiter does not replicate messages generated by nodes other than itself, it still spends time in receiving messages of others to drop them subsequently. Therefore, isolators, who directly deliver only their respective messages, make relative gains in this scenario.

---

[14]Note that an exploiter does not receive nor forward (replicate) messages created by other exploiters in the OMN.

- Finally, when an OMN consists of isolators in larger proportion, cooperators benefit the most by helping one another; a certain fraction of such gains made by cooperators is reaped by exploiters as well.

Now in each of the above discussed scenarios, motivated by the performance of one group of nodes (for example, exploiters) over a time period, the other group of nodes (for example, cooperators and isolators in this case), too, try to take similar advantage by adapting their strategy (behavior). As a consequence, after certain intervals of time, the size of one group of nodes become larger than the others, which lowers the payoffs, on an average, of the groups having the higher count of nodes [259, 260]. Because of these dynamics, an oscillatory effect among the size of different groups of nodes is observed, as shown in Fig. 6.1. For example, if initially the size of group of cooperators is greater, then, after certain time duration, the number of exploiters in the network increases and their group size becomes largest. Similarly, again, after a certain period of time, the number of isolators increase, and their group size becomes largest. This oscillation among the different group of nodes with respect to time form an RSP cycle. In this work, rocks resemble *cooperators*, scissors resemble *exploiters*, and papers resemble *isolators*.

In the following, we shall take a deeper look at the relationship among these three strategies, and investigate the impact of such behaviors on the performance of message delivery in OMNs. The precise definition of payoff for our RSP game of cooperation would also be provided. In the spirit of game theory, we assume that all the nodes in an OMN are *rational*. This means that any node would always prefer a strategy that delivers it greater payoff.

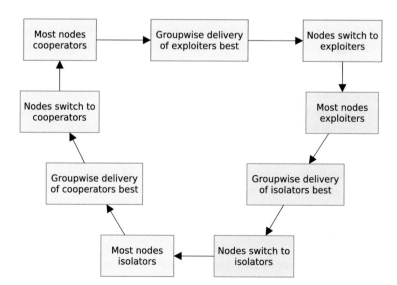

Fig. 6.1 Cyclical switch in the strategy of the nodes

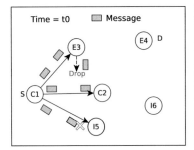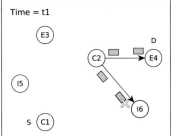

Fig. 6.2 Actions of the different types of nodes

6.3.1 Action of the Nodes

In this game, *populations* are the set of nodes belonging to the same group. We consider a finite set of nodes, N, and three groups of populations—*cooperators* (G_C), *exploiters* (G_E), and *isolators* (G_I). Thus, $N = G_C \cup G_E \cup G_I$. Note that these groups are mutually exclusive. In other words, a node cannot be both cooperator and exploiter at the same time. However, a particular node's behavior may change over time.

A set of strategies, $S = \{S_C, S_E, S_I\}$, is available to each node, where S_C, S_E, and S_I, respectively, denote the strategies cooperate, exploit, and isolate. Based on their contemporary strategy, nodes take different actions from the set of actions, $\mathscr{A} = \{Receive, Forward^{15}, Drop\}$, on any available message. When a message is sent to a node, the latter can either receive it or not[16] do so. On the other hand, a node having a message may choose either to forward or drop it. Let the receiving and forwarding actions be denoted by rc and fd, respectively. The payoff of a node is determined by its chosen actions.

Figure 6.2 illustrates the action of nodes belonging to the three groups. As shown in the figure, at time instant t_0, nodes $C2$ (cooperator), $E3$ (exploiter), and $I5$ (isolator) are in contact with the node $C1$ (cooperator). Node $C1$ sends a message to its neighboring nodes. Node $E3$ receives the message and drops it silently as the message is not destined for it. On the other hand, node $I5$ does not receive the message as the message is not destined for it. Node $C2$, being a cooperator, carries the message with itself, and at a later time instant t_1, it meets with the destination node $E4$ (exploiter) and delivers the message.

---

[15]As discussed earlier, only isolators directly *forward* a message to its corresponding destination; cooperators and exploiters *replicate* a message. However, here we use "forward" to represent both these kind of actions. The exact meaning should be evident based on the group of nodes that take the *Forward* action.

[16]To understand how a node "declines" to receive a message, let us recollect the summary vectors from Chap. 2. When a node sends a list of message IDs that it is (un)willing to receive in response to a summary vector, it can exclude(include) the specific message ID that it does not want to receive.

After a predetermined time interval, each node in the OMN collects the network-wide delivery ratio from a trusted authority. We assume this interval to be 12 h, and treat the corresponding time durations as one *generation* or round. The nodes try alternative strategies, if relevant, in different rounds of the game, and collect groupwise performance from a trusted authority after each generation.

Let us consider that the set of messages generated in the OMN be denoted by M. Each node may act as a source, sink (destination), or intermediate relay node. Let $m.src$ and $m.dst$, respectively, be the source and destination of any message $m \in M$, where $m.src, m.dst \in N$. Based on these, the actions of each type of nodes can be crisply defined.

The receiving action, $A_{rc}^C(m)$, of a *cooperator* $C \in G_C$ on a message $m \in M$, is abstracted as: $A_{rc}^C(m) = Receive, \forall m \in M$. Similarly, the forwarding action, $A_{fd}^C(m, j)$, taken by any node $C \in G_C$ on a message $m \in M$, while in contact with any other node $j \in N, C \neq j$, is abstracted as: $A_{fd}^C(m, j) = Forward, \forall m \in M$.

Representation of these actions for exploiters and isolators, however, is not as straight forward as above. The receiving action, $A_{rc}^E(m)$, of any *exploiter*, $E \in G_E$, for any message $m \in M$ is denoted by:

$$A_{rc}^E(m) = \begin{cases} Receive, \ if \ m.dst = E, \\ Receive \ and \ Drop, \ otherwise \end{cases}$$

The forwarding action, $A_{fd}^E(m, j)$, taken by any node $E \in G_E$ on a message $m \in M$ while in contact with any other node $j \in N, E \neq j$ is abstracted as: $A_{fd}^E(m, j) = Forward$, if $\forall E \in G_E$. Note that a node $E \in G_E$ possesses in its buffer only the messages generated by itself. Therefore, we do not need to include a condition to check the source of a message in the forwarding action of an exploiter. The receiving and forwarding actions of an *isolator* $I \in G_I$ on a message $m \in M$ are represented as follows: $A_{rc}^I(m) = Receive$, if $m.dst = I$, and $A_{fd}^I(m, j) = Forward$, if $m.dst = j$.

We assume that each node knows the groups to which the other nodes belong. Upon encountering another node, a node either receives or forwards the messages, or denies receiving them. Algorithm 6.1 describes the receiving action of a node x for an incoming message m. Note that if a node is cooperator or exploiter, i.e., it belongs to $G_C \cup G_E$, it always receives a message. However, an exploiter subsequently drops it. On the other hand, an isolator declines to receive a message unless it is the intended recipient (destination) of the concerned message.

Algorithm 6.2 describes the forwarding action of a node y having a message m. Similar to receiving, cooperators also forward (replicate) a message to other nodes.

Algorithm 6.1: The receiving action of a node x

Input:

- m: Incoming message

Output:

- Action taken by node

1 **if** $(x \in G_C \cup G_E$ *or* $m.dst \in G_I)$ **then**
2 | Receive the message.
3 | **if** $x \in G_E$ **then**
4 | | Drop the message.

5 **else**
6 | Decline receiving the message.

Algorithm 6.2: The forwarding action of a node y.

Input:

- m: Message in the buffer of y

Output:

- Action taken by node

1 $tempHosts = [\,]$ // An empty array
2 $tempHosts = neighbours(y)$ // Nodes that are within the communication range of y
3 $n = tempHosts.size()$
4 **if** $(y \in G_C$ *or* $m.src \in G_E$ *or* $m.src \in G_C)$ **then**
5 | Forward the message.
6 **else if** $(y \in G_E$ *and* $m.dst \notin G_E)$ **then**
7 | Drop the message.
8 **else**
9 | **for** i *from* 1 *to* n **do**
10 | | **if** $(m.src \in G_I$ *and* $m.dst == tempHosts[i])$ **then**
11 | | | Forward the message.

6.3.2 Analysis of Cooperation Strategies

In this section, we explore the impact of nodes' strategies on the message delivery ratio in OMNs. The approach presented here is derived from the general concept of *Simpson's paradox* discussed in [261, 262]. According to Simpson's paradox, the success of several groups is reversed when the groups are combined. We fuse together the concepts of RSP game, evolutionary game theory, and Simpson's paradox to study the aspects of cooperation in OMNs.

Simpson's paradox arise when inferences are drawn from improperly aggregated data. There are several real-life examples illustrating this paradox. Based on the admission statistics of graduate students in 1973, the University of California, Berkely, was sued for alleged bias against women in the admission process. It was revealed that while 44 % men were admitted, only 35 % of the women applicants got admission. On detailed investigation, Bickel et al. [263] discovered that, in reality, when per-department statistics were considered, there actually was a little bias in favor of admitting women students! The alleged bias against women arose due to improper aggregation of data that lacked consideration of several aspects, such as autonomy of a department and difficulty to get admission to certain departments. Interested readers may go look at an interactive visualization (http://vudlab.com/simpsons/) of this paradox for better understanding.

We consider three diverse scenarios—*Most Cooperators*, *Most Exploiters*, and *Most Isolators*. In the "most cooperators" scenario, most of the users choose the strategy cooperate, and the rest of the users choose exploit and isolate. Similarly, in the "most exploiters" and "most isolators" scenarios, most of the users choose the strategies exploit and isolate, respectively. Let $M_D \subseteq M$ be the set of messages delivered in the network. For each group, G_C, G_E, and G_I, we define the groupwise delivery ratio of messages denoted by α_C, α_E, and α_I, respectively, as follows:

$$\alpha_C = \frac{|\{m_D \in M_D | m_D.src \in G_C\}|}{|\{m \in M | m.src \in G_C\}|}. \tag{6.7}$$

The definitions of α_C and α_I follow likewise. The reader may recollect from Chap. 2 that the (network-wide) delivery ratio of messages is determined as $|M_D|/|M|$. Based on the work of Hauert et al. [259], we make the following observation in context of the scenario considered here.

Proposition 6.1 *When most of the nodes in a network adopt the strategy S_C, the exploiters do the best in terms of groupwise delivery ratio of the messages. Likewise, when most of the nodes in a network adopt the strategy S_E, the isolators gain the most. Also, when most of the nodes choose S_I, the cooperators do the best among the three strategies.*

The physical implications of the above Proposition is as follows. Let us consider the case when most of the nodes cooperate and they find the exploiters to benefit. This attracts the rational nodes to adopt the strategy exploit, in which case, the isolators benefit. The nodes are then lured to behave as isolators, which maximizes the payoffs of the cooperators. Thus, the strategy of any majority group fails to dominate the network performance.

In the following, we present a few theoretical characterizations of the RSP-based cooperation game considered here.

Theorem 6.1 *Consider an OMN, where all the nodes have equal chances of meeting one another and the number of isolators remains constant. The probability of delivery of the messages over multiple hops is higher when the number of cooperators is greater than the number of exploiters.*

Proof Let n_C, n_E, and n_I, respectively, be the number of cooperators, exploiters, and isolators such that, $n = n_C + n_E + n_I$. Let P_D be the probability of delivery of a single message in the network. We prove the theorem for two cases—single copy forwarding and multi-copy Spray and Wait (SnW) [42] forwarding in non-binary mode.

Case 1—*Single copy forwarding*: Let us consider the case of a message passing through K nodes, on an average, before reaching its destination. Thus, a message that is already delivered, must have passed through K cooperators. Therefore, the probability of delivery (P_D) of a message is computed as:

$$P_D = \Pi_{i=1}^K \left(\frac{n_C}{n}\right)^K = \left(\frac{n_C}{n_C + n_E + n_I}\right)^K = \left(\frac{1}{1 + \frac{n_E}{n_C} + \frac{n_I}{n_C}}\right)^K. \tag{6.8}$$

The value of P_D will be higher, when $n_C > n_E$.

Case 2—*SnW in non-binary mode*: Let us consider that a message is replicated $L - 1$ times in the OMN, where L is the maximum limit of replication imposed on the messages by SnW. Thus, the message is delivered if either the source node, or any of the other $L - 1$ nodes, meet the destination. Then,

$$P_D = \frac{1}{n-1} + \sum_1^{L-1} \frac{n_C}{n-2} \times \frac{1}{n-2} = \frac{1}{n-1} + \sum_1^{L-1} \frac{n_C}{n_C + n_E + n_I - 2} \times \frac{1}{n-2} \tag{6.9}$$

$$= \frac{1}{n-1} + \sum_1^{L-1} \frac{1}{1 + \frac{n_E}{n_C} + \frac{n_I}{n_C} - 2} \times \frac{1}{n-2}.$$

Once again, we observe that P_D increases when $n_C > n_E$ assuming n_I is constant.

Corollary 6.1 *The delivery rates of the messages in the case when all the nodes are cooperators will be greater than the same when all the nodes are isolators.*

Proof From Theorem 6.1, we have

$$P_D = \frac{1}{n-1} + \sum_1^{L-1} \frac{1}{n-2} \times \frac{1}{n-2} = \frac{1}{n-1} + \sum_1^{L-1} \frac{1}{(n-2)^2} \qquad (6.10)$$

$$\Rightarrow P_D > \frac{1}{n-1}, \; \forall n > 2.$$

For the *isolators*, $P_D = \frac{1}{n-1}$, since the messages are directly delivered, when the source node meets the destination node.

Corollary 6.2 *The delivery rates of the messages in a network with only exploiters degenerates to the case when all the nodes are isolators.*

Proof If the network consists of only exploiters, no intermediate node forwards the messages of the other nodes. Therefore, each exploiter directly delivers its messages to the destination, which is the same as the case when all the nodes are isolators.

Theorem 6.1 formally proves that having more number of cooperators in an OMN is beneficial for the network performance. Corollaries 6.1 and 6.2 draw related conclusions when cooperators and isolators, and exploiters and isolators are considered.

6.3.3 Relationship Among the Strategies

We conclude this chapter by taking a look at the impact of different behavior of nodes on the message delivery ratio in OMNs. For having specific insights, we evaluated the effects of different strategies chosen by the nodes using the ONE simulator [51] by considering a realistic synthetic mobility scenario, the *Helsinki city map* (HCM). We executed the simulations for 12 h with 100 mobile nodes deployed in a terrain of size 4500 × 3400 sq. km. Messages were generated uniformly at random with two randomly chosen nodes as their source-destination pair. The size of a message was chosen to be 25 KB and the TTL was set to 5 h. Each node had a buffer size of 5 MB; the transmission speed was taken to be 2 Mbps, and the transmission range was 10 m.

The RSP cycle of cooperation in OMNs discussed here leads to a dynamic system. However, in this chapter, we consider only *static* scenarios, where the nodes in the OMN do not change their strategies. In particular, we consider three independent scenarios with initially large proportions of cooperators, exploiters, and isolators in each of them, and investigate their respective groupwise delivery ratios. In the following chapter, we shall look at dynamic scenarios, where the nodes in the network *actually* change their strategies based on the payoff values.

The three cooperative strategies—cooperate, exploit, and isolate—were implemented atop the SnW routing protocol. In the the "most cooperators" scenario, we considered 80 % of the nodes using the strategy cooperate, whereas 10 % each of the remaining nodes used the strategies exploit and isolate, respectively. The percentage of nodes were similarly divided in the "most exploiters" and "most isolators" scenarios.

Figure 6.3 shows the groupwise delivery ratio of the messages obtained using the HCM mobility scenario. From the figure, it can be observed that when most of the nodes in the OMN adopted the strategy cooperate, the exploiters performed the best in terms of groupwise message delivery ratio. Similarly, when most of the nodes in the network adopted the strategy exploit, the isolators did the best, whereas the performance of cooperators aced the others when most of the nodes were isolators.

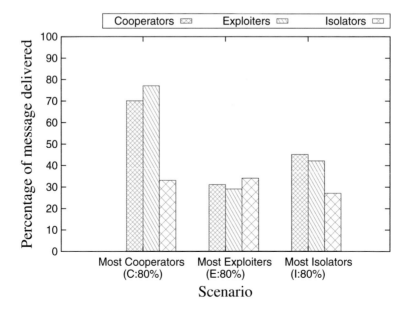

Fig. 6.3 Groupwise delivery ratio of the messages obtained using the HCM mobility scenario

To look at some specific figures, in the first scenario, the exploiters were the gainers with 77 % of messages created by exploiters being delivered. In the second scenario, the isolators were the gainers (34 % delivery of the messages created by isolators), and in the last scenario, the cooperators were the gainers (45 % of the messages created by cooperators being delivered). These results indicate a cyclical relationship similar to the RSP game, where each strategy cyclically dominates the other strategies chosen by the nodes.

We also observe that the percentage of delivered messages of the exploiters and the isolators were greater than those of the cooperators in any given scenario. However, the percentage of delivered messages of cooperators would be higher when averaged over all groups than that of the exploiters and the isolators, which is a well-known instance of Simpson's paradox. So, cooperation in the OMN would persist and increase.

Next, let us take a look at the evolution of network performance with time. Figure 6.4 shows the overall message delivery ratio of the OMN obtained at different instants of time by considering different percentages of the strategies. The figure indicates an increasing trend in the message delivery ratio with respect to time. As shown in the figure, when most of the nodes chose the strategy exploit, the delivery ratio reached up to 32 % at the end of 12 h time duration considered in the simulation. The performance slightly increased beyond 32 % when most of the nodes were isolators. However, the delivery ratio was more than double (68 %) when most of the nodes behaved as cooperators. The message delivery ratio of messages obtained in the "most cooperators" scenario was greater than the "most exploiters" and "most isolators" scenarios because the cooperators contributed toward delivery of the messages created by other (exploiter) nodes in the OMN as well.

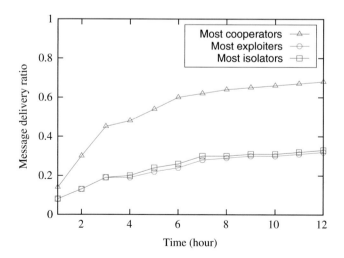

Fig. 6.4 Variation in message delivery ratio with respect to time

6.4 Summary

Since its inception, game theory has found wide applications across diverse streams. Unlike the *classical* game theory, its *evolutionary* counterpart is focused toward biology and evolutionary processes. In this chapter, we presented a quick tour of some of the fundamental concepts in game theory. In particular, we looked at the different components of a game formulation as well as analysis of Nash equilibrium. We also looked at different aspects of cooperative and non-cooperative games. Thereafter, we presented a brief overview of EGT, and discussed about ESS—a solution strategy for EGT. Subsequently, we looked at the RSP game, and several applications of EGT to diverse areas, such as biology, economics, and wireless networks.

A major portion of this chapter was devoted to the study of cooperation in OMNs using RSP game. In this regard, we considered three strategies for message forwarding—*cooperate*, *exploit*, and *isolate*. With analogy to the RSP game, here, the cooperators resemble rocks, exploiters resemble scissors, and isolators resemble papers. Thereafter, we looked into the impact of such behaviors upon the performance of message delivery in OMNs both conceptually and mathematically. Our study reveals that individual nodes gain higher payoffs if they defect rather than cooperate in message forwarding. This attracts other nodes to become exploiters, which, however, results in degradation of their performance. This, in turn, motivates exploiters to become isolators, and after certain time, isolators find cooperation to be a better strategy. As a consequence, similar to the RSP game, a cyclic transition among the three different strategies is set into motion.

In this chapter, we, however, looked at static scenarios where the nodes do not change their strategies dynamically. We shall continue this study in the following chapter by taking a comprehensive look at several cooperation enforcement schemes proposed for wireless ad hoc networks, in general. In addition to that, we shall consider the scenarios where nodes in OMNs switch to the most successful strategy over time. Such a scheme would be based on realistic constraints, and suitable for use in real-life OMNs.

6.5 Review Terms

- Classical game theory
- Core
- Rock-scissors-paper game
- Evolutionary stable strategy
- Neutrally stable strategy
- Cognitive radio networks
- Simpson's paradox
- Evolutionary game theory
- Shapley Value
- Replicator dynamics
- Nash equilibrium
- Backhaul networks
- Cooperative spectrum sensing
- Unstable and stable fixed points

Table 6.4 Payoff matrix

		A		
		C	E	I
B	C	2, 2	−3, 1	2, −1
	E	1, −3	1, 1	−1, 1
	I	−1, 2	1, −1	0, 0

6.6 Exercises

6.1 How evolutionary game theory is different from classical game theory?

6.2 What are the differences with respect to players in CGT and EGT?

6.3 What is replicator dynamics?

6.4 When does a strategy become an evolutionary stable strategy?

6.5 What is Simpson's paradox? Illustrate with an example.

6.6 Consider a pairwise population game among players with action set $A = \{F, R\}$ and payoffs $\pi(F, F) = 0, \pi(F, R) = 2, \pi(R, F) = -1$, and $\pi(R, R) = 1$. Deduce the replicator dynamics equations for this game and find all the fixed points.

6.7 Consider a payoff matrix as shown in Table 6.4. Find out whether the fixed point is asymptotically stable or unstable.

Chapter 7
Enforcing Cooperation in OMNs

Cooperation among nodes is a fundamental necessity in OMNs, where the messages are transferred using the store-carry-and-forward mechanism due to sporadic inter-node wireless connectivity. Multiple routing protocols in OMNs have been proposed in the literature based on the assumption that nodes are fully cooperative. However, this may not be always true when each node independently decides whether or not to accept messages of the other nodes. If the intermediate nodes do not cooperate in the communications process, the network is directly affected. So, the intermediate/relay nodes play key role of cooperation in OMNs. Unsurprisingly, a large volume of work is dedicated toward addressing this crucial aspect.

A quick walk-through of the existing literature reveals that diverse approaches have been taken toward ensuring cooperation in OMNs. Among the multitude of schemes, one of the well-known approaches is based on incentives. In this approach, nodes are provided with some credit to help deliver messages of other nodes. Alternatively, they gain reputation by helping others. This chapter presents a detailed discussion of a spectrum of such proposed in the literature for OMNs. On the other hand, the problem—whether a node should cooperate or not in the presence of other non-cooperative nodes—has been explored from the game theoretic perspective as well. A discussion on these topics is subsequently presented. An interesting question is, to what extent cooperation (or lack thereof) affects the performance of OMNs? We find the answer to this question as we look deeper into these schemes.

The latter portion of this chapter is aimed at detailed study of the design of a distributed scheme for enforcing cooperation in OMNs. Unlike many other existing schemes, the one concerned here is inspired by evolutionary theory and works in the lack of any centralized infrastructure. More specifically, we talk about different information collected by the nodes and how they are processed in order to approximate the overall network performance. Subsequently, we present theoretical characterization of the proposed scheme, and look at how closely it shadows the performance of an ideal alternative. Our hope in this chapter is not only to make the reader familiar with different approaches for inducing cooperation, but also to stimulate critical thinking toward any protocol development.

© Springer International Publishing Switzerland 2016
S. Misra et al., *Opportunistic Mobile Networks*,
Computer Communications and Networks, DOI 10.1007/978-3-319-29031-7_7

7.1   Cooperation Enforcement Schemes

Message delivery[1] in OMNs largely relies upon cooperation by the intermediate nodes. In the extreme case, when no node helps in delivering messages created by others, the routing process degenerates into direct delivery. Such a situation is clearly undesirable. However, in reality, due to resource limitations, the intermediate nodes may not always want to help the other nodes. Moreover, some nodes attempt to fulfill their own goal by utilizing resources of the others. Therefore, to promote cooperation among the nodes in such type of networks, different cooperation-enforcing schemes have been proposed. We broadly categorize them into two categories—(1) incentive- and reputation-based and (2) game theory-based schemes. An incentive-based scheme provides some credit (reputation) to the nodes to motivate them to cooperate. On the other hand, in game theory-based schemes, a node takes strategic decision depending on the individual (or network-wide) gains.

7.1.1   Incentive-Based Schemes

Many routing protocols for OMNs are based on the assumption that the intermediate nodes in the network are ready to forward messages received from the other nodes. However, when the owners of the nodes are people or parties, the nodes might behave selfishly or maliciously. Presence of these types of non-cooperative nodes forcefully degrades the performance of a network. Therefore, to overcome the effects of non-cooperative behavior, many incentive-based schemes have been proposed for OMNs, which indirectly enforce cooperation among the nodes. As the name suggests, incentive-based schemes provide some form of incentive to the nodes to motivate them to cooperate and refrain from selfish/malicious behavior. The nodes can subsequently "encash" the incentive so received to get their own messages forwarded. In general, incentives can be provided in two ways—(1) credit-based and (2) reputation-based. In credit-based incentive schemes, virtual currency or pricing acts as the credit. In reputation-based schemes, reputation of the nodes is calculated by their neighbors based on the message forwarding actions. Thus, if node X finds that the reputation of node Y is less than some chosen threshold, X would refuse to accept messages from Y.

7.1.1.1   Credit-Based Schemes

To encourage message forwarding among the nodes, credit is given to each of the forwarding nodes in the form of virtual currency. The nodes receive payments when

---

[1]Portions of Sect. 7.1 are reproduced with kind permission from Springer Science+Business Media: "Next-Generation Wireless Technologies, Cooperation in Delay Tolerant Networks, 2013, 15–35, Sudip Misra, Sujata Pal, Barun Kumar Saha".

they help in forwarding messages created by the other nodes. The nodes can use these credits to forward their own messages. Thus, if node A knows (for example, through acknowledgments) that node B has forwarded some message(s), it provides appropriate credit to node B. Node B, in turn, uses its accumulated currency to ask the other nodes to forward its messages.

Shevade et al. [264] proposed pairwise Tit-for-Tat (TFT) incentive mechanism for DTNs. The authors described two constraints—*generosity* and *contrition*—to maximize cooperation among the nodes. The basic TFT scheme prevents relaying messages when two nodes meet for the first time since no message has been relayed by the other node until that time. To overcome this drawback, the *generous TFT* scheme was proposed. In a generous TFT mechanism, a node must relay at least ε number of messages. Generous TFT helps to overcome asymmetric traffic demands. So, it is possible that a node may act selfishly after it has achieved this criterion. To prevent such selfish behavior, another constraint, contrition, is considered. Contrition prevents nodes from indulging into selfishness by getting a valid counterattack to its own mistakes.

Zhu et al. [265] proposed secure credit-based incentive scheme (SCI), a framework for enforcing cooperation among selfish nodes in DTNs. In addition to cooperation, SCI also adds a security mechanism for *layered coin*, a virtual electronic credit system where credits are distributed by the intermediate nodes without involvement of the senders. A layered coin consists of a *base layer* originated at the source node and multiple *endorsed layers* generated by the intermediate nodes. The base layer expresses different rewarding policies, such as payment rate, remuneration conditions, and class of service requirement. When any intermediate node receives a layered coin, it verifies the message's lifetime, likability of layers chain, sender's certificate, supporting signatures of the base layer, and intermediate nodes certificates. After performing these verifications, the intermediate node determines the next-hop forwarding node and creates an additional endorsed layer. The endorsed layers append an unforgettable digital signature based on the previous layers. The authors introduced a *concatenated layer* technique, which prevents malicious users for cheating credits and concatenates different layers with one other by injecting the generator information of the next layer into the previous layer. The authors further proposed two performance optimization techniques to minimize the transmission overhead in the network.

In [266], the authors proposed SMART, a secure multilayer credit-based incentive scheme for DTNs. SMART not only stimulates cooperation among the nodes, but also prevents malicious users from cheating credits by introducing a secure incentive scheme based on layered coins. The propagation path of a message can be easily tracked by checking the digital signature of an endorsed layer. Digital signature provides the security and the propagation path of the end-to-end connection from source to destination. Each intermediate node periodically submits the layered coin to a virtual bank (VB), which takes charge of credit clearance. The last intermediate node submits the collected layered coins to the VB for clearance. The VB checks the deposited layered coins and share credits among each forwarding nodes according to a predefined reward policy.

MobiCent [267] is a credit-based incentive scheme suitable for mobile social networks [11], where people carry low power mobile devices. MobiCent is compatible with replication-based routing protocol in DTNs. In this scheme, the selfish nodes in DTNs are not detected. Rather, the nodes are motivated into cooperation by providing them incentives for message relays. A mobile device operates in two modes—(1) long-range low-bandwidth radio and (2) short-range high-bandwidth link. The first mode ensures that a mobile device can maintain connectivity with distant devices; in the second mode, devices can exchange huge amount of data opportunistically with their closely located peers. Due to these two modes of operation, a mobile device in OMNs faces challenges in forwarding data since disconnection among the nodes is the general norm. The authors in [267] described two types of attacks—edge[2] insertion and hiding. These two attacks are addressed by an incentive-compatible payment scheme. *Edge insertion* attacks are launched by selfish nodes by inserting a sybil[3] in the earliest path. Adding a sybil on the latter path does not change the eligible path set. In *edge hiding* attacks, the selfish nodes hide the paths and can hold the message instead of forwarding it to other nodes. To counter edge insertion attacks, the multiplicative decreasing reward (MDR) algorithm is used, which prevents the relay and clients from gaining in edge insertion attack. To prevent edge hiding attacks, minimum cost and minimum delay selection algorithms are used to determine an incentive-compatible relay set by examining a sufficient subset of paths ever revealed before the deadline. These two algorithms prevent the selfish nodes from launching edge insertion and hiding attacks, and thereby, ensure cooperation among the nodes.

Lu et al. [139] proposed the practical incentive (Pi) protocol in which the selfish nodes are stimulated to cooperate by forwarding messages created by others. Pi utilizes both credit-based and reputation-based incentive schemes, and considers a single-copy data forwarding algorithm. Each source node sends a message by attaching some incentive with it. This incentive scheme is *fair* to the source node. An intermediate node may or may not participate in message forwarding. If an intermediate node forwards a message, which finally reaches its destination, then all the intermediates node get credits, which leads to increased reputation from the source node. However, if any forwarded message fails to reach the destination, then each of the intermediate nodes is assigned higher reputation values from a trusted authority. Unlike the previous case, here, the source node is not required to provide any credit to the intermediate forwarding nodes. Thus, each intermediate node in the network is stimulated for cooperation. The Pi protocol also guarantees security to the incentive scheme by using a layered coin model and verifiable encrypted signature techniques. Pi provides fair incentives and prevents free riding attacks, i.e., the layer removal and addition attacks. However, this method requires a Trusted authority

---

[2] A DTN/OMN can be represented with a *contact graph*, which consists of multiple edges observed at different time instants. In such a graph, an edge indicates a contact between two given nodes at a specified time instant. We shall discuss more about graph-based representation of DTNs/OMNs in Chap. 8.

[3] Sybil attacks [268] are launched by nodes in a network by forging multiple identities.

(TA) for storing associated credit and reputation for each node. On the other hand, "iTrust" [269] encourages cooperation among the forwarding intermediate nodes by providing incentives and by detecting and punishing the misbehaving nodes. "iTrust" introduces the concept of a periodically accessible trusted authority, which detects the misbehaving nodes based on the forwarding history evidence.

7.1.1.2 Reputation-Based Schemes

In reputation-based incentive schemes, each node is evaluated by its neighboring nodes and assigned a reputation value based on its cooperating behavior with the other nodes. If a node cooperates, its reputation value is high and the node receives better services.[4] The reputation value of a node is, however, reduced when it does not cooperate with others in forwarding messages created by the other nodes. Therefore, in order to avoid low reputation values, all nodes try to cooperate with one another in message forwarding.

Wei et al. [270] proposed MobiID, a user-centric and social-aware incentive scheme based on reputation to mitigate the effects of dishonest nodes in the network. MobiID assumes the presence of an offline system manager where each node registers to join the network. This scheme uses self-check and community-check of either the previous- or next-hop nodes or their community before forwarding a message. A reputation value is assigned to a node when it forwards a message to another node. Nodes are ostracized when their corresponding reputations fall below a given threshold. Therefore, each node exerts itself to cooperate in forwarding messages originated at other nodes in order to earn increased levels of reputation.

Zhang et al. [271] noted that even when credit-based incentive schemes are used, the selfish nodes could engage in other types of misbehavior. The authors observed that there are three potential reasons for isolating the selfish nodes in DTNs:

1. Improve the network performance (in terms of message delivery and associated delay) by not involving them in the routing process,
2. By preventing the selfish nodes to join the network, they are indirectly forced not to behave selfishly, and
3. Negative impact of selfishness in the network is minimized.

In contrast to the credit-based schemes, reputation-based schemes take into account the opinions (*reputation*) of the other nodes in the network before taking any routing decision. To discourage selfishness in DTNs, Zhang et al. [271] proposed Pri, a practical reputation-based incentive scheme. Pri assumes the use of Offline Security Manager (OSM), which is in charge of key distribution. Nodes register in OSM before joining the network. Pri is comprised of three modules—(1) monitoring behavior of the nodes, (2) computing reputation of the nodes, and (3) responding to the other

---

[4]In this context, service refers to message replication/forwarding, which is the primary operation in OMNs and other communication networks.

nodes' requests. Successful forwarding credentials (SFCs) are used as positive evidences of the nodes' successful forwarding of the messages.[5] Authentication and integrity of such SFCs are ensured using digital signatures. When a node receives a message, it sends out an SFC to the forwarding node. Any intermediate node, N_{i+1}, also sends out the SFC received from the immediate downstream node, N_{i+2}, to the node N_i that had originally forwarded to itself. While the SFCs serve as first-hand evidence, second-hand evidence is gathered through reputation propagation by the nodes.

Watchdogs (or guard dogs) are frequently used to guard premises or properties, and they alert the presence of any intruder by barking. Inspired by this, "watchdog"-based mechanism have been widely used in MANETs [272–274] for detecting the presence of selfish or malicious nodes in the network. Dias et al. [275] took a similar approach and proposed a cooperative watchdog system (CWS) for detecting the presence of selfish nodes in VDTNs. CWS assigns and updates reputation score for every node in the network. Any node say, X, assigns a reputation score (ranging between 0 and 100) to another node say, Y, by considering three aspects:

1. Reputation of Y as observed by X itself,
2. Reputation of Y observed by the neighboring nodes, and
3. The cooperative value assigned by the watchdog at X.

CWS consists of three different modules, classification, neighbors evaluation, and decision that are responsible for evaluating these values. The classification module categorizes a node into a particular type,[6] and a reputation score is assigned accordingly. Finally, the decision value combines the individual values to determine the final reputation score. CWS informs the nodes about their updated reputation scores. Based on these evaluations, a node is said to be selfish node when its reputation score is below 20. If a selfish node is detected by another node, the latter's CWS alarms and spreads the news to the entire network. In this way the CWS detect the presence of a selfish node and saves the resources (for example, buffer space) of cooperative nodes by not sharing them with the selfish nodes.

7.1.2 Game Theory-Based Schemes

In game theory, each rational user in the network takes individual decisions based on the fact that its actions affect others. In the following, we briefly take a look at some of the game theoretic approaches for inducing cooperation in DTNs/OMNs.

---

[5] A malicious node may falsely claim that it has forwarded messages to several nodes. Verifying such claim, however, is difficult in OMNs since frequent disconnections result in delay in information arrival at the nodes. Unforgeable token of proof, such as SFCs, help in ensuring that no false claim can be made.

[6] Cooperative, partial cooperative, neutral, suspected, and selfish.

To avoid selfish behavior and stimulate cooperation, Buttyan et al. [276, 277] proposed the concept of barter and developed a non-cooperative game theoretic model. The authors considered delay-tolerant personal wireless networks in a touristic city that distribute information among the tourists. It may happen that selfish users download interesting messages from others, but they do not store and distribute messages for the other users. Their model prevents this selfish behavior using the concept of barter. When two nodes A and B come in contact with each other, they first send the description of their messages. Then, they send a list of messages that they want to download from each other. Two types of messages were considered in the system—primary and secondary. A message is primary if the mobile node has interest in it, and secondary if the mobile node has no direct interest in it. Each message has some barter value for each node and the value is determined by its age and type. Using the game theoretic model, the mobile nodes A and B decide the messages that they want to download from each other to maximize their own benefit. The game model described consists of the players (P), strategy space (S_i), and the payoffs. Each individual node acts as a player. The node behavior depends on two parameters, (s, h), where s is the $\frac{secondary}{primary}$ ratio of the nodes represented by the players and h is the threshold value below which the nodes do not download the secondary messages from the other players. The payoff of the players is the average total score of the nodes in the groups. Each node receives a score after each interaction, i.e., $Score = Gain - Loss$, where $Gain$ is the total value of primary messages downloaded in the interaction and $Loss$ is total number of exchanged messages in the interaction. Using the concept of best response and Nash equilibrium, it was observed that message delivery rate increases if the mobile nodes follow the Nash equilibrium.

Yin et al. [278] proposed a model named pay-for-gain (PFG). In this mechanism, each node keeps record of the identification of encountered nodes, the total amount of buffer space it has lent from the encountered nodes, and the amount of buffer space it will lend to the encountered nodes. The authors defined a Loan Feedback (LFB) function, which helps in computing the amount of buffer space lent to users. When each pair of node offers equal-sized buffer space that they borrow from each other, the strategy PFG is similar to TFT, and according to the Nash equilibrium, it is the maximum optimization possible. When the buffer space of a node is smaller than the expected buffer size of another node, then the node offers smaller buffer space between them, which is the Nash equilibrium of maximum optimization. So, both the conditions lead to the maximal interest.

Abdelkader et al. [279] observed that message forwarding has two perspectives—that of the node and network—and raised an interesting questions in the context of DTNs—"how much should a node cooperate?" Such a threshold (a fair level of cooperation) would encourage a node to help others in the network; at the same time, it would not be labeled as a selfish node. To this end, the authors designed a new utility function to capture the fairness in node cooperation. In particular, each node maintains a received–offered table (ROT), where various metadata about received (data and acknowledgment) messages are stored. Moreover, the system imposes three constraints:

1. An acknowledgment message contains a list of all the intermediate nodes along the delivery path. Only the nodes included in that list receive a reward.
2. Hop count of any message can be H at most.
3. Since multiple copies of any message exist in the network, the first few copies of any message reaching to its destination would be eligible for a reward.

Accordingly, the utility function is defined for each node by considering the number of messages received/forwarded, cost of forwarding, and benefit received by the node. The overall network utility is obtained by adding up the utility values of the individual nodes. The authors noted that by tuning different parameters of the utility function, the nodes can achieve fair level of cooperation while reducing the cost in message delivery.

7.1.3 Other Approaches of Cooperation

Apart from the mechanisms discussed above, there are other different approaches to address the issue of cooperation in OMNs. In this section, we briefly review some of those schemes.

Panagakis et al. [280] studied the effect of nodes' cooperation (or lack thereof) on three routing protocols—Epidemic [18], two-hop routing, and binary SnW [42] using the random direction mobility model. The performance of these algorithms was studied in terms of delivery delay of the messages and the corresponding transmission overhead. Cooperation was categorized and measured in terms of a node's probability to

1. Drop a received message due to misbehavior (or possible resource constraints) or exhibit normal behavior of forwarding (this is referred to as Type I cooperation), or
2. Carry the message, but probabilistically forward it upon encounter with another node (which is said to be Type I cooperation).

On receiving a message, a non-cooperative node either drops it with probability P_{drop}, or forwards it with probability $P_{forward}$. So, the degree of cooperation in the first case is $1 - P_{drop}$, whereas in the second case it is $P_{forward}$. Based on these, the authors investigated the sensitivity of the aforementioned three routing protocols with the degree of cooperation in terms of average message delivery delay and transmission cost. When Type I cooperation was considered, it was found that Epidemic protocol far more sensitive to the degree of cooperation in terms of normalized transmission cost. However, when normalized message delivery delay was taken into account, SnW exhibited the highest sensitivity.

Resta and Santi [281, 282] addressed a similar problem as [280]. However, they took it a step forward with the detailed characterization of the performance of routing protocols under different scenarios. In particular, Resta and Santi [281] considered four different behaviors of the nodes—*coop*, *def*, *rand*, and *tft*. In *coop*, the nodes behave normally according to the underlying routing protocol. In other words, they are fully cooperative. In *def*, the nodes discard messages from the other nodes and send only their own messages. In *rand*, nodes forward messages created by other nodes with a probability p; whereas with the *tft* strategy, a node forwards other nodes' messages with a probability p, where p depends upon the contemporary network conditions. Specifically, a node sets p to the delivery ratio of messages created by it. Based on the results of performance evaluations, Resta and Santi made an interesting observation—even a moderate level of cooperation (around 25 %) was found to provide much better performance as compared to the scenario with fully non-cooperative nodes.

Keranen et al. [283] considered forwarding, non-forwarding, and partly forwarding behaviors of the nodes and investigated their impact upon message delivery ratio for three mobility scenarios (Random Waypoint, map-based model, and the real-life KAIST traces [225]) and three routing protocols (Epidemic [18], SnW [42], and PRoPHET [19]). In [280, 281], the nodes accept and silently drop the messages with some probability, which waste energy for the reception of the messages. On the contrary, in [283], the nodes do not accept the messages in the first place, but exhibit a coherent behavior. This work does not consider any incentive mechanism for cooperation. Instead, the authors proposed two types of behavior for cooperation—static and dynamic. Under the static cooperation scheme, behavior of a node remains the same throughout the network operation. However, under the dynamic scheme of cooperation, the nodes acclimate their behavior on the basis of time/energy constraints. Forwarding nodes are cooperative nodes, which help in storing and forwarding the messages. Non-forwarding nodes are non-cooperative, which do not help in storing/forwarding the messages of others, but receive messages destined to them using other nodes as free riders. Partly forwarding nodes receive others' messages and forward if they come in direct contact with the corresponding destination nodes. Therefore, with such nodes, a message travels at most two hops for delivery to its destination. The authors evaluated the network performance by considering the message delivery ratio achieved under the different routing protocols. The relative performance of SnW was less than that of Epidemic and PRoPHET protocols. However, all the three routing protocols offered good performance in the presence of 20–60 % of non-cooperating nodes. This observation concurs with [281], and suggests that OMNs are, in general, robust against selfishness.

7.2 Distributed Cooperation Enforcement

In Chap. 6, we considered three different behaviors of nodes namely, cooperate, exploit, and isolate. To briefly recapitulate, cooperators help others in delivering messages; exploiters replicate messages created only by themselves, whereas isolators directly deliver their messages to their intended destinations. In general, these three strategies capture a wide spectrum of behaviors observed in reality. In Chap. 6, we studied a scenario where behavior of the nodes remained unaltered. This, however, does not generally hold true in practice.

Evolutionary theory [284] postulates that each individual in a population periodically checks the alternative strategies and selects the best one for survival. Inspired by this, we extend our study from the previous chapter by considering an OMN in which the nodes periodically compare their individual performance with the locally available network-wide performance (in terms of delivery ratio of messages) and switch to the strategy of the most successful group, if relevant. Our objective in this is to devise a fully distributed scheme, named *distributed information-based cooperation ushering scheme (DISCUSS)*, that works solely based on interactions among the nodes. In particular, use of no infrastructure or centralized authority is assumed, since their presence in an OMN cannot be guaranteed, in general.

We represent an OMN as (N, M, S), where N denotes the set of nodes in the network, M denotes the set of messages generated by the nodes, and S denotes the set of strategies selected by any node in forwarding the messages; $S = \{Cooperate, Exploit, Isolate\}$. Accordingly, we consider three groups of nodes, *cooperators*, *exploiters*, and *isolators*, as explained earlier in Sect. 6.3. In the following, we formally present our objective in such networks.

Definition 7.1 An OMN is said to be a *Strategy Defined-Opportunistic Mobile Network (SD-OMN)* if all the nodes in the OMN follow any message forwarding strategy described by the set $S = \{Cooperate, Exploit, Isolate\}$. At any point of time, a given node chooses only one strategy from S.

DISCUSS is inspired by the existing research in *evolutionary theory* [284], which shows that the size of groups in a population (set of nodes in our case) changes with the various competing strategies. In particular, the following presents a mapping of the concepts from Darwin's Natural Selection Theory into our context involving SD-OMN.

- Nodes in SD-OMN exhibit variation in their message forwarding actions.
- Such variation occurs as the nodes follow different kinds of strategies as defined by the set S.
- The most successful strategy in the set S survives in the population and is propagated to the other nodes over time.

Let n_C, n_E, and n_I, respectively, denote the number of cooperators, exploiters, and isolators in an SD-OMN at any time instant t. Therefore, we have $|N| = n_C + n_E + n_I$. The objective of SD-OMN is to motivate the non-cooperators in the network into cooperation in order to optimize the performance of the network.

Therefore, the goal function can be written as

$$\max_{t \in \mathbb{R}} n_C(t),$$
$$\text{s.t. } 0 < t \leq \Gamma, \ 0 < n_C(t) \leq |N|. \tag{7.1}$$

Here, Γ is the lifetime of the system. When the concerned performance objective is the delivery ratio of messages (fraction of the created messages that are delivered to their respective destinations), the performance function should be maximized. Therefore, in this context, the goal function, can be alternatively represented as

$$\max_{t \in \mathbb{R}} \sum_{s \in S} \alpha_s(t),$$
$$\text{s.t. } 0 < t \leq \Gamma, \ 0 < \alpha_s(t) \leq 1. \tag{7.2}$$

Here, $\alpha_s(t)$ denotes the delivery ratio of the group of nodes with strategy $s \in S$ at any time instant t.

Initially, each node begins with a message forwarding strategy. Each node distributively shares information with the other nodes without involving a CA. Therefore, as a result of intermittent connectivity, each node has limited information about the effect of its actions. When two nodes mutually encounter with one another, they exchange their past history (own delivered message list and nodes' delivery probability list of the node) between themselves. Based on the local knowledge of past history, each node updates its own behavior, if relevant. The nodes select their own communication mode—whether to *cooperate*, *exploit*, or *isolate* in message forwarding—based on their past history as well as the performance of the other nodes known until then. Note that since our focus is on the aspects of cooperative communication in OMNs, we assume that the nodes reliably exchange information with one another. Consideration of scenarios where nodes provide false information would be a subject of an independent study on its own. Therefore, we do not consider any such issue in the remainder of this chapter.

7.3 A Detailed Look at DISCUSS

DISCUSS consists of two phases—*acquiring information* and *strategy adaptation*—that are repeated in every generation. The first phase requires information on—(1) messages created (*CM*) by the nodes, (2) delivered messages (*DM*), and (3) message delivery probabilities (*DP*) of the nodes. The second phase uses this information to decide the best paying strategy. We discuss the two phases in details below.

7.3.1 Information Acquisition

Let i and j be two nodes that come in contact with each other at time t. Let $CM^i(t)$ be the list of messages generated by node i at time t. Then,

$$CM^i(t) = \{(m^i_r, t^i_r)\},$$

where m^i_r is the rth message created by the node i, at time $t^i_r, r \geq 0$. In other words, $CM^i(t)$ consists of the list of messages together with time stamps created by node i till time t. Similarly, we define $DM^i(t)$ as:

$$DM^i(t) = \{(m^i_q, t^i_q)\},$$

where m^i_q is the qth delivered message of the node i at time t^i_q. Therefore, $DM^i(t) \subseteq CM^i(t)$.

Each node i maintains its own delivery probability as well as its view of the other nodes' delivery probabilities known till time t in DP^i_t as:

$$DP^i_t = \{(k, p^i_k, t^i_k)\}, \ k \in N, \ t \geq t^i_k, \forall k \tag{7.3}$$

Here, p^i_k is the last updated information stored at node i on the delivery probability of the node k known at time t^i_k. As usual, by message "delivery", we refer to the first event of reception of a message by its corresponding destination node. Since OMNs use multi-copy routing, a destination node can receive the same message more than once, but discards it in all instances but the former. As a consequence, the delivery ratio defined above always lies between 0 and 1. The delivery probability of any node $k, \forall k \in N$, at time instant t is determined as:

$$p_k(t) = \frac{|DM^k(t)|}{|CM^k(t)|}. \tag{7.4}$$

For simplicity, we write $p_k(t)$ as p_k by dropping the time notation. Information is updated between pairs of nodes when they come within their respective communication ranges. More specifically, nodes i and j, $i, j \in N$ maintain, and exchange, a list of delivery probabilities for every node they have met before. Let p^i_k and p^j_k, respectively, be the values of p_k known by nodes i and j. Then, node i updates the delivery probability value p^i_k in its list DP^i_t according to the following rule:

$$p^i_k = \begin{cases} p^j_k, & \text{if } t^j_k > t^i_k, \\ p^i_k, & \text{otherwise.} \end{cases} \tag{7.5}$$

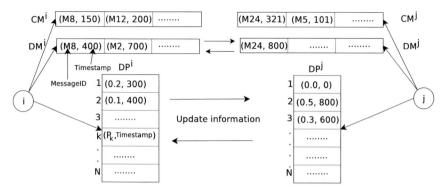

Fig. 7.1 Information update between nodes i and j during a contact

Figure 7.1 illustrates an example of comparison and update process between two nodes i and j. On every contact, node i updates its information with node j, as discussed below.

- Node i has three lists, CM^i, DM^i, and DP^i, at time t. Similarly, node j has CM^j, DM^j, and DP^j.
- Node i generates two messages, $M8$ and $M12$, at $t = 150$ and 200, respectively. Similarly, node j has messages $M24$ and $M5$ created at $t = 321$ and 101, respectively.
- Nodes i and j list the past delivered messages of nodes. Node i checks node j's list of delivered message, and updates its own list of delivered messages. Initially, nodes i and j have $DM^i = \{(M8, 400), (M2, 700)\}$ and $DM^j = \{(M24, 800)\}$.
- Node i finds its delivery probability at current time t as $p_i = |CM^i|/|DM^i|$, and updates its delivery probability list.
- Node i compares its own DP^i list with that of node j and updates its own list, if necessary.
- Let node i chooses the kth node time stamp (t_k^i) from the DP^i list, and compares it with the kth node's time stamp (t_k^j) of node j's DP^j list. If $t_k^i < t_k^j$, node i updates its kth node's delivery probability by dp_k^j. With reference to Fig. 7.1, node i checks the 1st node's time stamp, which is $300 > 0$, and consequently makes no change. Again, it checks the 2nd node's time stamp. In this case, $400 < 800$. So, the message delivery probability of this node is updated to 0.5. This process is repeated for the rest of the nodes maintained by i. After this, the updated delivery probabilities list of node i looks like $DP_t^i = \{(0.2, 300), (0.5, 800), (0.3, 600)\}$.

Algorithm 7.1 describes the steps required for updating node i's lists on encountering node j, $\forall i, j$.

Algorithm 7.1: Delivery probability update by node i on contact with node j

---

Inputs : $CM_t^i, CM_t^j, DM_t^i, DM_t^j, DP_t^i, DP_t^j$
Output: Updated values of DM_t^i and DP_t^i

```
1  C ← |CM_t^i|;                              // Number of messages in CM^i
2  D ← |DM_t^j|;                              // Number of messages in DM^j
3       /* Check, whether DM^j list contain any created messages of
   node i. If so, add it to DM^i. */
4  for u ← 1, C do
5   |  for v ← 1, D do
6   |   |  if (m_u = m_v) then
7   |   |   |_ Add m_v to DM_t^i;

8  // Find delivery probability of i and j
9  P_i ← |DM_t^i|/|CM_t^i|; P_j ← |DM_t^j|/|CM_t^j|;
10      /* Node i compare its own DP list with node j's DP list */
11 for u ← 1, P_i do
12  |  for v ← 1, P_j do
13  |   |  if t_k^v > t_k^u then     /* Compare time stamp of kth node at node i
   and the same kth node at node j. */
14  |   |   |_ p_k^u ← p_k^v, t_k^u ← t_k^v;  /* Update delivery probability and time
   stamp of kth node at node i */
```

Further, each node i, $i \in N$, maintains a record of the *group weights*,[7] $\Psi_g^i(t)$, $g \in \{C, E, I\}$. Initially, at time $t = 0$, $\Psi_g^i(t) = 0$. At any time t, when a node i comes in contact with another node j belonging to a group g, where $g \in \{C, E, I\}$, node i updates its value of the corresponding group weight as:

$$\Psi_g^i(t) = \Psi_g^i(t - 1) + \mathbb{I}_g^i(t), \tag{7.6}$$

where, $\mathbb{I}_g^i(t)$ is the indicator decision variable, which, in turn, is defined as follows:

$$\mathbb{I}_g^i(t) = \begin{cases} 1, & \text{if node } i \text{ receives a message from group } g, \\ 0, & \text{otherwise.} \end{cases}$$

For example, $\Psi^i(t) = \{\Psi_C^i, \Psi_E^i, \Psi_I^i\}$ indicates that until the time instant t, node i received Ψ_C^i, Ψ_E^i and Ψ_I^i messages, respectively, from the cooperators, exploiters, and isolators. Thus, Ψ_g denotes the relative importance of each group over the others.

---

[7]Based on the three strategies described earlier, the nodes can be divided into three groups. The set of all these groups is $\{C, E, I\}$.

7.3.2 Strategy Adaptation

Cooperation is necessary and important in OMNs for increasing the network performance. So, for enhancing cooperation among nodes, we propose a dynamic *strategy adaptation* mechanism. This mechanism bridges the nodes' self-recommendation approach based on the local information gathered during contact with other nodes. The objective of the proposed approach is to maximize the expected delivery probability of messages in the Kth time-slots by motivating the exploiters and isolators to shift their strategies to cooperation.

We evaluate the proposed approach using two scenarios: *static* and *dynamic*. In the *static* scenario, the number of cooperators (n_C), exploiters (n_E), and isolators (n_I) remain invariant over time t, i.e.,

$$\left. \begin{array}{l} n_C(t) = n_C(0), \\ n_E(t) = n_E(0), \\ n_I(t) = n_I(0), \end{array} \right\} \qquad (7.7)$$

and $n_C + n_E + n_I = |N|$, $\forall t$.

In the *dynamic* scenario, the number of cooperators (n_C), exploiters (n_E), and isolators (n_I) vary with time. A *generation* is defined as a stage in the life cycle of a node. The strategies of individuals (nodes) remain same in a *generation*. The *generation interval* (τ), where $\tau = \{1, 2, \ldots, K\}$, is the time required to complete a generation. In Fig. 7.2, initially, the size of cooperators, exploiters, and isolators are the same.

At the end of a generation, every node compares its individual performance with that of the other groups and dynamically changes its strategy, if required, accordingly. Such a self-evaluation can be performed by each node individually during the run-time based on information exchanged with the other nodes, as discussed below.

Each node i has a delivery probability list (DP_t^i), where it stores delivery probabilities of the other nodes based on the most recent information available, together with

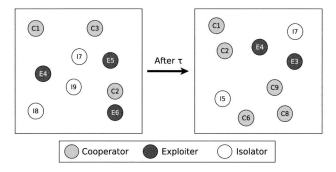

Fig. 7.2 Strategy adaptation by the nodes in the SD-OMN after a generation interval (τ)

the corresponding time when the information was obtained, and then time stamps it. Any node i, $i \in N$, computes the average delivery ratio α_g, $g \in \{C, E, I\}$, of cooperators, exploiters, and isolators, as follows:

$$\left.\begin{aligned}
\alpha_C^i &= \frac{\sum_{k \in N_C} p_k^i}{|N_C|} \\
\alpha_E^i &= \frac{\sum_{k \in N_E} p_k^i}{|N_E|} \\
\alpha_I^i &= \frac{\sum_{k \in N_I} p_k^i}{|N_I|}.
\end{aligned}\right\} \tag{7.8}$$

where, α_C, α_E, and α_I, respectively, denote the average delivery ratio of the cooperators, exploiters, and isolators in the OMN. Here, N_C, N_E, and N_I denote the set of cooperators, exploiters, and isolators that node i, $i \in N$, met within time interval τ.

Definition 7.2 The weighted factor, ω, of a group g, $g \in \{C, E, I\}$, is defined as:

$$\omega_g = \frac{\Psi_g}{\sum_{s \in \{C, E, I\}} \Psi_s}.$$

Node i determines the *weighted factor* of each group from $\Psi^i(t)$. The weighted factor of cooperators (ω_C^i), exploiters (ω_E^i), and isolators (ω_I^i) of node i are calculated using Definition 7.2, as follows:

$$\left.\begin{aligned}
\omega_C^i &= \frac{\Psi_C^i}{\Psi_C^i + \Psi_E^i + \Psi_I^i} \\
\omega_E^i &= \frac{\Psi_E^i}{\Psi_C^i + \Psi_E^i + \Psi_I^i} \\
\omega_I^i &= \frac{\Psi_I^i}{\Psi_C^i + \Psi_E^i + \Psi_I^i}.
\end{aligned}\right\} \tag{7.9}$$

Definition 7.3 The weighted average message delivery ratio γ_g of a group g is defined as

$$\gamma_g = \alpha_g \times \omega_g. \tag{7.10}$$

From Definition 7.3, the *weighted average delivery ratio* of cooperators (γ_C), exploiters (γ_E), and isolators (γ_I) of node i is calculated as follows:

$$\left.\begin{aligned}
\gamma_C^i &= \alpha_C^i \times \omega_C^i \\
\gamma_E^i &= \alpha_E^i \times \omega_E^i \\
\gamma_I^i &= \alpha_I^i \times \omega_I^i.
\end{aligned}\right\} \tag{7.11}$$

Finally, node i determines the most successful group, g_{succ}, as follows:

$$g_{succ} = \begin{cases} C, & \gamma_C > \gamma_E, \ \gamma_C > \gamma_I \\ E, & \gamma_E > \gamma_C, \ \gamma_E > \gamma_I \\ I, & \gamma_I > \gamma_E, \ \gamma_I > \gamma_C. \end{cases} \qquad (7.12)$$

Now, node i changes its group to g_{succ}, if its delivery probability, computed using (7.4), is less than the maximum of the weighted average delivery ratio of the three groups, i.e.,

$$p_i < max(\gamma_C^i, \gamma_E^i, \gamma_I^i). \qquad (7.13)$$

The strategy adaptation mechanism is repeated exactly in the same way for every node after each *generation interval* for switching to a successful strategy. In particular, we note that if a node fails to perform self-evaluation due to one reason or another (for example, when a node does not have information about the other nodes available yet), it continues with its own previous strategy. The flowchart presented in Fig. 7.3 shows the sequence of steps performed by a node executing the DISCUSS algorithm in an OMN.

7.4   Characteristics of DISCUSS

This section presents the complexity analysis and theoretical analysis on the characteristic of DISCUSS.

7.4.1   Theoretical Analysis

Definition 7.4 A SD-OMN with N nodes is said to have reached an *equilibrium*, when $(1 - \delta)N$ (where $0 \le \delta < 1$) nodes have a common strategy $s \in S$, and the remaining δN nodes have different strategy/strategies. The common strategy of the $(1 - \delta)N$ nodes is said to be the *equilibrium strategy*.

Definition 7.5 A SD-OMN is said to have reached (K, δ) *convergence* if for K successive generations, where $K \ge 1$, at least the $(1 - \delta)N$ nodes have the same common strategy $s \in S$, while the remaining δN nodes could possibly have different strategy/strategies.

Theorem 7.1 (Necessary criteria for (K, δ) convergence) *Let s_i be the equilibrium strategy at the ith generation. If $s_i = s_j, \forall i, j \in \{1, 2, \dots, K\}$, SD-OMN has attained (K, δ) convergence.*

Proof The proof follows from Definition 7.5.

Fig. 7.3 Flowchart
depicting the steps in
DISCUSS

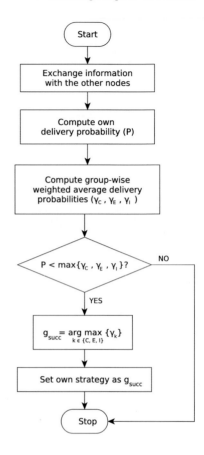

Theorem 7.2 (Sufficient criteria for (K, δ) convergence) *Let s_i be the equilibrium strategy for the ith generation. It is sufficient to say that SD-OMN moves toward (K, δ) convergence with $\delta \to 0$ when*

$$\gamma_{s_i} \gg \max_{s_i}\{\gamma_{s_i'}\}, \ s_i' \in S - \{s_i\}.$$

Proof Let N_{s_i} be the set of nodes with strategy s_i, and $N_{s_i'}$ be the set with strategies other than s_i. Using Eq. (7.10), we have,

$$\gamma_{s_i} = \sum_{n \in N_{s_i}} \alpha_{s_i}^n \omega_{s_i}^n.$$

Assuming, $\gamma_{s_i} \gg \gamma_{s_i'}$, we have

$$\sum_{n \in N_{s_i}} \alpha_{s_i}^n \omega_{s_i}^n \gg \sum_{n \in N_{s_i'}} \alpha_{s_i'}^n \omega_{s_i'}^n$$

$$\Rightarrow \sum_{n \in N_{s_i}} \alpha_{s_i}^n \Psi_{s_i}^n \gg \sum_{n \in N_{s_i'}} \alpha_{s_i'}^n \Psi_{s_i'}^n$$

$$\Rightarrow \sum_{n \in N_{s_i}} \frac{p_{s_i}^n}{|N_{s_i}|} \Psi_{s_i}^n \gg \sum_{n \in N_{s_i'}} \frac{p_{s_i'}^n}{|N_{s_i'}|} \Psi_{s_i'}^n.$$

The above holds true if any one of the following conditions holds: (1) $|N_{s_i}| \gg |N_{s_i'}|$, (2) individual $\alpha_{s_i} \Psi_{s_i} \gg \alpha_{s_i'} \Psi_{s_i'}$, and (3) both (1) and (2). Let us look at the individual cases.

Case 1:

$$|N_{s_i}| \gg |N_{s_i'}|$$

$$\Rightarrow \quad \frac{|N_{s_i}|}{|N_{s_i'}|} \gg 1$$

$$\Rightarrow \quad \frac{|N_{s_i}| + |N_{s_i'}|}{|N_{s_i'}|} \gg 1$$

$$\Rightarrow \frac{1}{\delta} \gg 1 \Rightarrow \delta \ll 1 \Rightarrow \delta \to 0.$$

This implies that SD-OMN moves toward convergence.

Case 2: $\alpha_{s_i} \Psi_{s_i} \gg \alpha_{s_i'} \Psi_{s_i'}$ implies that most of the nodes from $N_{s_i'}$ switch to N_{s_i}, which, in turn, implies that in the next generation, SD-OMN will attain convergence.

Case 3: If both Cases (1) and (2) occur, SD-OMN will attain convergence. Hence, the proof.

Theorems 7.1 and 7.2 bear significance in the characterizing the convergence time of strategies in the concerned SD-OMN. In other words, for given K and δ, one can determine if and when the OMN converges to an equilibrium strategy.

Theorem 7.3 *The message delivery ratio of the SD-OMN increases with the increasing number of cooperators. Mathematically,*

$$lim_{|N_C| \to |N|} \sum_{s \in S} \alpha_s(t) = 1.$$

Proof Let $|N_C| \gg |N_{s'}|$, where $s' \in S - \{C\}$. The chances of message delivery for a group g increases with the increase in the corresponding Ψ_g. So, we can approximate α_s as $\alpha_s(t) \approx \frac{\Psi_s}{\sum_{s \in S} \Psi_s}$. As the number of cooperators is high, Ψ_C is also high, since the cooperators forward the messages of exploiters as well. So, we can write $\Psi_C \gg \Psi_{s'}$.

Therefore,

$$\alpha_C(t) = \frac{\Psi_C}{\sum_{s \in S} \Psi_s} \approx \frac{\Psi_C}{\Psi_C} = 1.$$

Moreover, as $\Psi_{s'} \to 0$, $\alpha_{s'} \to 0$, $s' \in S - \{C\}$. Therefore,

$$\sum_{s \in S} \alpha_s(t) = \alpha_C(t) + \sum_{s' \in S - \{C\}} \alpha_{s'}(t) = 1.$$

Hence, the proof.

Theorem 7.3 holds true in a SD-OMN when one of the following conditions is satisfied.

1. The traffic generation rate is low, but messages are generated throughout the lifetime of the system considered.
2. The traffic rate is high and messages are generated for a short period of time (possibly during the initial generation of the SD-OMN).

In the first case, due to low traffic rate, a less number of messages are generated in the time period considered. Theoretically, almost all of them can be delivered to the respective destination if most of the nodes are cooperators, so that $lim_{|N_C| \to |N|} \sum_{s \in S} \alpha_s(t) = 1$. In reality, depending on the contact patterns among the nodes, the same may be less than unity. In the second case, although more messages are created in the system, a comparatively longer lifetime allows the delivery of most of them. Again, when most of the nodes are cooperators, $lim_{|N_C| \to |N|} \sum_{s \in S} \alpha_s(t) = 1$. Moreover, this theorem verifies the goal function presented in (7.2).

The case of high traffic rate, however, deserves further discussion. In real life, there are practical constraints on the buffer capacity of the nodes and/or lifetime of the messages. Even if such constraints are relaxed, continuous generation of messages throughout the lifetime of the system does not provide sufficient time to the nodes to deliver them using the store-carry-and-forward approach that is typical of OMNs. This can be realized from the fact that in OMNs, the average message delivery latency is typically a few thousand seconds, or may be more. Therefore, in such cases, the claim of Theorem 7.3 can be suboptimal.

7.4.2 Complexity Analysis

We complete our characterization of DISCUSS by considering its space and time complexities.

Space Complexity

Each node caches the *CM*, *DM*, and *DP* lists in its buffer. Let σ, φ, λ, and μ bytes be required to store a node's address, message identity, time stamp, and delivery probability, respectively. Let us further assume that a node generates X number of messages out of which Y messages are delivered. Let N be the total number of nodes in the SD-OMN. We compute the space required for storing X messages in CM and Y messages in DM. The space complexity for storing CM, DM, and DP at the nodes, respectively, are

$$\mathscr{S}(CM) = X \times (\varphi + \lambda) \in O(X), X \gg (\varphi + \lambda)$$
$$\mathscr{S}(DM) = Y \times (\varphi + \lambda) \in O(Y), Y \gg (\varphi + \lambda)$$
$$\mathscr{S}(DP) = N \times (\sigma + \mu + \lambda) \in O(N), N \gg \sigma + \mu + \lambda$$

So, the space complexity for storing these lists in a node is $O(n)$, where $n = X + Y + N$. As an illustration, let $\sigma = 8$ byte (typical for Bluetooth hardware address), $\varphi = \lambda = \mu = 4$ byte, and $N = 100$. So, the size of the list DP is, $|DP| = 100 \times (8+4+4) = 1600$ byte. Similarly, $|CM| = 8X$ byte, and $|DM| = 8Y$ byte. So, the total buffer overhead, $B = 1600 + 8(X+Y)$ byte. With $X = Y = 100$, $B = 3200$ byte. Figure 7.4 shows the buffer requirement versus the number of messages for storing the lists, when $X = Y$. This is insignificant, given the fact that today's smart phones have storage capacities ranging upto a few gigabytes.

Time Complexity

The time complexity of Algorithm 7.1 is $\max\{O(XY), O(N^2)\} = O(m^2)$, where $m = N$ or $m =$ number of messages created or delivered. Thus, the time complexity is quadratic in terms of the number of nodes in the network, or the number of messages created and delivered—whichever is maximum. The most time consuming part of DISCUSS algorithm (as shown in Fig. 7.3) is the phase of acquiring information, which is described in Algorithm 7.1. The remaining steps of the DISCUSS

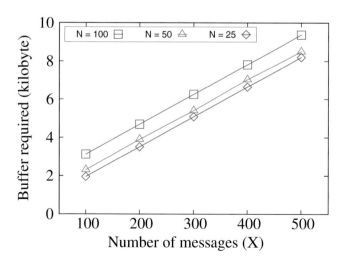

Fig. 7.4 Storage overhead of DISCUSS as a function of the number of messages

212 7 Enforcing Cooperation in OMNs

algorithm are linear or constant, and do not affect the order of complexity. So, the time complexity of DISCUSS is $O(m^2)$.

In DISCUSS, we assume that each node correctly informs its contemporary strategy to any node that it comes in contact with. This is essential in computing the groupwise delivery ratio of the messages, which, in turn, is used to determine the most successful strategy. However, in practice, such an assumption may not hold true. An exploiter is not expected to inform others that it is exploiting them. Rather, exploiters may wrongfully claim themselves to be cooperators. This assumption is relaxed in [285], where a *distributed strategy identification scheme* (DISIDE) for SD-OMNs is proposed. In DISIDE, nodes do not inform one another about their strategies. In contrary, each node estimates the strategy of others based on past observations of message replication. It has been shown in [285] that DISIDE is highly efficient. Moreover, the use of DISIDE with DISCUSS can result in close performance as compared to the original version of DISCUSS.

7.5 Performance Insights

In this section, let us take a quick look at how DISCUSS performs in practical scenarios. We evaluated DISCUSS using the ONE simulator [51] considering two synthetic mobility models—*Helsinki City Map* (HCM) and *Random WayPoint* (RWP), and four real-life traces—Infocom'06 [80], PMTR [83], Sassy [86], and KAIST [85]. At the beginning of a simulation, an initial strategy from the strategy set S was assigned to each node in such a way that the percentage of the nodes having each strategy was equal. Based on the information available at the end of every generation interval (τ), each node determined the most successful strategy and switched to it if the node was not already using the same strategy. Unless otherwise specified, the generation interval was taken to be 1 h. Apart from typical performance metrics, we also considered the *group composition (size)*, which is the percentage of cooperators, exploiters, and isolators in each generation with respect to time. We compared the results of performance evaluation obtained for three variants of the DISCUSS algorithm: (a) DISCUSS in a dynamic[8] scenario (referred to as "dynamic" in the plots), (b) DISCUSS in a static scenario (referred to as "static" in the plots), and (c) DISCUSS with global knowledge (referred to as "global" in the plots), which we discuss below.

---

[8]This actually is our proposed scheme where the nodes adapted their strategies with time. See Sect. 7.3.2 for further details.

7.5.1 DISCUSS with Global Knowledge

For the sake of completeness and evaluating the effectiveness, we also consider a variant of DISCUSS, where the nodes are assumed to possess *global knowledge* about the SD-OMN—every node is assumed to have precise information about the messages created and delivered by any node. In this case, we assume the presence of a Central Authority (CA) in the network with which the nodes can communicate instantaneously. Whenever a new message is created (or delivered), the CA is informed by the concerned node. Based on these information, the CA computes the *DP* of each node. At the end of each generation interval, each node retrieves the *DP* information of the other nodes in the network from the CA. Accordingly, each node adapts its strategies to the most successful one, if required. Note that this latter variant is purely of theoretical interest and diverges from reality, but also serves as a useful benchmark at the same time.

In the following, we take a look at the performance of DISCUSS in different scenarios.

7.5.2 Effects of Generation Interval

Let us begin by looking at the effects of generation interval upon the network performance. Figure 7.5 shows the delivery ratio of messages obtained using different mobility scenarios for different generation intervals (τ). In the KAIST and RWP mobility scenarios, the message delivery ratio was higher when $\tau = 1$ h, and it decreased as τ increased. This is because, as τ increased, the number of exploiters and isolators remained the same in respective strategies for τ time. So, the message drops increased in the OMN. It was observed that in 1 h, almost all nodes had information about the other nodes. So, they attempted to switch their respective strategies. The sooner they changed their strategies, the better they earned in terms of delivery ratio. So, the overall network-wide message delivery increased, as the generation interval was less. Similarly, using the Infocom'06 trace as well, the message delivery ratio increased when $\tau = 1$ h. But in the HCM scenario, the delivery ratio improved marginally when $\tau = 2$ h because in this case, in the 1st hour, the number of direct contacts were more than the indirect contacts, which implies that almost all messages of nodes are delivered. So, less number of nodes changed their strategies. This is ascribed to the fact that delivery ratio was high when $\tau \approx 1$ h. Based on these empirical observations, we considered $\tau = 1$ h for all the subsequent simulations.

7.5.3 Similarity Measurement

Similarity index is used to determine the similarity between two documents or entities. Using this metric, we establish the effectiveness of DISCUSS by comparing

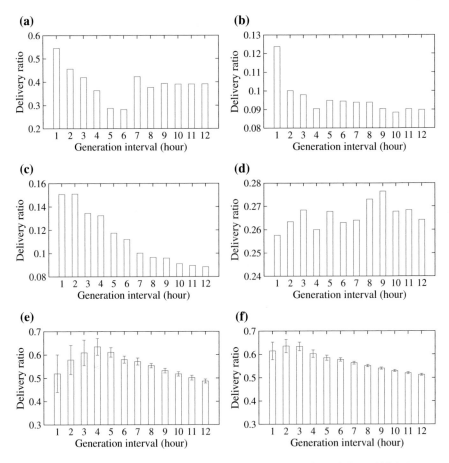

Fig. 7.5 Message delivery ratio for different generation intervals using **a** Infocom'06, **b** Sassy, **c** KAIST, and **d** PMTR traces, and **e** RWP and **f** HCM synthetic mobility scenarios

it with its variant in which global knowledge is available. Jaccard similarity index [286] has been widely used in several works related to DTNs (such as [287]) and wireless networks [288], in general. The application of Jaccard similarity index is suitable to evaluate the similarity of two sets, as considered here, unlike many other similarity indices, such as cosine similarity index, which operates upon two vectors.

Let, after a generation interval, N_D and N_G, respectively, be the set of nodes using DISCUSS and its variant with global knowledge. In particular, $|N_G| = |N_D| = |N|$; N_G and N_D differ only in terms of the strategies adapted by the different nodes, for example, node i may have strategy C in N_G and E in N_D. Then, the Jaccard similarity index [286] is defined as

$$J_s = \frac{|N_G \cap N_D|}{|N_G \cup N_D|}. \tag{7.14}$$

A measure of $J_s \in [0, 1]$ indicates the closeness in performance of DISCUSS with local and global knowledge.

Example 7.1 Let us consider two sets, $A = \{1, 2, 3, 4, 5\}$ and $B = \{2, 4, 6\}$. Then, $A \cup B = \{1, 2, 3, 4, 5, 6\}$ and $A \cap B = \{2, 4\}$. Therefore, the Jaccard similarity index for these two sets is $J_s = |A \cap B|/|A \cup B| = 2/6 = 1/3$. In other words, the two sets are about 33 % similar. $\square$

Figure 7.6 shows the similarities between DISCUSS with global and local knowledge for different time intervals. It can be observed that the value of Jaccard similarity index increased with time, since the nodes obtained more accurate information as time increased. For the Infocom'06, HCM, KAIST, and RWP mobility scenarios the Jaccard similarity index was observed to be above 85 % at the end of two days. However, for Sassy and PMTR, the similarities varied between 55–70 %. In RWP, initially, in the first hour, very less number of nodes changed their strategies. So, the similarity is higher at the beginning. In RWP, the nodes met less frequently and for short durations. Therefore, using the DISCUSS scheme, the nodes did not exhibit accurate possessing information about other groups. So, the similarity decreased significantly (30 %) between 1–10 h using RWP. Overall, DISCUSS, which is distributed and works based on locally available information, bears close resemblance to the scenario where global knowledge is used. This establishes the efficiency of DISCUSS.

7.5.4 Variation in Group Composition

Figure 7.7 shows the variation in group composition with time. The reader may recall that our objective in designing DISCUSS is to motivate increasing number of nodes in an OMN into cooperation. In other words, we expect the percentage of cooperators to dominate the other types of strategies; in the best case, all nodes should become and stay as cooperators.

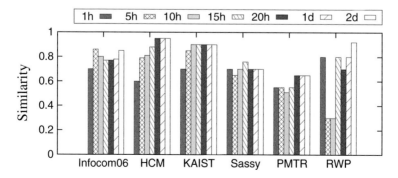

Fig. 7.6 Measure of similarity between DISCUSS and its variant with global knowledge using Jaccard Index

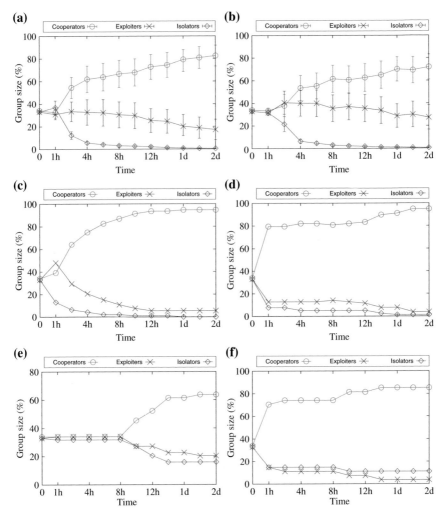

Fig. 7.7 Variation in group composition with time under different mobility scenarios: **a** HCM, **b** RWP, **c** KAIST, **d** Infocom'06, **e** PMTR, and **f** Sassy

As mentioned earlier, initially, in all the simulation scenarios, the percentages of cooperators, exploiters, and isolators were the same. Figure 7.7 shows that in all the scenarios, the residual number of cooperators varied between 85–100 %, and the number of exploiters and isolators varied between 0–10 %, except for the PMTR case. Using the PMTR trace, the percentage of cooperators, exploiters, and isolators were 78 %, 18 %, and 4 %, respectively. We observe from Fig. 7.7 that the group size of the cooperators increased with time. This is due to the reason that the weighted delivery ratio, as shown in (7.11), of the cooperators, in general, is higher because they help in forwarding messages created by other nodes. So, after each generation interval, more nodes switched their strategy to cooperate rather than to exploit or isolate. Another interesting phenomenon that can be observed for all the scenarios is that, approximately after one day, about 80 % nodes became cooperators because cooperate was the most successful strategy. In the HCM, RWP, KAIST, and Infocom'06 scenarios, the number of isolators reached almost 0 % after a day, and number of exploiters reached 2–3 %, except in the case of the HCM and PMTR traces.

(K, δ) Convergence

Since DISCUSS is based on the principles of evolutionary theory, it is not sufficient to ensure that all (most of) the nodes in an OMN choose the optimum strategy and reach equilibrium, but it also equally important to guarantee that they continue with that equilibrium strategy. The reader may recall from Chap. 6 that an RSP cycle of strategy shift was suggested based on the percentages of individual strategies and their performance. Therefore, we look at the "stability" of such an equilibrium strategy, which here is cooperation, using the notion of (K, δ) convergence (Definition 7.5).

From Fig. 7.7d, it can be observed that beyond 24 h, the percentage of cooperators was at least 90 %. Thus, $\delta = 10$ % of the nodes had strategies other than cooperation. Let us consider the time duration between 24 and 36 h consisting of 12 generations, with $\tau = 1$ h. We can infer from Fig. 7.7d that the SD-OMN attained (K, δ) convergence, with $K = 12$ and $\delta = 0.1$. Similar inferences can be drawn from others as well.

7.5.5 Delivery of Messages

Finally, let us take a look at the message delivery ratio obtained using DISCUSS under diverse scenarios. As mentioned earlier, here, we consider three different variants of DISCUSS—*static*, *dynamic*, and *global* scenarios. In the first scenario, as the reader would recall, the composition of the three groups of nodes do not change.

Figure 7.8 shows the delivery ratio of messages obtained under different mobility scenarios as a function of time.[9] It can be observed from the figure that the dynamic version of DISCUSS, where the nodes adapted their strategies, exhibited higher delivery ratio than in the static version. For example, in the Infocom'06 and HCM scenarios, the dynamic version outperformed the static one by 31 % and 17 %, respectively, at the end of two days. This is due to the reason that in the static scenario, the groupwise percentage of nodes remained the same throughout the simulation. As discussed earlier, exploiters and isolators do not cooperate in forwarding other nodes' messages. So, the message drop probability increased over time and the delivery probability decreased. On the other hand, in the dynamic scenario, the groupwise percentage of nodes changed over time and most of the nodes switched their strategy to cooperate. Hence, the message delivery ratio increased over time, in general.

---

[9]Note that this is not necessarily a cumulative—monotonic to be specific—function. For example, the delivery ratio of messages after first 20 h may be less than that achieved in the first 10 h.

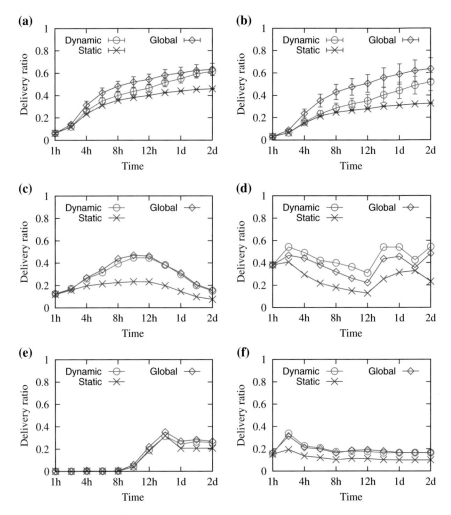

Fig. 7.8 Message delivery ratio using DISCUSS in different scenarios: **a** HCM, **b** RWP, **c** KAIST, **d** Infocom'06, **e** PMTR, and **f** Sassy

The reader may observe that the message delivery ratio using the Infocom'06 trace, as depicted in Fig. 7.8d, was low between 2–12 h. This is due to the reason that during this time, the number of contacts per hour was quite less, as shown in Fig. 7.9b. So, during this time, a less number of messages were delivered to the destination nodes.

However, in the HCM mobility scenario, as shown in Fig. 7.9a, the number of contacts varied in 350–550 range, which is nearly similar throughout the simulation duration. So, the probability of message delivery increased over time when the cooperators increased with time. In the KAIST mobility trace, the message delivery

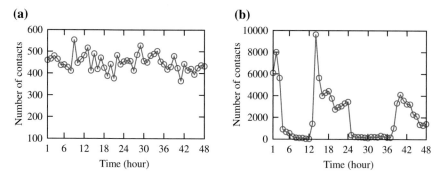

Fig. 7.9 Number of contacts in the network in the **a** HCM and **b** Infocom'06 scenarios

ratio decreased after 12 h, because the number of contacts was observed to be less among nodes in between 12–48 h and message generation rate was uniform throughout the simulation duration. The delivery ratio of RWP, KAIST, PMTR, and Sassy were 51 %, 15 %, 25 %, and 16 %, respectively, due to assorted number of contacts on different mobility scenarios.

Synthetic mobility models, such as RWP and HCM, are based on a set of equations and, therefore, offer "smooth" results. For example, in Fig. 7.9a, the contacts distribution is more or less uniform. However, real-life processes are not governed by simple formulae. As such, observations from real-life mobility often tend to show spikes and jumps, as can be seen in Fig. 7.9b.

Finally, let us take a look at how DISCUSS with local knowledge (indicated by "dynamic" in Fig. 7.8) fares in comparison to its variant with global knowledge (indicated by "global" in the figure). For the global scenario, accurate information was obtained by all the nodes and each node took decision accordingly. This implies that the performance profile for the global case is more accurate than the distributed case in the dynamic scenario. In the KAIST, PMTR, and Sassy scenarios, the delivery ratio was almost similar for both the global and dynamic scenarios as the number of contacts between nodes was higher and repetitive. In RWP, the delivery ratio of DISCUSS in the global scenario was more than that in the dynamic scenario, because in the latter case, the nodes did not get the complete information about the other groups of nodes due to less number of contacts. However, in the HCM scenario, the contacts were more and repetitive. Initially, the nodes had partial information about network performance. As time passed, they came in contact with more number of nodes. So, the delivery ratio in the case of the dynamic scenario was almost the same as that of the global scenario after one day. Similarly, in the Infocom'06 scenario, the delivery probability of the dynamic scenario was approximately the same as that in the global scenario. As a whole, Fig. 7.8 indicates that DISCUSS with only local knowledge closely shadowed the performance of that with global knowledge. This, once again established the efficiency of DISCUSS and feasibility of its use in practice.

7.6 Summary

Design of robust and versatile routing protocol for efficient communication in OMNs is, of course, a primary requirement. But at the same time, cooperative communication among the nodes demands an equal, if not more, attention. Network performance degrades as more and more nodes drift toward non-cooperative behavior, where they do not help other nodes in the OMN by replicating their messages. Over the years, researchers have proposed several mechanisms to mitigate such scenarios. One of the most common approaches is based on providing incentives to the nodes in return of their help; many other schemes rely on game theory. One of the fundamental requirements in this case is that the schemes should operate in distributed fashion. While certain form of infrastructure may be available in some scenarios, their versatility is not guaranteed in general.

DISCUSS differs from many of the existing schemes in this respect. In DISCUSS, the nodes do not rely upon a CA or any other infrastructure. Rather, the nodes exchange information with one another when they come in contact. Moreover, DISCUSS is inspired by evolutionary theory, where the nodes always try to adapt the most successful strategy in the OMN. In this chapter, we minutely looked at different aspects of DISCUSS using both theoretical and practical approaches. We observed that, after some period of time, DISCUSS can motivate a majority of nodes in an OMN into cooperation. This not only establishes the efficiency of DISCUSS, but also hints at its adoption in real world.

7.7 Review Terms

- Incentive
- Reputation
- Sybil
- Edge insertion
- Watchdog
- Population
- Generation interval
- Weighted factor of groups
- Jaccard similarity index

- Credit
- Layered coins
- Contact graph
- Edge hiding
- SD-OMN
- Generation
- Group weight
- (K, δ) convergence
- Equilibrium strategy

7.8 Exercises

7.1 Let us consider four broad areas in OMNs—multicasting, cooperation, trust, and privacy. Suppose that, in the year 2014, the number of research works focused on these four areas were 10, 75, 40, and 10, respectively; whereas, in 2015, these numbers were 5, 60, 25, and 45, respectively. What is the cosine similarity of research focus between these 2 years?

7.2 What is sybil attack?

7.3 What is a contact graph? What are edge insertion and removal attacks?

7.4 Discuss one of the challenges of using incentive-based schemes in OMNs.

7.5 What is a watchdog? How is this concept used in the context of network cooperation?

7.6 Let us focus on a single node say, X, in an OMN, which is using DISCUSS. Suppose that $\Psi(t) = \{\Psi_C = 15, \Psi_E = 10, \Psi_I = 5\}$. What are the weighted factors for each group? Also, consider that $\alpha_C = 0.5$, $\alpha_E = 0.8$, and $\alpha_I = 0.3$. What is the most successful strategy of the OMN as determined by X? If the strategy of X at the beginning of the current generation was cooperate, would there be any change in its strategy after this generation? Explain.

7.9 Programming Exercises

7.7 Implement the DISCUSS scheme as discussed here in the ONE simulator. You can use three strings for identifying the three strategies. In particular, you need to think about how to enforce message reception and transmission rules for the three different behaviors. What action should be taken, if any, after a message has been received? Moreover, check the methods in the routing classes to identify how to delete a message. What conditions should be taken care of before deleting any message? Assume that all kinds of metadata are exchanged instantaneously without incurring any overhead similar to the implementation of PRoPHET.

Part IV
Advanced Topics

Chapter 8
Heterogeneity in OMNs

Heterogeneity in real-life networks is perhaps inevitable. This is an aspect that affects every different kind of network—be it MANETs [289], WSNs [290], or mobile cloud computing environments [291]. Indeed, in today's world, we have several kinds of devices that together form communication networks. On one end of the spectrum, we have hand-held devices, such as smartphones and tablets. On the other end, we have core networking switches and routers. In fact, we not only have heterogeneous networks, but in many cases, the services offered are heterogeneous, too. In this context, the Internet of Things (IoT) should also be mentioned that aims to interconnect all sorts of heterogeneous devices [292].

In this chapter, we look at different aspects of heterogeneity in conventional networks, such as MANETs, as well as, in OMNs. We begin this chapter by looking at the fundamental causes of heterogeneity in communication networks. Interestingly, heterogeneity in a network can be manifested by as simple as storage capacity and energy levels of the devices, to much more complex aspects related to hardware and software. Moreover, the underlying architecture for communication can also lead toward a heterogeneous system. Subsequently, we look at the different forms of heterogeneity that arises in the link and network layers of the protocol stack. Moreover, we also look at different techniques in the existing literature to mitigate such issues. Subsequently, we take a look at heterogeneous contact patterns in OMNs, which results in diverse inter-contact times among the devices.

Next, we focus on several aspects of heterogeneity that are observed in OMNs. In this context, the reader might recall the fundamental characteristics of DTNs, which inherently indicate heterogeneity. In particular, we discuss about the heterogeneity in connection dynamics in such networks. Subsequently, we discuss about the diversity in hardware and their impact upon the network. However, heterogeneity is not only related to hardware, but software as well. Therefore, we investigate the causes of incompatibility in routing protocols, and summarize the adverse effects that heterogeneity has upon the performance of OMNs.

© Springer International Publishing Switzerland 2016
S. Misra et al., *Opportunistic Mobile Networks*,
Computer Communications and Networks, DOI 10.1007/978-3-319-29031-7_8

The latter portion of this chapter deals with an interesting topic—the representation of DTNs/OMNs as graphs. The topic is interesting, since traditional graph theoretic approaches fall short in DTNs due to its unique characteristics. To this end, we briefly study the temporal graph model proposed by Kostakos. We also touch upon the concepts of random graphs and evolving graphs. Then, we look at time-varying graphs, which attempts to unify different related formalisms existing in the literature. Subsequently, we discuss about the use of time-varying graphs to represent a heterogeneous OMN.

While an understanding of the root causes heterogeneity is fine, one really needs insights to tackle them. In this chapter, we look at some approaches to overcome the adverse effects of heterogeneity related to hardware. We also propose the concept of protocol translation units in order to bridge the functionality of two or more incompatible routing protocols in OMNs. Finally, we conclude this chapter with key performance insights.

> Although the aspects of heterogeneity in this chapter are discussed in the context of DTNs/OMNs, they are equally valid for the MOONs as well. In fact, in the previous chapters in Part *II*, we have looked at the performance degradation of MOONs when heterogeneity is considered.

8.1 Heterogeneity in Communication Networks

We begin this chapter by looking at different forms—and sources—of heterogeneity observed in traditional networks. Subsequently, we would study some heterogeneity aspects in the link layer and network layer of the protocol stack. We also look at a few diversities exhibited in DTNs in this section. However, a detailed study of extensive forms of heterogeneity in, and their impact upon, DTNs would be left for the latter sections of this chapter.

8.1.1 Overview of Heterogeneity

The issue of heterogeneity in the context of ubiquitous and pervasive computing has been widely addressed in the literature. da Rocha and Endler [293] discussed context management in heterogeneous mobile computing environments. The authors observed that heterogeneity arises in pervasive systems due to three aspects.

- **Hardware heterogeneity**: This arises due to incompatibility among different hardware, for example, laptops, smartphones, and PDAs. Diversity in networking technologies integrating these devices also give rise to such heterogeneity.

- **Software heterogeneity**: When devices have different software or applications, software heterogeneity arises. Thus, even if homogeneity in hardware is assumed, that still may not guarantee software heterogeneity. For example, laptops with similar specifications may have different operating systems (OSs) so that a video game that executes in one OS may not execute in the other.
- **Architecture heterogeneity**: Such heterogeneity arises when the devices do not follow any common network architecture for interconnection among themselves.

Schmohl and Baumgarten [294] discussed about these heterogeneity in detail. Hardware heterogeneity can arise due to diversity in various hardware that are equipped to the devices. For example, these could be diverse networking interfaces—a device with only Bluetooth cannot communicate with another device having only Wi-Fi. User interfaces and capacity of the devices in terms of power, computing, and storage also give rise to such heterogeneity. Software heterogeneity, on the other hand, are manifested due to difference in OS, applications, application domains (broad category of the application), and middleware[1] APIs. Finally, aspects, such as network topology and available (and utilized) services, give rise to architectural heterogeneity.

Unless such heterogeneity are handled appropriately, interoperability among different devices is challenged. In other words, the goal of pervasive and ubiquitous computing remains largely unfulfilled. Schmohl and Baumgarten [294] noted that existing research has often adopted middleware to respond to such heterogeneity. In such scenarios, the middleware acts as an abstraction layer residing in between the applications layer and the heterogeneous layer of hardware, software, OS, and network. The authors noted the following modes of communication in the context of interoperability.

- *Direct* communication among the mobile devices. For interoperability, such devices should follow common communication paradigm.
- Communication *brokered* among the devices by one or more centralized system. Such a system should be able to interact with heterogeneous devices and thus, enable interoperability among them.
- A user can activate a *unidirectional workflow* from his/her device to the server.
- An extension of the above is *service provisioning*, where, on triggering a workflow to the server, an appropriate response is obtained by the device.

Schmohl and Baumgarten also discussed different ways for representing heterogeneity (such as device capability databases and ontologies) and broad techniques for handling such heterogeneity. One common approach is to consider *bridges* that can interconnect two or more heterogeneous entities. Bridges can be hardware or software component. Subsequently, the authors presented a general architecture for implementing a middleware to overcome the heterogeneity.

---

[1]The online Oxford dictionary defines "middleware" as "Software that acts as a bridge between an operating system or database and applications, especially on a network." (Source: https://www.oxforddictionaries.com/definition/english/middleware).

Fig. 8.1 Neighbor database and priority tables maintained by the devices using virtual interface [295]

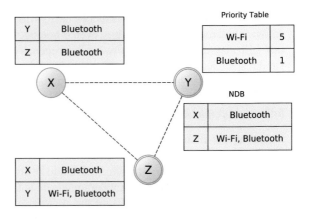

8.1.2 Heterogeneity at Link Layer

Stuedi and Alonso [295] noted that the assumption of devices in a MANET or WSN having a common communication interface has its limitations. In practice, such an assumption is not often true. For example, devices of the students in a university campus may be equipped Bluetooth or Wi-Fi (IEEE 802.11) interfaces, or may be both. On the other hand, a smart home may have altogether different and heterogeneous devices from different manufacturers. Accordingly, the authors explored the integration of heterogeneous MAC protocols in MANETs, with specific focus on 802.11 and Bluetooth.[2] The authors envisioned that such an integrated environment would allow transparent end-to-end communications, independent addressing, easy configuration and support mobility of the nodes. To this end, the authors proposed the use of software-based virtual interface to integrate the devices with different MAC layers inspired by the Linux Ethernet bridge.

Every node using the virtual interface mechanism maintains a neighborhood database (NDB). In NDB, a node stores a list of (its own) network interface(s) through which it can communicate with its neighbors. For example, let us consider that device X has only a Bluetooth interface, while its neighbors, devices Y and Z, have both Bluetooth and Wi-Fi interfaces. Figure 8.1 shows the NDB for each node. The entries X, Y, and Z in NDB are actually the MAC addresses of the respective devices, whereas "Bluetooth" and "Wi-Fi" in the table indicate the identifiers of the physical interfaces.

---

[2]A Bluetooth network consists of piconets and scatternets. A piconet is composed of up to 8 nodes, where one of the nodes acts as *master*, while the remaining act as *slaves*. A master node centrally coordinates data transmission to other nodes in the piconet. A scatternet is formed when one or more nodes (master, slave, or both) become members of other piconet(s). In such a scenario, two new roles of the nodes are introduced—master-slave and slave-slave. A master-slave node acts as master in one piconet, while behaves as slave in another piconet. On the other hand, a slave-slave node acts as slave in both the piconets.

Additionally, each node also maintains a priority table. When a node has multiple communication interfaces available, it assigns a priority value to each such interface. The interface with the least priority is subsequently used for communication. For example, the use of Bluetooth—having lower power consumption—can be prioritized for communication. Figure 8.1 also shows the priority table for the node Y.

Since the interface with the least priority value is always set to be used, Stuedi and Alonso introduced a threshold time, *maxdiff*, that helps in retaining or updating the priorities. The authors developed a command line interface to configure the virtual interfaces in Linux. The proposed system was evaluated using a MANET routing protocol. Experimental results indicated that the hand over times and throughput obtained using the virtual interfaces were quite close to those obtained using the actual physical interfaces.

8.1.3 Heterogeneity at Network Layer

Sadok et al. [289] presented the heterogeneous technology routing (HTR) framework for MANETs. Similar to other protocols, the link layer of HTR allows a device to scan for nearby compatible network nodes (also running HTR) and join such networks. HTR supports three modes of connection setup—a device can join as a *client* to an existing network, it can create its own network as a *server*, or can act in both the modes. The routing module used by HTR is based on OLSR. It uses two control messages—a HELLO message as heartbeat and a topology control (TC) message. HTR innovates by piggybacking service discovery messages into these control messages and thus, reduces the control overhead.

To ensure longer operational periods of the MANET, HTR introduces a metric called HTR score. In general, a heterogeneous MANET can have nodes with different link layer technologies, diverse energy levels, different number of network interfaces, and so on. The HTR score is aimed to calculate the difference in cost between a given node's network interface and that of its neighbor. In other words, the purpose of the metric is to help in deciding an optimum route for a packet. In particular, HTR adapts Dijkstra's shortest path algorithm with HTR score and link stability indicators to determine the shortest paths toward other nodes in the network. Sadok et al. [289] discussed the results of real-life experiments using HTR involving a heterogeneous network. It was observed that the presence of Bluetooth and Wi-Fi network interfaces in a network often creates bottleneck and thus, affects the network throughput. However, HTR was found to fare well as compared to a homogeneous network. The authors also discussed about the multipath routing extension for HTR in order to provide better performance guarantees.

Scott et al. [89] noted that emerging mobile networking solutions, such as PSNs, have different data sharing use cases, where the assumption of fixed source and destination nodes does not hold good. Moreover, efficient resource usage is a key aspect in mobile computing environments, unlike the traditional Internet. The authors, therefore, observed that the IP-based (or rather, infrastructure-based) architecture of the

Internet does not necessarily become suitable for PSNs. Interestingly, data transmission in the Internet (or even MANETs) is synchronous involving end-to-end communication paths. To this end, the authors discussed a set of guiding patterns for efficient communication in PSNs or similar environments.

Deviating from the traditional consideration of layer-based protocol stack, Scott et al. [89] proposed the *unlayered* architecture of Haggle for PSNs. Haggle is located in between the applications running in a device and its networking interfaces. Haggle is composed of four interrelated modules

- **User data**: Haggle proposes forwarding data as Application Data Units (ADUs), which consists of multiple key-value attributes. The "key" is a string; the "value" can be a string or binary stream. For example, metadata in the form of the following key-value pair stores the name of a file: {'file_name': 'song.mp3'}. The contents of the file can also be passed in a similar way: {'file_contents': 0xA0B1C2...}.
- **Delivery**: To deliver (or forward) a message, neighbors of a node must be known. When Haggle discovers nodes in the neighborhood, it labels the corresponding addresses as "nearby", while the other addresses remain unlabeled. Moreover, the "benefit" of forwarding a message is also estimated, based on which the forwarding decision is taken.
- **Protocols**: Addressing is required to transfer a message from one point to another. Haggle considers any identifier as an address if the identifier has an associated protocol to transfer the data. For example, in the infrastructure mode, an email address can be used to send a message using SMTP. On the other hand, in the infrastructure-less mode, Bluetooth and/or Wi-Fi MAC addresses are useful. Mappings of these addresses are contained in the ADUs.
- **Resource management**: This module is central to the operations of Haggle. Unlike traditional scenarios, where a user decides how much storage and CPU should be allocated for a task, such decisions are taken by this module in Haggle. Moreover, the cost-benefit analysis of message forwarding is also affected by the contemporary state of resources of a device.

Thus, Haggle enables communication in different modes (infrastructure-based or otherwise) with a complete abstraction to the underlying network. Moreover, prioritized tasks and optimized resource usage makes Haggle suitable for use in resource constrained mobile computing environments.

In a sense, one can regard DTNs as an extreme form of MANETs—they not only lack in infrastructure, but also end-to-end communication paths, in general. However, as we have discussed until now in this book, the characteristics of DTNs are widely different from the ones usually considered in the context of MANETs. It may be recalled that DTNs were also aimed at interconnecting heterogeneous networks. For example, in reality there could be many isolated MANETs (*regions*), which can communicate with one another using DTN protocols, but otherwise lack in inter-MANET communications. In such scenarios, a primary requirement would be seamless switching between the MANET and DTN protocol stack without any perceivable application-level glitches. Motivated by this, Petz et al. [296] presented MaDMAN, a middleware for dynamic switching between MANET and DTN protocols. In their proposed architecture, the network stack consisted of a collection of different possible transport, network, and link layer protocols. The network stack can be switched with another even while the application is running, which enables communications with asymmetric protocols.

MaDMAN consists of several modules, such as the context aggregator, session manager, and application interfaces. The context aggregator collects information about the network environment, for example, mobility, connectivity, and network statistics. Supplied with such information, the session manager decides upon the relevant communication mechanism. The application interfaces, on the other hand, enable different forms of inter-process communications. They also optionally share limited information that can be used to optimize the use of network resources. Results of experiments using MaDMAN indicated that as compared to a TCP-only protocol stack, an adaptive protocol stack performed well in scenarios involving long communication delays.

Le et al. [297] considered heterogeneous networks with both MANET- and DTN-based routing. In the model considered, mobile nodes on the surface are located in different, isolated islands of connected networks. While the nodes in such an island can communicate with one another using MANET routing protocols, communication between any two island is infeasible. To overcome such challenges, Le et al. leveraged Unmanned Aerial Vehicles (UAVs) to form a DTN, which can enable communication between two nodes on any isolated surface network via the UAVs. The authors noted that the main purpose of having UAVs here is not to form the above mentioned DTN. Rather, those UAVs already exist for some reason, and their mobility is utilized to improve communication.

8.1.4   Heterogeneous Contact Patterns

The reader might recall from Sect. 1.3.2, Chap. 1, that the empirical distribution of the aggregated ICT in DTNs exhibit power law characteristics. Additionally, multiple works, such as [298–300], have also studied the diversity in the pairwise contact patterns among the mobile nodes in DTNs. Conan et al. [298] observed that different real-life connection traces exhibit diversity in the number of contacts per node. Moreover, the mean ICT, as well as the mean contact time, varied widely across different traces. Further, in the three data sets that were considered, it was observed that mean contact duration was much less than the mean ICT. Thus, ICTs play an important role in message routing in DTNs. Conan et al. however, reported an interesting finding. The log-normal distribution was found to better fit the pairwise ICT distribution than

Fig. 8.2 Illustration of aggregated and pairwise ICTs

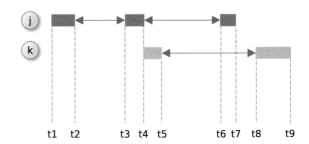

the Pareto and exponential distributions. Moreover, unlike power law curves (with power less than 2), log-normal distribution results in finite mean and variance. This led the authors to suggest the *power law paradox* [298]—while the aggregate ICT distribution was found to resemble power law distribution [11], the pairwise ICT distribution did not exhibit similar power law characteristics.

> Pareto distribution is an example of power law distribution. The probability density function (PDF) of a random variable X following Pareto distribution is given by $f_X(x) = a x_m^a / x^{a+1}$, where $x \geq x_m$. Here, x_m is the scale parameter and a is the shape parameter of the distribution. The cumulative density function (CDF) of the distribution is given by $F_X(x) = 1 - (x_m/x)^a$, where $x \geq x_m$.
>
> The PDF of an exponentially distributed random variable X is given by $f_X(x) = \lambda e^{-\lambda x}$, where $\lambda > 0, x \geq 0$. Its CDF is given by $F_X(x) = 1 - e^{-\lambda x}$.

Example 8.1 Figure 8.2 shows the contact events for any node say, i, with two other nodes, j and k. The contacts are shown in colored boxes. For example, a contact between nodes i and j begins at the time instant t_1, which continues up to t_2. Subsequently, at t_3, the two nodes again come in contact and stay until t_4. The final contacts between these two nodes occur at the time instant t_6. On the other hand, nodes i and k have two contacts beginning at time instants t_4 and t_8, and ending at t_5 and t_9, respectively. Note that the inter-contact times (shown with arrows in the figure) are usually larger than the contact durations.

Based on the figure, the aggregated ICTs of node i are $\{t_3 - t_2, t_6 - t_4, t_8 - t_5\}$. However, the pairwise ICTs between i and j are $\{t_3 - t_2, t_6 - t_4\}$; between i and k it is $\{t_8 - t_5\}$. $\qquad\qquad\square$

Tian and Li [299] considered three real-life connection traces and sorted the pairs of nodes in ascending order according to their respective mean ICT values. Subsequently, the contact pairs were distributed into ten percentile groups based on the means ICTs. Heterogeneity in the percentage of contacts among the 10 different groups were observed across all the data sets. Moreover, the authors further observed that the empirical distribution of ICTs of the pairs of nodes in lower percentile groups is more like power law, whereas the same for the higher percentile groups tend to be exponential. Further statistical analysis revealed that the distribution of pairwise ICTs can be divided into three groups—(1) those that follow exponential distribution, (2) those that follow Pareto distribution, and (3) those that follow none of the previous two distributions. Tian and Li noted that since exponential distribution is memoryless, such ICTs are derived from independent motion of the nodes. Thus, the underlying contacts were with "familiar strangers". On the other hand, Pareto-like ICTs can be derived from human priority, interest, or urgency. Therefore, the corresponding contacts in this case were with "friends".

Lee and Eun [300] analyzed a similar problem in further depth by considering the effects of heterogeneous contact process on the performance of DTNs. The authors assumed that the pairwise ICT distribution of the nodes is exponential, but different pairs having different parameters (mean ICT). However, it may be noted that, apart from the diversities in the mobility patterns and contact dynamics, there are several other factors that lead to a heterogeneous network. Besides, while works in [299, 300] focus on the reduction of communication opportunities in the network, heterogeneity of certain aspects (for example, incompatible network devices—in absence of any bridging [295], and routing protocols) turn available communication opportunities useless. In the remainder of this chapter, we would look at such heterogeneous aspects of DTNs. Moreover, we would also look at some quantitative figures to assess how adverse the effects of heterogeneity could be.

8.2 Aspects of Heterogeneity in OMNs

Until now in this chapter, we have looked at different aspects of heterogeneity in different kinds of networks. One may recall from the earlier chapters that the fundamental characteristics of DTNs/OMNs are widely different from that of the traditional networks. As such, when heterogeneity is imposed upon such networks, the performance degrades by a large extent.

In this section, we look at the various aspects that contribute to heterogeneity in an OMN.[3] These include the heterogeneity in the connection establishment events, and diversities in both hardware and software of the participating devices in the network. Some of these aspects might seem to bear similarity with other works described in the previous sections, but their consideration in challenged networks like OMNs is important.

---

[3]Sections 8.2, 8.3.3, 8.4, 8.5, and 8.6 have been reproduced by permission of the Institution of Engineering and Technology.

B.K. Saha, S. Misra, "Effects of heterogeneity on the performance of pocket switched networks", *IET Networks*, 2013, vol. 3, no. 2, pp. 110–118.

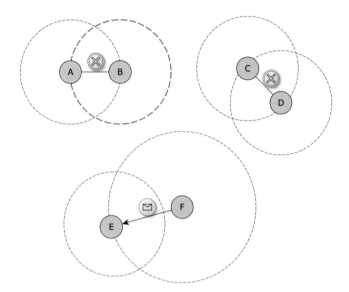

Fig. 8.3 Communication impairments in OMNs due to heterogeneous network interfaces (*A*–*B*), incompatible routing protocols (*C*–*D*), and diverse transmission ranges (*E*–*F*)

Figure 8.3 presents an overview of the effects of heterogeneous node capacities in OMNs. Although nodes *A* and *B* are within the physical transmission ranges of one another, they still cannot communicate due to their heterogeneous network interfaces. On the other hand, the nodes *C* and *D* fail to communicate due to incompatibilities arising due to their routing protocols. Finally, node *F*, which has a larger transmission range, can send messages to node *E*, but the reverse is not possible.

8.2.1   Heterogeneity in Connection Dynamics

Asymmetry in connections in an OMN (or any other wireless network, in general) need not be limited to the data rates, but manifested in the entire connection establishment process. Heterogeneity in the connection dynamics of the devices arises due to one or more of the following reasons:

- Asymmetric transmission ranges and/or speeds of the network interfaces,
- Diversity in the link-layer protocols of the devices,
- Asymmetric device scanning intervals.

Diverse transmission ranges can result in one-way connectivity between a pair of devices. Difference in data rates implies that, all other factors remaining the same, some nodes can transfer less data in a given time than the others. Further, each device scans for its neighbors periodically after a certain time interval. Devices with variable scan intervals would affect the frequency of neighbor discovery, and, hence, possibly decrease the number of connection establishment events. Such issues, however, can be induced by the underlying link-layer protocol of the devices, and their further consideration have been scoped out in this chapter.

8.2.2 Diverse Hardware of the Devices

The participating devices in OMNs can be of different models of smartphones and PDAs from different manufacturers, and, therefore, have different specifications. A particular aspect is buffer size of the devices, which determines how many messages could be stored by a device at any given time. Users' devices have a fixed buffer size, which determines how many messages can be stored by a device at any given time. In particular, the assumption of infinite buffer space is invalid in reality. The reader might argue here that in today's age, smartphones come equipped with several gigabytes of internal storage, or that users often attach external memory cards to their devices. This is true. However, at the same time, it should also be remembered that with increased storage capacity, our desire to store more and more data (such as images, music files, and videos) is also increasing. Therefore, an efficient application for OMNs should respect the storage constraints. The effects of buffer size on the performance of different routing protocols have been presented earlier in Sect. 2.3.2.

Another potential hardware related issue is the presence of devices with incompatible network interface. For example, let us consider two devices where the first has a Bluetooth interface, while the other has a Wi-Fi interface. Such devices, although may be within the transmission range of one another, cannot communicate due to their differences in the network interfaces. Thus, communication opportunities in the network are wasted. Since we discussed about similar aspects in Sect. 8.1.2, we do not further elaborate on it here.

Yet another relevant aspect is the energy capacity, which varies from device to device. Similar to the assumption of infinite buffer space, the assumption that devices have infinite energies is impractical. While a stand-alone smartphone can take a long time to discharge from its full battery level, in practice, one or more applications keep running in our smartphones, which accelerates the battery discharge rate. Such effects are more pronounced when communication interfaces (such as Bluetooth, Wi-Fi, and GPRS) are turned on. This might not be too much of a problem indoors—where one can keep the device plugged in to the power supply—but is of concern outdoors where one may not easily find a charging point. Therefore, it is desirable that application development for OMNs also take into consideration the available energy level of the devices.

In the following, we take a look at the impact of energy levels on the performance of OMNs. The typical energy consumption rates of the nodes were considered similar

Fig. 8.4 Delivery ratio versus delivery latencies of the delivered messages **a** without and **b** with energy constraints

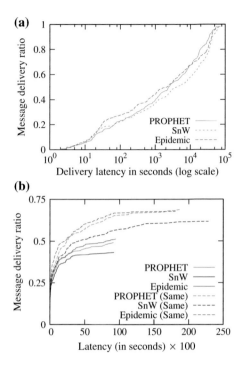

to that of Motorola Milestone.[4] In particular, the initial energy of each node was taken to be 1400 mAh, 3.5 V, and transmission and scanning energies as 0.7 and 2 J, respectively. The devices had 200 MB of buffer space, and all the nodes used either PRoPHET, SnW, or Epidemic routing protocol.

Figure 8.4 shows a semi-log plot of the delivery ratio of messages versus the delivery latencies. Figure 8.4a presents the base case when devices had *unlimited* energy, which although not practical, is useful for comparison. The Fig. 8.4 shows that, in case of the Epidemic routing protocol, about 30 % of the delivered messages had delivery latencies up to 1000 s. In other words, 60 % of the messages cannot be delivered within 1000 s time duration. Moreover, using the Epidemic routing protocol, a slightly higher percentage of the delivered messages had latencies in the range 25–2500 s.

Figure 8.4b illustrates the performance obtained when the devices had limited energy. In this figure, however, both the axes are linear in scale. It can be observed that limited energy budgets significantly degraded the performance in, otherwise, homogeneous OMNs. The plots corresponding to the performance obtained when the nodes had same initial energy are indicated by "Same" in the graph. The remaining

---

[4]http://www.gsmarena.com/motorola_milestone-3001.php, accessed 10 December 2013.

plots correspond to the cases when the nodes had varying levels of initial energy. In this case, it can be observed that using SnW resulted in a comparatively less delivery ratio, while the delivery latencies of the messages were quite high and variable. For example, beyond 100×100 s, the curves for Epidemic and PRopHET almost saturated. However, in the same range, a relatively steeper slope for SnW can be observed. This can be accounted to the reason that SnW is neither as aggressive as Epidemic, nor has any intelligence as PRoPHET.

Moreover, diversity in the initial energy levels of the devices further worsened the delivery ratio compared to the scenario when all the nodes had the same initial energies. In this case, we find that the delivery ratio of messages barely reached up to 50 %, and, once again, it was low for SnW as compared to the other routing protocols.

8.2.3 (In)Compatibility of Routing Protocols in OMNs

To discount the effects of intermittent connectivity, several routing schemes based on message replication have been proposed. In such schemes, multiple copies of a message are forwarded to different nodes with the hope that at least one copy would reaches the destination. Although we have studied these protocols in details in Chap. 2, in the following, we, once again, take a brief look at them to understand their semantics.

One of the simplest routing scheme is Epidemic [18], where every node replicates the messages they are carrying to the other nodes not having a copy of those messages. Thus, apart from summary vectors, which are likely used by other routing protocols, too, the Epidemic protocol does not involve any routing metadata exchange. Similarly, SnW [42] also follows a simple scheme. For each message generated, SnW assigns a count $L > 1$. Any node having a copy of the message forwards a copy to another node as long as $L > 1$. After forwarding, it reduces its own count of copies to $L/2$ or $L - 1$, depending on whether the protocol is run under binary mode or not. Accordingly, the receiving node would have $L/2$ or a single copy of the concerned message.

In PRoPHET [19], however, a node forwards a replica of a message to another node only if the other node has greater chances of encountering the destination of the message than itself. This, therefore, necessitates the exchange of delivery predictabilities when two nodes come in contact. In fact, there are many other routing protocols that assume the exchange of similar metadata for their successful operation. In general, although these routing protocols help in enhancing the message delivery ratio, often they are not *compatible* with one another, primarily due to two reasons:

- The protocol-specific headers added to the messages while they are created, and
- The operation modes of the protocols, for example, single- or multi-copy routing and message forwarding/replication criteria.

All messages exchanged in the network have headers containing general information such as, the source and destination of the message, and TTL. Apart from these, the headers often contain information specific to the underlying routing protocol. For example, the SnW messages have a header that specifies the number of copies of a given message is present with the current node. The header of a routing protocol can typically be interpreted by another node using the same protocol. This, therefore, limits the interoperability of the routing protocols in many scenarios.

Even if it is assumed that all the routing protocols use a standard header format, which can be interpreted by all, interoperability among them is further affected by the way each protocol works. To illustrate, consider the PRoPHET protocol. For instance, when two nodes running PRoPHET come in contact with each other, they exchange (and subsequently update) their respective delivery predictabilities of the other nodes. This sequence of interaction would fail if, for example, a PRoPHET router and a SnW router comes in contact. In fact, this is one of the adverse effects that we would attempt to mitigate later in this chapter.

8.2.4   Effects of Incompatibilities

So far, we have looked at different forms of heterogeneity and incompatibilities in OMNs and other networks. Let us now summarize the specific (adverse) effects that lack of interoperability among the devices results in OMNs:

- **Loss of communication opportunities**: As mentioned on multiple occasions, OMNs (or DTNs, in general) typically lack in end-to-end communication path. In OMN, a node often guesses[5] the next node to whom it should forward a message so that the message is likely to reach its corresponding destination. Such communication is possible only when a pair of devices are within the transmission range of one another. However, in the presence of incompatibilities, two devices cannot communicate even when they are near to each other, i.e., communication opportunities are lost.
- **Undelivered messages**: The reader might argue here that, since the nodes are mobile, contact durations are always finite and perhaps small. In other words, there are no infinite communication opportunities. So, there would be undelivered messages anyhow!

  The anomaly can be resolved as follows. If the messages have sufficiently large lifetime, if the traffic injection in the OMN stops at some point of time, and if enough time duration is considered, then the messages would *eventually* be delivered[6] to their respective destination nodes.

---

[5]If the movement of the nodes, however, is periodic, then a fixed time schedule can be utilized for routing. For example, if it is known that two nodes, A and B, come in contact daily at around 11 AM, message forwarding rules can be optimized using such information.

[6]Delivery latencies, however, would likely be high.

The scenario, however, becomes different in the presence of heterogeneity or incompatibilities in the OMN. Certain messages in the network possibly would always remain undelivered, no matter what resource and time are provided. For example, if S and D, respectively, are the source and destination of a message, and if the two nodes have incompatible network interfaces, S would not be able to deliver the message to D irrespective of how many times they come in contact or how long they stay in contact. The only way of delivery of the said message would be through a node that is compatible with both the network interface types.

- **Increased delivery latencies**: This follows from our above discussion. With the nodes in OMNs effectively having less communication opportunities, the time required to deliver messages, on an average, rises up. Even the "most" optimum routing protocol would likely suffer from degradation in a heterogeneous OMN.

Thus, it is desirable that such heterogeneity issues are addressed to achieve better performance in real-life OMNs. It may be noted that, in such heterogeneous OMNs, overhead ratio of message delivery would likely decrease. However, such achievement is of little help. This is because, such decrement comes at a heavy cost of reduced message delivery and increased delivery latency.

8.3 OMNs as Graphs

In this section, we would look at graph-based representation of DTNs, which holds for OMNs as well. Such representations are often useful for studying different aspects of network as well as the individual nodes. For example, in a graph-based representation of a social network, one may wish to measure the *centrality* of a node. An understanding of such aspects can be useful in many ways, for example, formulating a better routing protocol.

However, due to unique characteristics of DTNs, unlike other networks, we need to consider a bit different kinds of graphs for their representation. In the following, we discuss about different kinds of graphs that are used for representing dynamic systems. We look in detail at two particular types—temporal graphs and time-varying graphs. Subsequently, we discuss how an OMN with heterogeneous routing protocols can suitably be represented with such a graph.

8.3.1 Temporal Graphs

Traditionally, a network is often abstracted as a graph. The nodes/devices in a network are represented as the vertices of a graph, whereas the connections/links among the nodes are represented as edges of the concerned graph. With such abstraction, the problem of routing reduces to determining the shortest path (or, the most fault tolerant path, or any other path optimal with respect to a given metric) from a source node to a destination node. For example, one can use Dijkstra's algorithm to find the single source-all destination shortest paths, or Floyd-Warshall's algorithm to find out all pair shortest paths in a graph (network).

Life, however, is not easy. The above presented abstraction has some implicit assumptions—either all links in the network have same (edge) weight, or all the weights are known. It is hard to satisfy such assumptions in practice. In large networks where dynamic routing is used, a node often knows the weights of links connecting its neighbors, but not of nodes beyond the 1-hop neighbors. In some cases, the weights of the links farther in the network are typically learned over time. In other cases, a node leaves it up to its neighbor to determine the remainder of the best path to the destination. Such dynamic routing schemes work both in wired networks and MANETs.

The situation, however, is different in OMNs. Not only the nodes in an OMN have intermittent connectivity, but they often lack in end-to-end communication paths. For example, there might be communication link between two nodes, i and j, at time instant t, but the link may disappear at $t + \Delta t$, where Δt is small. This is not just for a single link, but, in general, holds true most of the time. As a consequence, a graph-based representation of an OMN is largely different from the related models described above. For example, let us look at an OMN at time instant t from a "global" viewpoint. In other words, we are able to see all the nodes and all the existing communication links. Now, an attempt to represent the OMN as a graph based on our observation (see Fig. 8.5) would reveal two general characteristics:

- **Sparse connectivity**: Such a graph would be sparsely connected. A large fraction of the nodes would not be connected with the other nodes. For example, in Fig. 8.5, we can find only three pairs of connected nodes. Moreover, finding a continuous chain of connected nodes, with chain length greater than 1, if often a rarity. This is especially true is the transmission range of the nodes is low.
- **Frequently changing**: Now, let us assume that we look at the same network once again at time instant t', where $t' > t$. What would we expect to see? First of all, due to mobility of the nodes, positions of many nodes would change. Moreover, due to change in position of the nodes, some of the previously existing links would break, some new connections would be established, and perhaps a fraction of the previously existing links would continue to exist. From the point of view of a graph, the nodes would remain the same, but the edges would vary likely by a large extent. In other words, the shape of the corresponding graph would continuously keep changing with respect to time.

Therefore, it is difficult to apply traditional graph-based approaches for OMNs. To address such issues, the concept of temporal graphs has been proposed. Simply stated, temporal graphs take into account different snapshots of the network taken at different instants of time, such as $t_1 < t_2 < \cdots < t_n$.

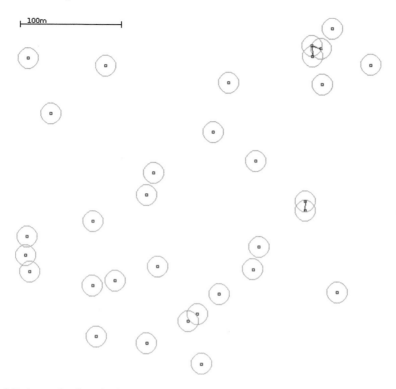

Fig. 8.5 A snapshot from the ONE simulator showing a partial view of 50 nodes moving according to TLW. The transmission range (shown with *green circles*) of the nodes was taken to be 10 m

Table 8.1 Contact events between pairs of nodes in an OMN

Node	Node	Contact time
A	B	1
A	D	3
C	E	4
B	E	9

To illustrate, let us consider the set of contact events in an OMN, as shown in Table 8.1.

Using a traditional approach, the above set of relationships can be represented as a static graph, as shown in Fig. 8.6. However, with such a representation, we loss all the *temporal information*—information with time relevance. In other words, we cannot state when those particular events took place by looking at the figure. Temporal graphs, in contrast, are able to capture all such underlying temporal information.

Fig. 8.6 A static graph
representation of contact
events from Table 8.1

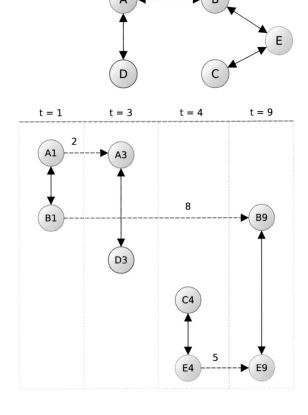

Fig. 8.7 A temporal graph
representation of contact
events from Table 8.1

    Kostakos [301] suggested a three step-approach to construct a temporal graph from
the given data set. The first step is to create a node at every time instants when it is
active. The node labels are appended with the time instant of the corresponding event.
Next, each such node is joined with one another in sequence with directed edges. The
difference between two successive time instants gives the weight of the corresponding
edge. Finally, directed, but unweighted, edges are connected between pair of nodes
to indicate a relationship (event) from the data set. Figure 8.7 shows the temporal
graph representation of the set of events presented earlier in Table 8.1. In Figs. 8.6 and
8.7, solid edges with bidirectional arrows indicate that the communication between
each pair of nodes has been bidirectional. In other words, the transmission ranges
of all the concerned nodes were equal. In case it was not, bidirectional arrowheads
can be replaced with unidirectional ones wherever appropriate. The dashed edges in
the latter figure, however, are unidirectional since they indicate the flow of time. It
can be observed that all relevant information, including temporal, from Table 8.1 are
retained in Fig. 8.7.

8.3.2 Time-Varying Graphs

Prior to Kostakos' [301] work on temporal graph, several other related models have been put forward to capture the dynamic nature of graphs. For example, in 1959, Erdős and Rényi [302, 303] proposed the concept of random graphs. A random graph $\Gamma(n, N)$ consists of n nodes and N edges that are chosen uniformly at random from $\binom{n}{2}$ edges. In other words, if all possible $\binom{\binom{n}{2}}{N}$ graphs are generated, and one such graph is chosen uniformly at random, we would get $\Gamma(n, N)$. Gilbert [304] presented an alternative version of random graphs. In Gilbert's version, in a random graph $G(n, p)$ with n nodes, an edge is inserted between any pair of nodes with a probability p. Since there can be $\binom{n}{2}$ distinct edges each with probability p, the expected number of edges in such a random graph becomes $\binom{n}{2} p$.

Example 8.2 Let us consider a random graph $G(n, p)$, where $n = 50$ and $p = 0.4$. Then, the number of possible edges in G is given by $\binom{n}{2} = \binom{50}{2} = 1225$. Therefore, the expected number of edges in G becomes $1225 \times 0.4 = 490$. □

Ferreira [305] discussed about *evolving graphs*, a combinatorial model to capture dynamic characteristics of graphs. Given an ordered sequence of time, $S_T = t_1, t_2, \ldots, c_k$, and a sequence of subgraphs of a graph G taken at those time instants, $S_G = G_1, G_2, \ldots, G_k$, which were viewed as an ordered sequence of subgraphs of a graph G at increasing time instants, the tuple (G, S_T, S_G) is called an evolving graph, where $\cup G_j = G$. In other words, a sequence of static graphs gives rise to a dynamic graph.

Casteigts et al. [306] presented the concept of time-varying graphs (TVGs) by integrating different related models and formalisms existing in the literature. Let V be a set of nodes and E be a set of edges (relations) connecting the nodes. Moreover, let L be a set of label (alphabet). Then, $E \subseteq V \times V \times L$. Thus, a given pair of nodes can be connected with multiple edges each having a different label. Casteigts et al. noted that the label or alphabet so described may consists of multiple attributes, for example, bandwidth of a link, as well as its loss factor. Further, let $\tau \subseteq T$ be the lifetime of the system; T can be a set of integers or positive real numbers. Then, a TVG is represented with a quintuple, $\mathscr{G} = (V, E, \tau, \rho, \zeta)$. Here, ρ is called the *presence* function, indicating the existence of a particular edge at a given time, and is represented as $\rho : E \times \tau \rightarrow \{0, 1\}$. The function $\zeta : E \times \tau \rightarrow T$ represents the (possibly time-varying) latency involved to travel an edge from one end point to the other. The static graph $G = (V, E)$ is called the *underlying* graph (or footprint) of $\mathscr{G}$.

Casteigts et al. noted that it is possible to look at the evolution of a TVG from the point of view of an edge or node. For example, one can consider the union of all time intervals when a particular edge was active (i.e., $\rho(e, t) = 1$). The consecutive sequence of appearance and disappearance of an edge e is called the *characteristic date* of e. Based on this, one can define the *characteristic date* of $\mathscr{G}$, which indicates the time of appearance and disappearance of edges in the TVG. Such a representation allows one to view a TVG as a sequence of static snapshots of $\mathscr{G}$ taken at successive instants of time. Accordingly, various concepts, such as temporal distance, topological distance, temporal eccentricity, and so on, can be defined.

Example 8.3 Let us consider that the edges in a TVG are denoted by e_{ij}, where i and j are the two endpoints of the edge. Let the following set indicate the activeness of the edges over different instants of time: $\{(e_{12}, 1), (e_{12}, 2), (e_{45}, 4), (e_{71}, 9)\}$. Then, based on this information, we have $\rho(e_{12}, 1) = 1$. Also, $\rho(e_{45}, 4) = 1$. However, $\rho(e_{72}, 5) = 0$. □

Commonly, we talk about *spatial* distance measured by Euclidean geometry. TVGs and similar other dynamic graphs, however, introduce the notion of *temporal* distance [306]. Essentially, it is the time taken to travel from one node to another. A related terminology is *topological* distance, which is the number of hops required to travel to reach from one node to another. Note that both temporal and topological distances are functions of time.

8.3.3 Representation of Heterogeneous OMNs

The TVG model proposed by Casteigts et al. is useful for formal representation and analysis of OMNs and other similar dynamic systems. TVGs can also be used to represent a heterogeneous OMN. In particular, the presence function, ρ, accounts for multiple scenarios of heterogeneity, for example, devices with incompatible network interfaces and asymmetric connection events. However, since ρ indicates the temporal presence of the links, it cannot capture the scenarios when a link exists, but no communication is possible, such as when diverse routing protocols are used. It may be noted that the label set, L, is useful in this case to some extent. For example, the label of an edge can indicate that communication via that link is not possible. However, incompatible communication links arise due to heterogeneous properties of the nodes. Therefore, in the following we describe a representation to capture the presence of incompatible communication links arising due to heterogeneous routing protocols.

Let us assign a property, R_i, to each node $i \in V$ in the TVG-based representation of an OMN. Further, let $\equiv$ be the equivalent operator so that $R_i \equiv R_j$ if both the nodes $i, j \in V$ have compatible routing protocols. In general, two distinct routing protocols are usually incompatible.

Let E_τ be the set of the edges that exists over the entire network lifetime, i.e., $E_\tau = \{e \mid e \in E, \rho(e, t) = 1, t \in \tau\}$. In other words, theoretically, there can be a lot of edges among the nodes in a graph-based representation of an OMN. However, in reality, not all such edges may be present. For example, consider two devices in close proximity, but their Bluetooth radios are off. In such a scenario, there would be no communication link between the two devices.

Based on this, let us define a function to represent the compatibility of communication links (edges) in an OMN. This is required, since, as discussed earlier, the mere presence of a communication link does not indicate that communication is possible.

Definition 8.1 Homogeneous edge: The homogeneous edge function, $\phi : E_\tau \rightarrow \{0, 1\}$, for any edge $e_{ij} \in E_\tau$ is 1 if the nodes i and j at the two end points of the edge have compatible routing protocol (i.e., $R_i \equiv R_j$), and 0, otherwise (i.e., $R_i \not\equiv R_j$); $i \neq j$.

Therefore, $E_C = \cup_{e \in E_\tau} \phi(e) \subseteq E_\tau$ gives the set of potentially *communicable* edges in the OMN, i.e., the edges through which messages can be exchanged. Moreover, we can also measure the extent of homogeneity (and therefore, heterogeneity) in such an OMN resulting due to the diverse routing protocols using the *communication degree*, as defined below.

Definition 8.2 Communication degree: The communication degree of an OMN is defined as $\alpha = \frac{|E_C|}{|E_\tau|}$.

> The communication degree depends on the number of nodes in the OMN, routing protocols used by them, as well as their mobility patterns.

A logical question that arises here is—what role, if any, do the nodes play in a heterogeneous OMN when there is *apparently* a possibility of communication? By *apparently*, we mean that the link layer of a device indicates that it can communicate with another device. Even if such a link layer connectivity exists, several factors, for example, energy levels and routing protocols, can prevent the actual communication. Let us consider the scenario when the messages sent by one node cannot be interpreted by the other due to the difference in their protocols. In this case, however, both the nodes consume energy during transmission/reception of the messages. Such an issue can be circumvented if the link layers of the devices advertise the routing protocols used by them, and, therefore, do not engage in further communication if the other device is not found to use a compatible protocol.

For example, let us consider that the nodes maintain a protocol incompatibility database (PID)—a list of other nodes' addresses using any incompatible routing protocol. Whether another node is compatible, or not, can be learned after having multiple interactions with the latter. To illustrate, let us consider the node j comes in contact with the node i, $\forall i, j \in V$. Node i observes that in $\theta > 0$ successive contacts, it has either received no messages from j, or some transmissions in the form of a stream of bits that it (node i) cannot interpret as per its routing protocol. Under such a scenario, i assumes that j has an incompatible routing protocol, and stores j's address in its PID denoted by $\pi_i = \{(j, \delta_j) \mid j \in V, (i, j) \in E_\tau\}$. Here, δ_j represents the time duration for which a node's address is stored in the database. This is helpful when corruption in the message(s) received occur due to interference or channel error. In such cases, a node may remove an entry (j, δ_j) from its PID, if $\delta_j > \delta$, where $\delta > 0$ is a time threshold.

8.4   Overcoming the Adverse Effects of Heterogeneity

This section explores how the adverse affects of heterogeneity—both in terms of hardware and software—can be mitigated. The approach presented here derives from the general concept of *bridges* discussed in [294].

8.4.1   Hardware Incompatibility

The capacity of existing OMNs with different network interfaces can be easily augmented in the presence of devices that are accompanied with multiple types of network interfaces. For example, a group of devices having only Bluetooth adapters can be bridged to a group of devices having only Wi-Fi capabilities if they come in contact with one or more devices having both types of network interfaces.

Let us consider that devices are equipped with two types of network interfaces—*if1* and *if2*—that are assumed to be incompatible with one another. Moreover, let us consider that certain devices in the OMN have only either *if1* or *if2*, and the remaining have both. Any device having *if1* (*if2*) can communicate with other devices having *if1* (*if2*) or both. Communication is not possible otherwise. To understand how this helps, consider a device X equipped with both *if1* and *if2*, which comes in proximity of another device Y. If X discovers that Y has *if1*, it initiates communication. However, if Y does not have *if1*, X checks whether Y has *if2*. Assuming Y to have the latter, X and Y begins their communication. Any other device Z, with *if1*, would not be able to directly communicate with Y even if they are located close to each other.

8.4.2   Protocol Translation Units

Let us now look at how incompatibility arising due to heterogeneous routing protocols can be mitigated. We address the incompatibility issues between two specific routing protocols—PRoPHET and SnW. They are representative of two different categories of routing protocols used with OMNs/DTNs—routing with (a) Fixed number of and (b) Unlimited copies of the messages. Moreover, while SnW maintains the state of a message (L), PRoPHET considers the state of connectivity among the nodes ($P_{(a,b)}$). Thus, PRoPHET uses some "intelligence" in decision making. Although variations of these protocols have been proposed, the principles described here holds good for them as well.

To overcome the communication impairments caused due to heterogeneous routing protocols, we consider the use of Protocol Translation Units (PTUs). PTUs are "special devices" that can interact with two or more routing protocols both in

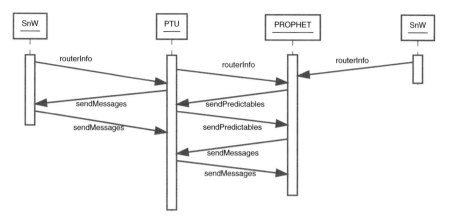

Fig. 8.8 Interactions among different types of routing protocols and PTUs

terms of interpretation of protocol-specific headers and sequence of interactions. The PTUs run a *hybrid* routing protocol encapsulating the syntax and semantics of both PRoPHET and SnW protocols. This enables a PTU to communicate with both types of routers. This can be further extended to encapsulate the logic of multiple other protocols.

8.4.2.1   The Use of PTUs

PTUs enable message exchanges among the heterogeneous devices that would have been infeasible, otherwise. To understand how the PTUs handle the dynamic scenarios arising in the OMNs, let us consider two devices X and Y using the routing protocols SnW and PRoPHET, respectively. Although X cannot successfully send a message to Y, it can do so to a PTU device. The latter, in turn, helps in forwarding the message to Y directly or through other intermediate nodes using the PRoPHET routing protocol.

It is considered that the devices periodically emit beacon signals, which also provides information[7] about the routing protocol used by the respective devices. The PTUs are assumed to advertise both the routing protocols in their beacon messages. Thus, any device that is running PRoPHET (SnW) initiates communication with the other devices if the received beacons advertise the use of PRoPHET (SnW). Figure 8.8 shows the interaction among the different routers and the PTUs. In the figure, the PTU identifies the routing protocol of the other device, and behaves accordingly. The figure also shows the failed interaction between a SnW and a PRoPHET router.

---

[7]For example, a specific bit pattern in the beacon header can indicate a particular routing protocol.

Algorithm 8.1: Interaction of the PTUs with PRoPHET routers

Input:

- All messages carried by the device

Output:

- Exchange new messages with the other device

1 **for** *msg in directly deliverable message* **do**
2   **if** *msg has SnW or PTU header* **then**
3   Remove the header
4   Add PRoPHET header
5   Forward the message

6 **for** *msg in remaining messages* **do**
7   `// PRoPHET messages`
8   **if** *msg does not have SnW header* **then**
9   Replicate and send according to (2.2)–(2.4)

10 **for** *msg in messages* **do**
11   `// SnW messages`
12   **if** *msg has SnW header with* $L > 1$ **then**
13   Update header with $L = L/2$
14   Replicate, remove header and send

Algorithm 8.1 presents the interaction logic between a PTU and any other device using the PRoPHET routing protocol. At the beginning, all the deliverable messages (i.e., the messages destined for the other device) are transferred. In case any such message was received from a SnW router, the corresponding SnW headers are removed, and the PRoPHET headers are added before forwarding. Replication of the remaining messages takes place in the following two phases:

1. In the first phase, all the messages received from the other PRoPHET routers are replicated depending on the delivery predictabilities, as shown in (2.2)–(2.4).
2. Subsequently, any message received from the SnW routers are replicated, provided $L > 1$. This ensures that the last copy of the message is directly delivered to the destination node.

Algorithm 8.2 presents a similar logic of interaction between the PTUs and the SnW routers. After the directly deliverable messages, if any, are sent, messages originating from other SnW routers are replicated depending on L. Finally, messages received from the other PRoPHET routers are replicated after heading a new header indicating L.

8.4.2.2 Complexity Analysis

Let us assume that n messages are generated in the concerned OMN. Thus, a PTU can have at most n messages in its buffer. So, the space complexity of the PTU turns out to be $O(n)$.

It may be noted that in Algorithm 8.1, a PTU can identify the directly deliverable messages in $O(n)$ time. Moreover, the actions such as removing/updating message header and replicating/forwarding a message can be performed in constant time. Therefore, the time complexity of the proposed algorithm becomes $O(n)$, which is true for the Algorithm 8.2 as well.

Algorithm 8.2: Interaction of the PTUs with SnW routers

Input:

- All messages carried by the device

Output:

- Exchange new messages with the other device

1 **for** *msg in directly deliverable message* **do**
2 **if** *msg has PRoPHET or PTU header* **then**
3   Remove the header
4   Add SnW header with $L = 1$
5 Forward the message
6 **for** *msg in remaining messages* **do**
7 `// SnW messages`
8 **if** *msg has SnW header with $L > 1$* **then**
9   Set $L = L/2$, replicate and send
10 **for** *msg in messages* **do**
11 `// PRoPHET messages`
12 **if** *msg does not have SnW header* **then**
13   Replicate *msg*, add SnW header with L, and send

8.5 Key Insights

Until now, in this chapter, we have explored several forms of heterogeneity. We have also looked at some mechanisms to alleviate the threat of incompatibility in OMNs. In this section, we look at some quantitative results that characterize the performance of OMNs in such scenarios. The effects of heterogeneity in OMNs were evaluated using the ONE simulator [51]. Real-life connection traces of 78 nodes from the Infocom'06 data set [80] were used for a duration of 12, 18, and 24 h starting from the first connection setup event. At first, we discuss in details the different scenarios that were considered for performance evaluation. Subsequently, we would look at some related comparative statistics.

The first scenario explored the possible impacts of asymmetry in the connection dynamics of the devices. The connection "Up" events, indicated by the device discovery times and contact durations in [80], were considered to be *unidirectional*—a connection was created only from the first node of the contact pair to the second node, but the reverse was not considered. This scenario was contrasted with the case when such events were considered to be bi-directional.[8]

We investigated the effects of incompatible network interface of the devices. A fraction of nodes with two network interfaces, *if1* and *if2*, were considered. Half of the other nodes used *if1*, while the remaining nodes had *if2*. These two types of network interfaces were considered to be incompatible with one another. The transmission speeds of both the interfaces were set to 2 Mbps. The effects of the presence of nodes with multiple network interfaces was then studied. The SnW routing protocol was used for this scenario.

Next, we explored the interactions of two different routing protocols—PRoPHET and SnW. We divided the nodes in the OMN into two groups. The ones in the first group used the SnW routing protocol. All the nodes in the other groups used either (1) SnW, (2) PRoPHET, or (3) were PTUs. Similarly, in the next scenario, we also considered all the nodes in the first group to use PRoPHET, while the composition of the second group varied as earlier. All the nodes were considered to have infinite buffer sizes in order to eliminate the effects of the buffer size, if any, on the evaluation metrics.

In the final scenario, the variation in communication degree was explored. In the first case, we divided the 78 nodes from the Infocom'06 trace into two groups. The first group contained the 10–50 % of the nodes incremented in steps of 10 %; the other group contained the remaining percentage of nodes. In the two other cases, we considered 5 and 10 nodes, respectively, to be the PTUs. The remaining 68 nodes were divided into two groups in a similar way.

We considered uniform message creation events within the first 5 h of the simulation duration. A total of 400 messages were created, with their sizes uniformly distributed between 50 KB to 1 MB. The TTL for the messages were set larger than the simulation durations. In all the scenarios, SnW was used in the binary mode with $L = 16$, unless otherwise specified. For the PRoPHET router, the settings mentioned in [19] were taken. Finally, except for the first scenario described before, all other connection events were considered to be symmetric (bi-directional).

In the remainder of this section, we look at the results of the simulations, and present related analysis.

---

[8]This is the only place in this book where we consider the connections to be unidirectional. Unless otherwise stated, all connection events considered elsewhere in the book are assumed to be bi-directional.

Fig. 8.9 Effects of (a)symmetric connection events on the delivery ratio of the messages

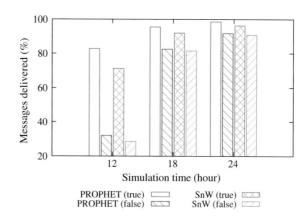

8.5.1 Heterogeneous Connection Events

Figure 8.9 shows the impact on message delivery ratio when (all) the nodes used either the SnW or PRoPHET routing protocols. The "true" and "false" cases, respectively, indicate the scenarios whether the connection events were considered to be symmetric or not. It can be observed that, for the lesser durations of simulation (or low message density per unit time), the asymmetry in connection among the devices reduces the delivery ratio of the messages by 30–40 %. When sufficient time is given (the 24-h case), the ratio improves significantly. This underscores the fact that, given enough time, the messages in the network are *eventually* delivered.

8.5.2 Incompatible Networking Devices

Figure 8.10a shows how the delivery ratio of messages in the OMN is affected in the presence of incompatible network interfaces. In particular, a fraction of the devices in the OMN—shown along the x-axis—had network interface *if2*, while the others used *if1*, which is incompatible with *if1*.

The 0 % case along the x-axis in Fig. 8.10a represents the scenario when all the nodes had *if1*, i.e., the network was homogeneous. Thus, the maximum performance in terms of message delivery ratio was obtained in this case. It can be observed that the delivery ratio steadily decreased as long as 20 % of the devices had incompatible network interfaces (*if2*). This can be explained by considering the fact that all the nodes were partitioned into two mutually exclusive groups based on their network interfaces. As the size of each such group increased, more number of nodes failed to exchange the messages among themselves, which reduced the delivery ratio.

Fig. 8.10 Effect of different networking interfaces when: **a** Different percentage of the nodes had incompatible network interface *if2*, and **b** The nodes had dual network interfaces

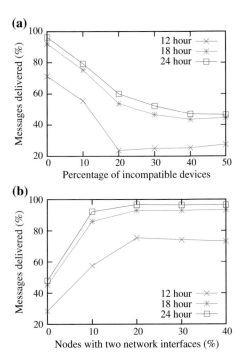

Moreover, the decreasing trend continued up to the 50 % case, when all the nodes in the OMN were equally partitioned into two groups of equal sizes. Any increase beyond the 50 % case would result in a reversed scenario, where the message delivery ratio would show an increasing trend. Thus, the performance curve is almost[9] symmetrical around the 50 % case.

The impact on the message delivery ratio in the presence of the nodes with dual network interfaces is shown in Fig. 8.10b. In this scenario, the percentage of nodes shown along the x-axis had dual network interfaces—both *if1* and *if2*. 50 % of the remaining nodes had only *if1*, while the other nodes were equipped with *if2*.

The figure suggests that the increasing participation of nodes with dual network interfaces in the network highly enhanced the delivery ratio of messages. A steady increase can be observed till 20 % presence of such nodes. This is due to the reason that the remaining nodes with either *if1* or *if2* got opportunities to transfer their messages to each other through the nodes with dual network interfaces. Beyond the 20 % limit, the delivery ratio stagnates and reaches the maximum level as determined by the routing protocol.

---

[9]The curve may not be *exactly* symmetrical, since the contact process in the OMN is not uniform. Some nodes come in contact more frequently than others. Therefore, depending on which node belongs to which group, there would be slight differences in the resulting curves.

Fig. 8.11 Percentage of messages delivered in presence of different types of routing protocols together with **a** SnW and **b** PRoPHET

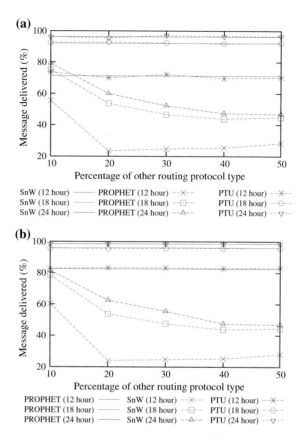

8.5.3   Heterogeneous Routing Protocols

Figure 8.11a, b present the delivery ratio of the messages when different routing protocols were considered. Figure 8.11a shows the case when certain fraction of the nodes—shown along the x-axis—used SnW, while the remaining nodes used some other routing protocol. In Fig. 8.11a, the plots labeled with "SnW (k hour)" denote the base case performances when all the nodes used the SnW routing protocol and the simulation duration was k hour.

The plots with labels "PRoPHET (k hour)" represent the scenarios when different fraction of the population (shown along the x-axis) used the PRoPHET protocol. It can be observed that, in comparison to the base cases, when the fraction of the nodes using PRoPHET protocol increased, the delivery ratio drastically decreased. Finally, the delivery ratio obtained with equivalent fraction of PTU nodes are shown. It can be observed that, while varying the fraction of PTUs from 10 to 50 % in the network, the delivery ratio obtained is almost the same as the best cases considered.

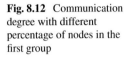

Fig. 8.12 Communication degree with different percentage of nodes in the first group

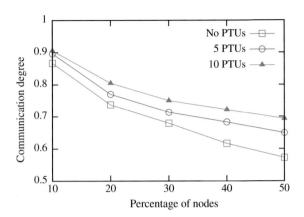

Figure 8.11b plots similar performance measurements when some of the nodes used the PRoPHET routing protocol, while the other nodes used SNW or was a PTU.

Finally, the variation in the communication degree (α) of the OMN is shown in Fig. 8.12. A certain fraction of the nodes (shown along the x-axis) used the PRoPHET routing protocol, while the remaining used SnW. It can be observed that with the increasing group sizes, α sharply decreased. This is due to the reason that with increasing group sizes, inter-group routing compatibility kept on decreasing. However, in the worst case, when both the groups had equal number of nodes, the presence of 10 PTUs enhanced α by 12 %. And as discussed earlier, any increase in communication degree potentially improves the chances of message delivery in the OMN.

8.6   Observations

By now, we have addressed the issues related to network heterogeneity both qualitatively and quantitatively. The observations from this chapter are summarized in the following points:

- Heterogeneous connection dynamics (the simplest case due to different transmission ranges) substantially reduces the delivery ratio of the messages.
- Hardware incompatibility arising due to incompatible network interfaces is hard to address particularly because, one may opt for software upgrade, but not for purchasing a new phone. Therefore, any contact opportunity with devices with multiple interfaces should be used to the best. This may require the routing protocols to use information available from the link-layer of the devices.
- For approaching reality, any new protocol or mechanism proposed should take energy consumption of the nodes into consideration.

- The performance degradation due to software-based incompatibility among the routing protocols is severe, but could be prevented. This does not require all the users to update their software. Rather, the presence of few "special" devices (for example, devices with middlewares or the PTUs as proposed here) could boost the performance.

Therefore, while targeting application and protocol development for real-life networks, it would be helpful to keep the different aspects of heterogeneity in mind.

8.7 Summary

In this chapter, we took a close look at different aspects of heterogeneity that arises in a real-life network. We looked at the root causes of such heterogeneity, and characterized them. While no network at a large scale, such as MANETs and WSNs, is immune to heterogeneity, the performance OMNs heavily degrades due to its unique characteristics. In particular, we looked at how various diversities manifested by the hardware and software of the devices lead to heterogeneity at different layers of the protocol stack. Moreover, even the contact process among the devices, too, is heterogeneous. For example, the distribution of movement to meet with "friends" differs from that with "strangers".

One of the important aspects covered in this chapter was the consideration of graph-based representation of DTNs. Various models, such as temporal graphs and TVGs, were discussed. As an application, we represented a heterogeneous OMN using TVG. Apart from heterogeneity in hardware, we also discussed why different routing protocols become incompatible with one another. To overcome such challenges, we considered the use of PTUs. Each PTU runs a hybrid routing protocol, which encapsulates the functionality of two or more routing protocols. This enables a PTU to communicate with a node using a routing protocol that the node understands. To address the question about the availability of the PTUs in real-life OMNs, we consider that a certain fraction of the users already posses such devices. This is possible either when the users purchase such devices, or are promoted by some person/organization.

Of course, it would be interesting to consider other forms of diversities, including the interactions among the diverse PTUs. Moreover, a comparison between the middleware-based and hybrid routing protocols-based approaches with respect to the performance overheads of the devices would be an interesting study. It may be noted that, in general, when we talk about heterogeneity, we only consider diversities related to the devices in a network. Since in OMNs—or rather MOONs—participation of human beings in the network operations play a key role, the prevalent situation is a bit different. Therefore, whether or not diversity in human actions and decision making affects the performance of a MOON remains a open question.

Before closing this chapter, we would like to observe that heterogeneity in a network is not necessarily always *evil*. In some cases, heterogeneity can be helpful.

As a simple example, consider that all the nodes in an OMN have uniform and very high ICT values. In such a scenarios, all the messages, on an average, would experience higher delivery latency. However, if the ICT distribution is nonuniform, then at least some of the messages can be delivered with lower latency. Of course, detailed evaluation is required to determine the extent of such gain, if any.

8.8   Review Terms

- Hardware heterogeneity
- Architecture heterogeneity
- Virtual interface
- Aggregated ICT graphs
- Temporal graphs
- Random graphs
- Presence function
- Communication degree
- Software heterogeneity
- Middleware
- Power law paradox
- Pairwise ICT
- Time-varying graphs
- Evolving graphs
- Homogeneous edge function
- Protocol translation unit

8.9   Exercises

8.1 Let ω and ω', respectively, be the message overhead ratio obtained using a specific routing protocol in a homogeneous and heterogeneous OMN. Then, $\omega' \leq \omega$. Explain why. Assume that all messages have infinite TTL and were generated within a certain time period, and the buffers of the nodes have unlimited storage capacity.

8.2 Apart from the different types of heterogeneity discussed in this chapter, what other aspects of heterogeneity can arise in a network? Which of them would affect the network performance? Which of them would not?

8.3 Inter-contact times in DTNs are sometimes found to be exponentially distributed, and sometimes follow Pareto distribution. Characterize the underlying human contact process that leads to such distributions.

8.4 How does the semi-log plot of the complementary CDF (CCDF) of an exponential distribution look like? How about the log-log plot of CCDF of a power law distribution? Explain using the expressions for their respective CCDFs.

8.5 Given a set of observations from a Pareto distribution, estimate its shape parameter, α, using the maximum likelihood method.

8.10 Programming Exercises

8.6 Simulate an OMN consisting of 100 nodes using the ONE simulator. In the first scenario, assume that all the nodes have a network interface of type 1, whose transmission range is 10 m and transmission speed is 1 Mbps. In the second scenario, assume that 50 nodes have an interface of type 1, whereas the remaining nodes have a network interface of type 2. The transmission range of the latter interface is 250 m and speed 10 Mbps. How does the performance of the two scenarios vary?

8.7 Write a report in the ONE simulator to measure the aggregate ICTs of an OMN. The output of the report should print the empirical CCDF of the measured values.

8.8 Use the above written report to get the CCDF of aggregate ICT for two scenarios—(1) using the Infocom'06 trace, and (2) using a random waypoint (RWP) mobility model. Plot the resulting CCDFs. What difference do you observe and why?

Chapter 9
Opportunistic Mobile Networks: Toward Reality

Having explored different dimensions of OMNs, in this chapter, we pause for a moment, and look at the reality. Where do DTNs/OMNs stand today? How well it has been adopted? What are its promises?

We begin by taking a look at some comprehensive statistics on research efforts in this domain. Subsequently, we discuss about Request for Comments (RFCs) and look at some of the RFCs published for DTNs. This is followed by a quick tour of a sample of current inventions based on DTNs/OMNs. In the latter part of this chapter, we discuss some of the recent trends in OMNs and different avenues along which it can evolve. Finally, this chapter ends with a few prospective project topics that can be undertaken by the reader.

9.1 Comprehensive Statistics

In this section, we present some high-level aggregate statistics on the research trends in DTNs in the recent past. In particular, we considered the time period between 2010 and 2014, both inclusive. We used the online search facility made available by Scopus[1] to gather the following statistics.[2] We listed the scholarly articles that contained the keywords "delay-tolerant networks" or "opportunistic networks" in their title, abstract, or keywords. Subsequently, we divided our data into two groups—conference and journal publications. Based on these data, we plotted the number of publications per year, as shown in Fig. 9.1. In general, a steady upward trend of publications count can be observed in the figure. This is a clear indicator of rising popularity of OMNs among the contemporary researchers.

Collectively, research publications in this domain were found to originate from as high as 70 countries. Figure 9.2 shows the top six countries and their percentage

---

[1] http://www.scopus.com/.

[2] The query URL is http://goo.gl/Ux3OrK.

© Springer International Publishing Switzerland 2016
S. Misra et al., *Opportunistic Mobile Networks*,
Computer Communications and Networks, DOI 10.1007/978-3-319-29031-7_9

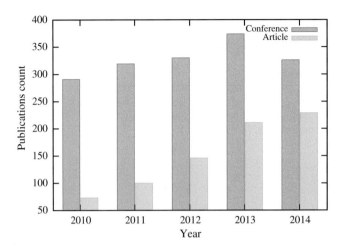

Fig. 9.1 Statistics on the number of conference and journal articles published in between 2010 and 2014

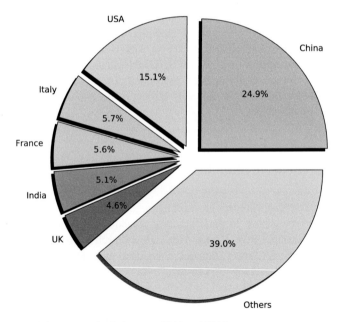

Fig. 9.2 Top publishing countries in between 2010 and 2014

contributions based on the number of scholarly articles published. China can be found leading the pack, with the United States of America at the second place. Contributions from Italy and France, which are at third and fourth positions, respectively, are almost similar. They are closely followed by India at the fifth position, and United Kingdom at the sixth place.

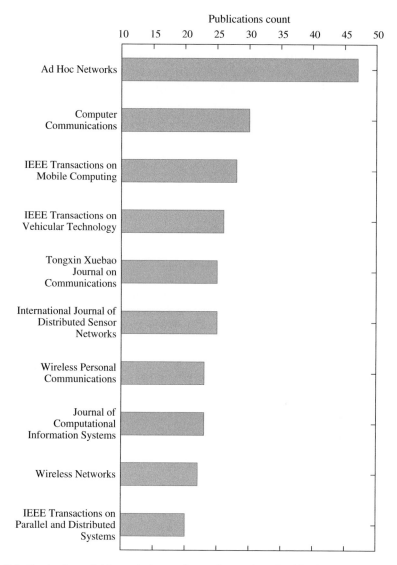

Fig. 9.3 Top ten journal titles ranked according to the number of publications in between 2010 and 2014

Finally, Fig. 9.3 shows the top ten journals where articles related to DTNs/OMNs have been published in the previously mentioned time period. It may be noted that the statistics presented herein are solely based on the publications indexed by Scopus. As such, the corresponding numbers are most likely to inflate in reality. Nevertheless, the information summarized in this section is helpful in obtaining a high-level overview of the contemporary state of research in DTNs/OMNs.

9.2 A Look at the Standards

In this section, we briefly talk about the Internet Engineering Task Force[3] (IETF), and more importantly, about RFCs. RFCs are peer-reviewed technical documents published by the IETF, which has often evolved into standards. Subsequently, we look at some of the RFCs that have been proposed in the context of DTNs. Finally, we take a look at a few patents that have been granted till date to inventions involving DTNs/OMNs.

9.2.1 Request for Comments

The IETF was formed in 1986 as a coordination platform for volunteers to develop and standardize protocols for the Internet. The mission of IETF is "to make the Internet work better" [307] with an emphasis on the engineering aspects. The IETF consists of several working groups, and decisions are made on a consensus basis. The outcomes of discussions in the working groups are published as Requests for Comments (RFCs), which document underlying concepts, protocols, and notes. Each RFC has a unique number. Moreover, once published, the contents of a RFC never change, although they can expire. Any changes or revisions must come as new RFC with newly assigned numbers. In such sense, a RFC is a static document.

However, all RFCs do not constitute as standards. The status of a RFC can be classified into one of the following categories:

1. **Informational**: RFCs under this category contain various types of information, such as tutorials, data, and proprietary protocols. Moreover, various RFCs under this category have been published, commonly known as *April Fools' Day RFCs* with publication date of 1st April, that contain humor. "A Standard for the Transmission of IP Datagrams on Avian Carriers" (RFC 1149), "The Address is the Message" (RFC 1776), "The Twelve Networking Truths" (RFC 1925), and "Electricity over IP" (RFC 3251) are few such examples.
2. **Historic**: RFCs, which have historical importance, but may not be of much importance currently, belong to this category.
3. **Experimental**: This category includes IETF documents and individual drafts submitted, which are likely in "work in progress" state. However, if such a proposal is widely adopted and found to work well, it (proposal) can evolve into a standard.
4. **Standards Track**: RFCs of this category are likely to become standards for the Internet.

While an algorithm gives an insight to a process, a protocol for practical use is often further more detailed and complex consisting of various parameters. Moreover, the behavior in the presence of several possible scenarios need to be clearly defined.

---

[3]https://www.ietf.org/.

RFCs serve the purpose of such detailed technical specifications. In the following, we take a look at some of the RFCs related to DTNs published till date. However, rather than providing a highly technical discussion, we summarize the objectives and essence of the corresponding RFCs. Readers interested in protocol development for real-life applications are suggested to go through the relevant RFCs in details.

RFC 4838: Delay-Tolerant Networking Architecture

RFC 4838 [14] is an informational RFC published in 2007. This RFC presents an overview of the architecture of DTN, which has evolved from the previously proposed Interplanetary Internet (IPN). The document describes in details about DTN endpoints and their identifiers, traffic priority classes, bundling and routing, congestion control, and security issues among others. The RFC talks about fragmenting DTN bundles in order to ensure higher utilization of the available contact durations. In particular, two types of fragmentation of bundles are proposed—proactive and reactive. In proactive fragmentation, a DTN node fragments the application data into smaller chunks, pack them in different bundles, and transmit them individually. The destination of the bundle is responsible to extracting each smaller chunks and reassembling them into the application data. On the other hand, when a DTN node detects that only partial transmission was successful, it can invoke reactive fragmentation to divide the larger data into smaller parts, and transmit the latter part in the next contact opportunity.

RFC 6256: Using Self-Delimiting Numeric Values in Protocols

RFC6256 [308], categorized as informational, describes the problems faced when fixed-width fields are used with protocols. For example, the receive window size in TCP is 16 bits wide, whereas IPv4 uses a 32-bit address. Although these values were sufficient when the protocols were originally designed, with advances in network speed and size, their respective value ranges need upgrade. Self-Delimiting Numeric Values (SDNVs) were proposed in earlier RFCs to allow sending just as much data as required while not using fixed-width fields at the same time. The RFC noted that since protocol handshaking and negotiation is ill suited for DTNs, the use of SDNVs is a potentially good choice. The SDNV-based encoding scheme works as follows. An integer of interest is represented with a bit string. If length of the bit string is not a multiple of 7, it is left padded with zeroes. The resulting 7-bit chunks are then transmitted. Thus, the need for fixed-length integer fields can be avoided.

RFC 6258: Delay-Tolerant Networking Metadata Extension Block

Published in 2011, RFC 6258 [309] belongs to the experimental category. This RFC proposes an extension block for DTN bundles to carry metadata about the payload, which can be used by to process the bundle at the DTN nodes. Such metadata is typically intended to be application-specific information. For example, keywords found in a HTML document, or album information of a song, and so on. A DTN node that generates a bundle can insert multiple metadata block(s) into it. An intermediate node that receives such a bundle can either insert new or delete existing metadata block(s) from the bundle. Finally, the recipient node of the bundle is expected to make use of the available metadata. SDNV allows optimal representation of nonnegative integer, which are commonly used in protocol headers, for example, TCP sequence number and message lengths.

RFC 6259: Delay-Tolerant Networking Previous Hop Insertion Block

Similar to RFC 6258, RFC 6259 [310] also defines an extension block for DTN bundles known as Previous Hop Insertion Block (PHIB). In essence, the RFC states that when a DTN endpoint forward a bundle to another endpoint, the identifier of the former is inserted into the bundle. Such information about the previous hop of a bundle can be useful in multiple ways. For example, in case of flooding-based routing protocols, retransmission of the bundle to the previous hop endpoint can be prevented. However, a PHIP is maintained only for a single hop. For example, if a DTN node A forwards a bundle to another node B, A inserts a PHIB with its identification in the bundle. However, when B transmits the bundle to another node, it removes the PHIB and inserts a new one into the bundle. If a bundle comes with a PHIB, the receiving node is made the previous hop information available for relevant decision making.

RFC 6693: Probabilistic Routing Protocol for Intermittently Connected Networks

The well-known PRoPHET [19, 20] routing algorithm has been extensively documented in the experimental RFC 6693 [311]. The RFC discusses PRoPHET's relation with DTN in terms of architecture, and describes the operation semantics and metrics in details. Additionally, it presents multiple optimization tips for practical purposes. For example, not exchanging predictability values that lower than a threshold. Moreover, the RFC suggests to maintain an exponentially weighted moving average of the predictability values for each known node. The RFC also addresses queuing policies, forwarding strategies, and security aspects of the protocol for use in the real life.

> A routing algorithm can be moderately simple. However, when translating it into a protocol for real life, several aspects need to be considered.

9.2.2   Patents

The potential of DTNs/OMNs can be easily apprehended if one looks at the rich collection of patents that have been applied by/granted to the inventors. A summary of a sample of such (granted) patents are listed[4] in Table 9.1. The diversity of these inventions range from deep scape communications to urban scenarios and health care. Apart from these, worldwide, there are several other patents that have been

---

[4]These information were collected using the Google Patents search interface.

Table 9.1 A summary of selected patents on DTNs and OMNs

Publication number	Title	Publication date	Original assignee	Overview
US7930379 B2 [312]	Interface for a delay-tolerant network	19 Apr 2011	Intel Corporation	The invention describes an interface based on file systems together with necessary locks as a mean to enable communication in DTNs. The file system has incoming and outgoing directories, among others, to manage such file transfers
US8032124 B2 [313]	Health-related opportunistic networking	4 Oct 2011	Microsoft Corporation	This patent presents a system for transmitting health-related (for example, basic diagnostic) data using OMNs. It also has a machine learning and reasoning-based component, which, based on the data type, can automatically select the destination and intermediate relay nodes
US8363565 B2 [314]	Routing method and node	29 Jan 2013	Nec Corporation	This invention addresses a problem in the PRoPHET routing protocol, wherein transitive predictability between two nodes, *A* and *C*, do not increase even if *A* and *B* are adjacent, and *B* and *C* are always adjacent
US8611213 B1 [315]	Transmitting delay-tolerant data with other network traffic	17 Dec 2013	AT&T Intellectual Property I, L.P.	The patent describes the methodology for staging delay-tolerant data (for example, emails, file backups, cloud off-loading, and data arising from participatory sensing), and transmitting them at a later time when the network exhibits low-activity rates

(continued)

Table 9.1   (continued)

Publication number	Title	Publication date	Original assignee	Overview
US8787157 B2 [316]	Multi-streaming multi-homing delay-tolerant network protocol	22 Jul 2014	Honeywell International Inc.	This invention presents a method for generating packets, assigning priority to them, and selecting one or more communication link(s) for their transmission in the context of deep space networking. Moreover, a method to decide whether or not discard packets in the absence of any link is discussed
US8842659 B2 [317]	Routing method intended for intermittently connected networks	23 Sep 2014	Thales	This invention describes an routing algorithm using a matrix of ICT values and hop limiting a message to delivery to k relays at most
US9015822 B2 [318]	Automatic invocation of DTN bundle protocol	21 Apr 2015	Raytheon Company	This patent describes the methodology of intercepting a packet and possibly enabling its transfer using DTN protocols (by encoding appropriately) to the legacy applications

applied for or, have already been granted. Interested readers are encouraged to go through them for further information.

The current and the previous sections of this chapter can be looked in conjunction. Since their inception, DTNs/OMNs have piqued intense research interest, which is evident from the sheer volume of published scholarly articles. However, this new paradigm of communication in the absence of end-to-end routes that DTN has unveiled is not merely of theoretical interest. Several RFCs have been published till date, which, although mostly belong to the experimental or informational categories, can still act as the cornerstones of future standards. In fact, if one looks at the length of RFC 6693 on PRoPHET, which runs more than a hundred pages, one would surely understand that translating a research concept into practically usable component requires a substantial effort. Concurrently, the growing space of inventions involving DTNs would perhaps be a further boost toward their realization in our daily life.

9.3 Promising Avenues

In this section, we discuss about some of the recent trends that are observed in the context of DTNs/OMNs. Moreover, we also look at different frontiers along which OMNs can evolve in the future. The topics mentioned here are indicative and by no means exhaustive. Several other innovations are expected to pour in as more time passes by.

9.3.1 Opportunistic Computing

Looking back at the past, we find that DTNs or OMNs were conceived to enable communication among the otherwise disconnected devices. However, in the recent years, handheld devices, such as smartphones, tablets, and PDAs, have witnessed giant technological advances. Such devices now are not only meant for communication, but come along with high computational power and resources (for example, high-resolution cameras, speakers, inbuilt sensors, microphone, communication interfaces, extensive internal and external memory, and various software). Motivated by this, Conti et al. [319–321] envisaged Opportunistic Computing (OC) as the next step of evolution of OMNs.

In OC, users can avail diverse and distributed resources—apart from those available with their own devices—through OMNs. The authors noted that in such a computational paradigm, users would be able to "compose the functionality of different resources" [320]. For example, a user can collect images from different users to stitch together into a video. However, the task of adding music to the video may be off-loaded to some other device(s) which is(are) equipped with such capabilities. In a sense, OC can be regarded as the melting pot of pervasive computing and OMNs. The heterogeneous composition of the network is, in fact, a boon here.

Conceptually, in OC, every resource of a device can be represented as a service offered by the device. Such services can be exposed to others via suitable application programming interfaces (APIs). Of course, it would be unwise to let everyone out in the world to access resources of a device. Therefore, the aspects of social relationships, human trust, and security play a central role in OC. Moreover, opportunistic task off-loading with service delay guarantees remain a research challenge. Nevertheless, the authors envisage that different areas, such as health care, transportation systems, and crisis management scenarios, would benefit from OC [320].

9.3.2 Remote Healthcare

Although the world has progressed a lot in the current century, universal and affordable health care still remains a challenging task. However, in the recent years, powered by the advances in sensor technologies and widespread use of WSNs, the healthcare sector has witnessed a boom. Today, a person can measure his/her own blood pressure sitting at his/her home without having to go visit a doctor. Telemedicine, which is now practiced to different extents around the world, would perhaps gain further momentum with such advances. In general, remote health care is no more a myth—related enabling technologies[5] are already in place.

As noted earlier in the context of OC, OMNs bear a promise of wide applicability in the remote health care sector. It is possible that OMNs can further extend the outreach to remote areas or difficult terrains where neither the Internet nor cellular networks exist. Moreover, in the immediate aftermath of a large-scale disaster, OMNs could be the only possible mean of communication, and, therefore, can be leveraged to provide medical care. The invention described in [313] worked toward similar directions. Unlike the prevalent telemedicine scenarios, OMNs do not require any investment in network infrastructure. Moreover, increasing number of people are owning at least a single smart device today. Considered together with daily human mobility, OMN-based remote health care is almost in a position to work. However, there remains a major challenge in addressing the end-user requirements. What kind of biophysical parameters should be monitored at the patient's side? Who should be the designated doctor? How frequently should doctors look at the acquired health statistics? Finally, what incentive does the doctors earn in this process? Moreover, OMNs, in general, are prone to high message delivery latencies. Thus, whether or not critical care can be delivered via OMNs remains to be explored.

9.3.3   5G and OMNs

With prospects of the fifth generation (5G) cellular technology looming large, OMNs can charter a new paradigm. In particular, the device-to-device (D2D) communication model in 5G promises a fertile field of application for OMNs. In 5G, four different communication dimensions [322] arise depending on whether a device is communicating with the base station or another device, and what kind of link establishment is in use. Security, interference management, efficient resource allocation, and appropriate pricing schemes are projected as major research challenges for this communication model. However, it has been noted that such D2D communication can lead to gains[6] in terms of operation capacity, data rate, and message delivery latency. The huge gap between the revenue of operators and devices that exist today can also be bridged [322]. D2D communication would also help in taking a further leap toward traffic off-loading technologies.

However, D2D itself is not entirely a new concept. MANETs and OMNs have operated for long have been based upon this, where two devices directly communicate in the absence of any network infrastructure. 5G, however, takes it a step forward

---

[5]http://www.nextservices.com/remote-healthcare-delivery-from-data-to-drones/.

[6]http://www.ericsson.com/research-blog/5g/device-device-communications/.

by considering two different types of link establishment in D2D communications. The integration of cellular network protocols together with OMNs/MANETs, nevertheless, would be interesting. Such a coexistence of 5G and OMNs would give rise to heterogeneous networks. However, in contrast to what we saw in Chap. 8, this kind of heterogeneity is expected to have positive influences upon the overall network performance.

9.3.4 Traffic Off-Loading

The surge in smartphone ownership and use of smartphone-based applications has resulted in the voluminous growth of traffic in cellular networks and infrastructure-based mobile networks. Since all these networks have limited capacity, such traffic explosion would lead toward congestion and suboptimal quality of service. To cope with such problems, traffic off-loading via DTNs/OMNs have been considered by the researchers [323–325]. The basic idea behind traffic off-loading is to let a portion of the data inflow pass through OMNs so that a chunk of bandwidth of the infrastructure-based network can be freed up. For example, in the traditional scenario, both an audio call and request for a video passes through the cellular network. However, with off-loading, the cellular network can be used for the audio call, whereas request for the video can be opportunistically searched in the neighborhood of the user. In this context, once again, several issues of heterogeneity arise, such as capacity of the devices, interests of the users, traffic type, and so on [323–324]. Moreover, as noted above, with the arrival of 5G, such traffic off-loading techniques would gain momentum.

9.3.5 OMNs and the Internet of Things

Internet of Things (IoT) is the current buzzword in the world of technology. Researchers envisage a smart world where every little device would be connected to the Internet. For example, a can of milk in the future can send a notification to its owner's smartphone that it is about to be empty. Similar to any other networking scenario where delay can be tolerated, OMNs can play a role in the context of IoT perhaps giving rise to an *opportunistic* IoT. As an example, if a person is hoping to place a quick order for a can of milk at the local grocery and planning to pick it up at a later point of time, such a preorder message can be opportunistically sent to the store via his/her neighbors. Taking it a further step forward, can OMNs be leveraged for *remote* payments similar to how near-field communication (NFC) has enabled mobile payments?

However, in the context of IoT, privacy concerns remain a big issue.[7] Such threats arise from two dimensions. First, the interconnected devices are themselves prone to be hacked. Second, a huge volume of owners' data would be stored in the cloud, which itself is prone to attack. Therefore, can OMNs help in plugging in the potential privacy leaks by reducing data storage that otherwise exist in the IoT ecology?

The above are merely a partial list of several future possibilities. Several other innovative directions, such as on-the-fly data centers [326] and Opportunistic Mobile Games [327], are also emerging based on the principles of OMNs. In the future, we hope to witness various other OMN-powered applications, as well as, further evolution of OMN itself.

9.4   Prospective Project Topics

We conclude this chapter with a few prospective project ideas. These not only include scope for contributing to the ONE simulator, but goes beyond to general contributions to the research community.[8] Readers with interest in large-scale software development might consider the following aspects:

Link layer implementation
The ONE simulator presently supports network layer and above. Thus, one can develop routing protocols and applications. However, exploiting the characteristics of the link layer and physical layer are not yet feasible. Therefore, one might consider adding one or more popular link layer protocols (for example, IEEE 802.11) to the ONE simulator.

Generalized framework for trace import
One of the greatest features of the ONE simulator is the integration of real-life traces (in form of externally loaded events) into simulations. A desirable feature would be the availability of a framework that provides a GUI-based approach for integration of any trace (for example, mobility and connection) file. This would include an intermediate step of preprocessing the actual trace files, wherever relevant, for using with the ONE simulator.

Parallel execution of simulations
Executing simulations often take long time depending upon the logic and operations that have been implemented. Moreover, one may wish to execute a large set of scenarios by running a simulation in batch mode. Possible scope for distributed and parallel execution of such process might be explored. To begin with, one may consider maintaining and version controlling the source code at a central repository, checking out the code at each (physical or virtual) machines, and executing them in parallel.

---

[7]http://www.politico.com/agenda/story/2015/06/internet-of-things-privacy-concerns-000107.

[8]A list of contributions by the community toward the ONE simulator can be found at the simulator's home page, http://www.netlab.tkk.fi/tutkimus/dtn/theone/.

Performance profile database[9]

Till date, several routing protocols and mobility models, among others, have been proposed for DTNs/OMNs. Moreover, implementations of a few protocols are already available in the public domain. However, due to limited space in a research article, it is often difficult to discuss in details about all the underlying parameters and effects of their variation. Therefore, apart from the scholarly articles, it would be nice to have a database that stores the performance of different protocols (say, in terms of message delivery ratio, delivery latency, and overhead ratio) under diverse scenarios. An input set can contain sufficient parameters to distinguish between two scenarios, for example, name of the routing protocol, number of nodes, simulation terrain size, mobility model, traffic intensity, and buffer size. Thus, when a new protocol is developed, one can simply pull data from this database, and compare their respective performances. In general, it is desirable that we move toward *reproducible research*[10] in the domain of OMNs.

9.5 Summary

In this chapter, we took a look at the OMNs from the perspective of current reality. We looked at some contemporary statistics that show a trend of increasing research interest in this domain. Moreover, effort has also been put forward to translate multiple protocols from the domain of research into technical specifications in the form of RFCs. This is complemented by several patents for inventions related to DTNs that have been filed (or granted) over the years. In general, when we talk about DTNs/OMNs, we not only speak about research based on purely theoretical or academic interest. Rather, the volume of RFCs and patents indicate the willingness of the community to push OMNs toward reality. We hope more and more applications would gradually emerge in the future.

In the latter portion of this chapter, we discussed in detail some of the emerging trends in OMNs such as, Opportunistic Computing (OC) and traffic off-loading. With the continued growth of resource rich smartphones, OC is expected to gain prominence in the coming days. On the other hand, opportunistic traffic off-loading is a promising technology to decongest existing network infrastructure, and is expected to achieve further momentum with the introduction of 5G. Apart from these, we also discussed about the possibilities of using OMNs for remote health care and together with the Internet of Things. We, finally, concluded this chapter with a set of sample project topics that would not only be an exercise for the reader, but also beneficial for the community as a whole.

---

[9]This idea is based on a query raised in the mailing list of the ONE simulator on April 01, 2015.
[10]http://reproducibleresearch.net/.

9.6   Review Terms

- Request for Comments
- Previous Hop Insertion Block
- 5G
- Traffic off-loading
- Self-delimiting Numerical Values
- Opportunistic Computing
- Device-to-device communication
- Internet of Things

9.7   Exercises

9.1 What is a RFC? What are its different categories?

9.2 What is traffic off-loading? Why is it increasingly becoming important in today's world?

9.3 Heterogeneity in a network is not always harmful. Discuss on this.

9.4 Describe the different communication paradigms in 5G involving D2D.

9.5 How do the privacy and security risks in IoT originate?

9.6 What is reproducible research?

Chapter 10
The Big Picture

This book presented a short tour of OMNs. It looked back at its origins in DTNs, its characteristics, advances made over the last two decades, and various applications. In this book, we looked at the aspects of routing in OMNs, issues of cooperation and heterogeneity, and a different dimension—aspects of human users in the network. In this final chapter of the book, we put our learning in perspective, and look at the big picture emerging out of all these discussions.

10.1 Challenges and Applications

Unlike traditional networks, end-to-end communication paths between a source and destination node are typically unavailable in OMNs. Thus, routing in such networks essentially reduces to the identification of a set of "best" nodes that can ensure movement of a message toward its destination node and subsequent delivery. In particular, such routing is replication-based—a transmitting node gives a copy of a message to the receiving node, and itself retains the original message. Such redundancy is helpful in improving chances of message delivery. Moreover, due to intermittent connectivity among the nodes, information about the global network topology is hardly available. Therefore, routing decisions are usually made on locally observable phenomena. Several signals, such as frequency and duration of encounters, chances of replication, and likelihood of meeting, are used to make intelligent decisions. Moreover, some routing algorithms also consider social network metrics, for example, centrality of the nodes, for this purpose. Decisions made in these ways are often found to result in better chances of message delivery. However, at the same time, we should also be careful that overhead involved in routing messages, on an average, remains low to the extent possible.

In general, we envisage that OMNs can be used in different scenarios—whether in the aftermath of a disaster that lacks of communication infrastructure, or in urban

© Springer International Publishing Switzerland 2016
S. Misra et al., *Opportunistic Mobile Networks*,
Computer Communications and Networks, DOI 10.1007/978-3-319-29031-7_10

settings where we do not wish to use available infrastructure.[1] In this book, we focused on scenarios where there is no centralized communication facility available. However, researchers have also considered scenarios with mutual co-existence of both technologies. In fact, if we look back at PSNs, by definition, they make use of global network connectivity, if available. An interesting example is that of (cellular or Wi-Fi) traffic offloading. In such scenarios, traffic from possibly congested infrastructure-based networks can be offloaded to OMNs for delivery. As we noted in the earlier chapter, with glimpses of 5G and D2D communication in the horizon, traffic offloading can prove as an useful application scenario. Additionally, other areas, too, for example, retail sector, can benefit from OMNs. A key fact to remember here is that data delivery in OMNs can involve long delays. Therefore, real-time or delay-sensitive applications may not gain much from OMNs. On the other hand, delay-tolerant applications—for example, ordering a can of milk from a nearby smart shop using Internet of Things—can perhaps benefit a lot by leveraging OMNs.

10.2   Human Aspects and Heterogeneity

The potential for real-life applications based on OMNs seem to be large. This is especially promoted by the fact that smartphones are becoming cheaper nowadays, and they come equipped with various network interfaces. Nevertheless, even if we go for real-life development, prior simulation-based studies can also be insightful. For example, the scalability of a solution can be easily verified—with certain assumptions and simplifications made—without having an actual deployment. Given the importance of such approaches, we studied key features of network simulation in details in this book.

In this context, a central aspect that comes into the picture is human mobility. Experiments have revealed that human mobility results in power law-distributed inter-contact time. Such observations are interesting. In reality, human beings do not move around randomly. Moreover, we meet with some people more often than others. The heavy tailed ICT distribution is able to explain such aspects. Various experimental traces pertaining to human mobility are available for public access. While going for simulations, one should make use of them, wherever possible, for realistic performance evaluation. On a related note, the truncated Levy walk mobility model was proposed to capture different characteristics typical to human mobility.

However, simplifications made by our models often do not hold in reality. A prominent example is that of heterogeneity. Since devices are manufactured by different vendors, they come with different capabilities and specifications. More often than not, one can expect diversities in communication interfaces, routing protocols, and other aspects related to devices' hardware and software. Therefore, it is essential that issues of heterogeneity are kept in mind while exploring the real world. However, heterogeneity is not necessarily evil, but can actually be useful in some scenarios.

---

[1]This could be due to various reasons, for example, congestion and privacy concerns.

For example, devices with longer transmission ranges and higher storage capacity may induce positive effects upon the network performance.

Talking about performance, we should also keep in consideration that, in most scenarios, the end users of all our applications developed are human beings. For example, when we talk about PSNs or MOONs, by definition, they involve human users. Perceptions of humans, however, differ from that of machines, which often can be approximated by rational behavior. In general, involvement of human beings with the network gives rise to a new dimensions as captured by MOONs. For instance, human beings have innate intelligence and preferences. In one of the chapters of this book, we observed that actions (movement to be specific) driven by such intuitive intelligence can, in general, help mission prospects by increasing the chances of communication.

On the other hand, emotions constitute a fundamental aspect of a human being's identity. Our actions are often motivated by contemporary emotions. Moreover, signals received from our environment also affect our emotions. Now, the aspects of emotions are not something that only psychologists should be concerned with. It would perhaps be helpful for application developers to keep certain considerations in their mind as well. For example, in a related chapter, we noted that the likelihood of visiting a website further depends on its first impression and that choice of a color palette can evoke different reactions. In the context of communications, we also noted that our emotions often dictate to which extent we would engage in communications with other. In particular, crisis scenarios, which reflect emotions, such as fear and anxiety, in high proportions, may lead to traffic explosion and congestion. Thus, as developers of OMN—or other infrastructureless network-based applications, some prior planning—in terms of storage, protocols, or otherwise—to mitigate such scenarios might prove helpful.

10.3 Issues of Cooperation

Undoubtedly, one of the key ingredients of the recipe for a fully functional network is inter-node cooperation. This is true for distributed networks, in general, and especially for DTNs/OMNs. Unlike the traditional Internet or MANET, OMNs lack in end-to-end communication paths between any pair of nodes. Therefore, while nodes in the Internet or MANET can possibly use an alternative path, such luxury is not available in OMNs. In OMNs, any node plays three distinct roles—(1) Source of a message, (2) Destination of a message, and (3) Intermediate relay node. It is the latter role that is critical for the health of an OMN. On one extreme, all intermediate relays (nodes who replicate messages of other nodes to other nodes) can be fully cooperative. While this is the ideal situation, in reality, we may not always reach there. On the other hand, some nodes can be malicious—they do not help others in the network, but rather try to cause harm. Somewhere in between are the selfish nodes, who do not intend to (sometimes or always) help others, but take help from others whenever available. Therefore, it is a fundamental necessity to prohibit selfish nodes—or better, to motivate them into cooperation. We addressed this issue to a great length in this book.

Several approaches toward enforcing cooperation in communication networks have been proposed in the past, for example, those based on incentives and game theory. However, one of the challenging issues is actually providing the incentive to a node due to lack of end-to-end paths. This is often alleviated by the assumption of centralized authorities—some well behaving nodes which take care of the incentives transaction. However, their presence in every scenario can perhaps not be guaranteed. Evolutionary theory seems to be an interesting prospect in this context. In general, when the nodes in OMN exhibit various message replication behaviors, the system can, as a whole, evolve and reach to an equilibrium. In other words, based on Darwin's theory, the "best" performing strategy would emerge out as the frontrunner, and other nodes would adapt the same behavior. In a previous chapter, we noted that when an OMN begins with equal proportions of different strategies, cooperative behavior triumphs eventually. In real-life, one, however, should be taking into consideration different other aspects that come into play in this context. Moreover, hybridization of incentive-, reputation-, and evolutionary theory-based approaches would be an interesting topic. Nevertheless, a few key points should be kept in our consideration while working toward a real-life solution. First, cooperation is necessary, but cannot be taken for granted. Second, the context of one's apparent non-cooperation should be evaluated. For example, with low battery power, it is impractical to expect someone to cooperate by replicating messages of other nodes. However, such behavior is clearly avoidable when ample resources are available. Finally, the overall goal should be that users get benefit from the OMN to the extent possible. In other words, rather than prohibiting selfish nodes, a desirable approach would be to motivate them into cooperation.

To summarize, in this book, we looked at recent advances in OMNs along several dimensions. We presented our discussion in the book in a way suitable for a diverse audience. To be honest, this is not the "ultimate" book on OMNs. While we discussed many topics in details, we also provided pointers to a lot many resources for further consultations. In such a sense, this is not the end of this book. This rather is the beginning of a new journey for the reader to undertake.

Author Biographies

Sudip Misra is an Associate Professor in the Department of Computer Science and Engineering at the Indian Institute of Technology Kharagpur. Prior to this he was associated with Cornell University (USA), Yale University (USA), Nortel Networks (Canada) and the Government of Ontario (Canada). He received his Ph.D. degree in Computer Science from Carleton University, in Ottawa, Canada, and the masters and bachelors degrees respectively from the University of New Brunswick, Fredericton, Canada, and the Indian Institute of Technology, Kharagpur, India.

Sudip has several years of experience working in the academia, government, and the private sectors in research, teaching, consulting, project management, architecture, software design and product engineering roles.

Sudip's current research interests include algorithm design for emerging communication networks. He is the author of over 260 scholarly research papers (including 160+ papers, of which 40+ papers are in ACM/IEEE Transactions, Journals, and Magazines). He has won nine research paper awards in different conferences. He was awarded the 3rd Prize in the Samsung Innovation Award (2014) at IIT Kharagpur, and also the IEEE ComSoc Asia Pacific Outstanding Young Researcher Award at IEEE GLOBECOM 2012, Anaheim, California, USA. He was also the recipient of several academic awards and fellowships such as the Young Scientist Award (National Academy of Sciences, India), Young Systems Scientist Award (Systems Society of India), Young Engineers Award (Institution of Engineers, India), (Canadian) Governor Generals Academic Gold Medal at Carleton University, the University Outstanding Graduate Student Award in the Doctoral level at Carleton University and the National Academy of Sciences, India—Swarna Jayanti Puraskar (Golden Jubilee Award). He was also awarded the Canadian Governments prestigious NSERC Post Doctoral Fellowship and the Humboldt Research Fellowship in Germany.

© Springer International Publishing Switzerland 2016
S. Misra et al., *Opportunistic Mobile Networks*,
Computer Communications and Networks, DOI 10.1007/978-3-319-29031-7

Sudip is the Editor-in-Chief of the International Journal of Communication Networks and Distributed Systems (IJCNDS), Inderscience, U.K.. He has also been serving as the Associate Editor of the Telecommunication Systems Journal (Springer), Security and Communication Networks Journal (Wiley), International Journal of Communication Systems (Wiley), and the EURASIP Journal of Wireless Communications and Networking. He is also an Editor/Editorial Board Member/Editorial Review Board Member of the IET Communications Journal, IET Wireless Sensor Systems, and Computers and Electrical Engineering Journal (Elsevier).

Sudip has published 8 books in the areas of wireless ad hoc networks, wireless sensor networks, wireless mesh networks, communication networks and distributed systems, network reliability and fault tolerance, and information and coding theory, published by reputed publishers such as Springer, Wiley, and World Scientific. He was invited to chair several international conference/workshop programs and sessions. He served in the program committees of several international conferences. Sudip was also invited to deliver keynote/invited lectures in over 30 international conferences in USA, Canada, Europe, Asia and Africa.

Barun Kumar Saha holds a B. Tech. degree (in Information Technology) from Haldia Institute of Technology, India, and an M.S. (by Research) degree from Indian Institute of Technology Kharagpur. Barun has worked with Wipro Technologies at different locations in India. In his role, he was responsible for supporting the critical daily operations of two global retail giants.

Barun's primary area of research is Opportunistic Mobile Networks. Additionally, he is also interested in mobile ad hoc networks and other general areas of wireless networks. Barun's research works have been published in several top notch journals and conference workshops including, but not limited to, IEEE Transactions on Computers, IEEE Wireless Communications Magazine, and ACM MobiCom workshop. He has also peer reviewed scholarly articles for many reputed journals and conferences, such as IEEE ANTS, Wireless Networks (Springer), International Journal of Communication Networks and Distributed Systems (Inderscience), EURASIP Journal on Wireless Communications and Networking (SpringerOpen), and Telecommunication Systems (Springer).

Barun is known in the research community for his widely acclaimed blog (http://delay-tolerant-networks.blogspot.com/) on DTNs and the ONE simulator. In his humble attempt of giving back to the community, he developed the ONE Knowledge Base (http://theonekb-barunsaha.rhcloud.com/) for benefit of the community members. Barun is also interested in the use of modern technology for enhancing the quality of education. Sponsored by the Ministry of Human Resource Development, India, Barun led the development of two virtual labs (http://vlssit.iitkgp.ernet.in/) on Software Engineering and Advanced Network Technologies, which are accessed by several people across the globe. His other open source contributions can be found in

GitHub. Apart from these, Barun has also penned a book on Bengali poems. Further details about Barun can be found in his personal website (http://barunsaha.me/).

Sujata Pal received her Ph.D. degree from Indian Institute of Technology Kharagpur, BE (in Computer Science) from North Orissa University, India, and MTech (in Multimedia and Software System) from West Bengal University of Technology, India. Sujata was awarded a Tata Consultancy Services (TCS) Research Scholarship for 4 years to enable her pursue the Ph.D. program. She was also the recipient of prestigious Schlumberger Faculty for the Future Fellowship for the year 2015. Currently, Sujata is a Post-Doctoral Fellow in Computer Engineering at the University of Waterloo, Canada. She has almost 6 years of teaching experience as an Assistant Professor in two different engineering colleges in their Information Technology and Computer Science departments. Sujata's publication record shows that she has well-rounded research experience comprising publications in high quality international journals, conferences, and a book chapter. Her current research interests include Delay Tolerant Networks, Mobile Ad hoc Networks, and Opportunistic Mobile Networks.

References

1. F. Warthman. (2003, March) Delay-tolerant networks (DTNs): A tutorial v1.1. Accessed: 08 Oct. 2012. [Online]. Available: http://www.dtnrg.org/docs/tutorials/warthman-1.1.pdf
2. B. Huffaker, M. Fomenkov, D. Plummer, D. Moore, and K. Claffy, "Distance metrics in the internet," in *IEEE International Telecommunications Symposium (ITS)*. Brazil: IEEE, Sep. 2002, pp. 200–202.
3. V. G. Cerf, "An Interplanetary Internet," *Space Operations Communicator*, vol. 5, no. 4, pp. 14–19, 2008.
4. Mars science laboratory: Data rates/returns. Accessed: 08 Oct. 2012. [Online]. Available: http://mars.jpl.nasa.gov/msl/mission/communicationwithearth/data/
5. Underwater acoustic modem models. Accessed: 08 Oct. 2012. [Online]. Available: http://www.link-quest.com/html/models1.htm
6. K. Fall, "A delay-tolerant network architecture for challenged internets," in *Proceedings of the 2003 Conference on Applications, Technologies, Architectures, and Protocols for Computer Communications (SIGCOMM '03)*. New York, NY, USA: ACM, 2003, pp. 27–34.
7. V. Cerf, S. Burleigh, A. Hooke, L. Torgerson, R. Durst, K. Scott, E. Travis, and H. Weiss. Interplanetary Internet (IPN): Architectural definition. Accessed: 31 Oct. 2012. [Online]. Available: https://tools.ietf.org/html/draft-irtf-ipnrg-arch-00
8. I. F. Akyildiz, D. Pompili, and T. Melodia, "Underwater acoustic sensor networks: research challenges," *Ad Hoc Networks*, vol. 3, no. 3, pp. 257–279, 2005.
9. M. Khabbaz, C. Assi, and W. Fawaz, "Disruption-tolerant networking: A comprehensive survey on recent developments and persisting challenges," *IEEE Communications Surveys Tutorials*, vol. 14, no. 2, pp. 607–640, 2012.
10. C.-M. Huang, K.-c. Lan, and C.-Z. Tsai, "A survey of opportunistic networks," in *Proceedings of the 22nd International Conference on Advanced Information Networking and Applications - Workshops (AINAW '08)*. Washington, DC, USA: IEEE Computer Society, 2008, pp. 1672–1677.
11. P. Hui, A. Chaintreau, J. Scott, R. Gass, J. Crowcroft, and C. Diot, "Pocket switched networks and human mobility in conference environments," in *Proceedings of the 2005 ACM SIG-COMM Workshop on Delay-tolerant networking (WDTN '05)*. New York, NY, USA: ACM, 2005, pp. 244–251.
12. D. B. Johnson and D. A. Maltz, "Dynamic Source Routing in Ad Hoc Wireless Networks," in *Mobile Computing*, Imielinski and Korth, Eds. Kluwer Academic Publishers, 1996, vol. 353. [Online]. Available: http://citeseer.ist.psu.edu/johnson96dynamic.html
13. C. T. Bhunia, S. Maity, S. Saha, S. Swanaz, and B. K. Saha, "Pre-emptive dynamic source routing: A repaired backup approach and stability based DSR with multiple routes," *CIT*, vol. 16, no. 2, pp. 91–99, 2008.

© Springer International Publishing Switzerland 2016
S. Misra et al., *Opportunistic Mobile Networks*,
Computer Communications and Networks, DOI 10.1007/978-3-319-29031-7

14. V. Cerf, S. Burleigh, A. Hooke, L. Torgerson, R. Durst, K. Scott, K. Fall, and H. Weiss, "Delay-tolerant networking architecture," Internet Requests for Comments, RFC Editor, RFC 4838, April 2007. [Online]. Available: https://www.rfc-editor.org/rfc/rfc4838.txt

15. A. Chaintreau, P. Hui, J. Crowcroft, C. Diot, R. Gass, and J. Scott, "Pocket switched networks: Real-world mobility and its consequences for opportunistic forwarding," University of Cambridge, Computer Laboratory, Technical Report UCAM-CL-TR-617, February 2005.

16. S. Phoha, "Guest editorial: Special section on mission-oriented sensor networks," *IEEE Transactions on Mobile Computing*, vol. 3, no. 3, p. 209, 2004.

17. B. K. Saha and S. Misra, "Could human intelligence enhance communication opportunities in mission-oriented opportunistic networks?" in *Proceedings of the 1st ACM MOBICOM Workshop on Mission-Oriented Wireless Sensor Networking (ACM MiSeNet '12)*. ACM, August 2012, pp. 15–20.

18. A. Vahdat and D. Becker, "Epidemic routing for partially-connected ad hoc networks," Duke University, Tech Report CS-2000-06, 2000. [Online]. Available: http://issg.cs.duke.edu/epidemic/epidemic.pdf

19. A. Lindgren, A. Doria, and O. Schelén, "Probabilistic routing in intermittently connected networks," in *Proceedings of the 1st International Workshop on Service Assurance with Partial and Intermittent Resources (SAPIR)*, ser. Lecture Notes in Computer Science, P. Dini, P. Lorenz, and J. Souza, Eds., vol. 3126. Berlin, Heidelberg: Springer Berlin / Heidelberg, Aug. 2004, pp. 239–254.

20. S. Grasic, E. Davies, A. Lindgren, and A. Doria, "The evolution of a DTN routing protocol - PRoPHETv2," in *Proceedings of the 6th ACM Workshop on Challenged Networks*, ser. CHANTS '11. New York, NY, USA: ACM, 2011, pp. 27–30.

21. P. Tournoux, J. Leguay, F. Benbadis, V. Conan, M. Dias de Amorim, and J. Whitbeck, "The accordion phenomenon: Analysis, characterization, and impact on dtn routing," in *IEEE INFOCOM 2009*, Brazil, April 2009, pp. 1116–1124.

22. I. Rhee, M. Shin, S. Hong, K. Lee, S. J. Kim, and S. Chong, "On the levy walk nature of human mobility: Do humans walk like monkeys?" *IEEE/ACM Transactions on Networking*, vol. 19, no. 3, pp. 630–643, Jun. 2011.

23. S. Trifunovic, F. Legendre, and C. Anastasiades, "Social trust in opportunistic networks," in *2010 INFOCOM IEEE Conference on Computer Communications Workshops*, San Diego, CA, March 2010, pp. 1–6.

24. I.-R. Chen, F. Bao, M. Chang, and J.-H. Cho, "Dynamic trust management for delay tolerant networks and its application to secure routing," *IEEE Transactions on Parallel and Distributed Systems*, vol. 25, no. 5, pp. 1200–1210, May 2014.

25. E. Ayday and F. Fekri, "An iterative algorithm for trust management and adversary detection for delay-tolerant networks," *IEEE Transactions on Mobile Computing*, vol. 11, no. 9, pp. 1514–1531, Sept. 2012.

26. S. C. Nelson and R. Kravets, "Achieving anycast in DTNs by enhancing existing unicast protocols," in *Proceedings of the 5th ACM Workshop on Challenged Networks*, ser. CHANTS '10. New York, NY, USA: ACM, 2010, pp. 63–70. [Online]. Available: http://doi.acm.org/10.1145/1859934.1859948

27. J. Cucurull, M. Asplund, S. Nadjm-Tehrani, and T. Santoro, "Surviving attacks in challenged networks," *IEEE Transactions on Dependable and Secure Computing*, vol. 9, no. 6, pp. 917–929, Nov 2012.

28. Y. Ma and A. Jamalipour, "Opportunistic geocast in disruption-tolerant networks," in *2011 IEEE Global Telecommunications Conference (GLOBECOM 2011)*, Dec 2011, pp. 1–5.

29. G. Costantino, F. Martinelli, and P. Santi, "Investigating the privacy versus forwarding accuracy tradeoff in opportunisticinterest-casting," *IEEE Transactions on Mobile Computing*, vol. 13, no. 4, pp. 824–837, April 2014.

30. U. Lee, S. Y. Oh, K.-W. Lee, and M. Gerla, "Relaycast: Scalable multicast routing in delay tolerant networks," in *IEEE International Conference on Network Protocols, 2008. ICNP 2008.*, Orlando, FL, Oct 2008, pp. 218–227.

31. M. Grossglauser and D. N. C. Tse, "Mobility increases the capacity of ad hoc wireless networks," *IEEE/ACM Trans. Netw.*, vol. 10, no. 4, pp. 477–486, Aug. 2002.
32. K. Scott and S. Burleigh, "Bundle protocol specification," Internet RFC 5050, Nov 2007.
33. I.-R. Chen, F. Bao, M. Chang, and J.-H. Cho, "Trust management for encounter-based routing in delay tolerant networks," in *2010 IEEE Global Telecommunications Conference (GLOBECOM 2010)*, Miami, FL, Dec 2010, pp. 1–6.
34. S. Misra, S. Pal, and B. K. Saha, "Distributed information-based cooperative strategy adaptation in opportunistic mobile networks," *IEEE Transactions on Parallel and Distributed Systems*, vol. 26, no. 3, pp. 724–737, 2015.
35. T. Karagiannis, J.-Y. Le Boudec, and M. Vojnović, "Power law and exponential decay of intercontact times between mobile devices," *IEEE Transactions on Mobile Computing*, vol. 9, no. 10, pp. 1377–1390, Oct 2010.
36. I. Parris and T. Henderson, "The impact of location privacy on opportunistic networks," in *2011 IEEE International Symposium on a World of Wireless, Mobile and Multimedia Networks (WoWMoM)*, Lucca, June 2011, pp. 1–6.
37. ——, "Privacy-enhanced social-network routing," *Computer Communications*, vol. 35, no. 1, pp. 62–74, 2012.
38. S. Zakhary, M. Radenkovic, and A. Benslimane, "Efficient location privacy-aware forwarding in opportunistic mobile networks," *IEEE Transactions on Vehicular Technology*, vol. 63, no. 2, pp. 893–906, Feb 2014.
39. C. Shi, X. Luo, P. Traynor, M. H. Ammar, and E. W. Zegura, "Arden: Anonymous networking in delay tolerant networks," *Ad Hoc Networks*, vol. 10, no. 6, pp. 918–930, 2012.
40. D. Goldschlag, M. Reed, and P. Syverson, "Onion routing," *Commun. ACM*, vol. 42, no. 2, pp. 39–41, Feb. 1999.
41. N. Thompson, S. Nelson, M. Bakht, T. Abdelzaher, and R. Kravets, "Retiring replicants: Congestion control for intermittently-connected networks," in *2010 Proceedings of IEEE INFOCOM*, San Diego, USA, March 2010, pp. 1–9.
42. T. Spyropoulos, K. Psounis, and C. S. Raghavendra, "Spray and wait: an efficient routing scheme for intermittently connected mobile networks," in *Proceedings of the 2005 ACM SIGCOMM workshop on Delay-tolerant networking (WDTN '05)*. New York, NY, USA: ACM, 2005, pp. 252–259.
43. M. Radenkovic and A. Grundy, "Congestion aware forwarding in delay tolerant and social opportunistic networks," in *2011 Eighth International Conference on Wireless On-Demand Network Systems and Services (WONS)*, Bardonecchia, Jan 2011, pp. 60–67.
44. J. Lakkakorpi, M. Pitkänen, and J. Ott, "Using buffer space advertisements to avoid congestion in mobile opportunistic DTNs," in *Proceedings of the 9th IFIP TC 6 International Conference on Wired/Wireless Internet Communications*, ser. WWIC'11. Berlin, Heidelberg: Springer-Verlag, 2011, pp. 386–397.
45. L. Leela-amornsin and H. Esaki, "Heuristic congestion control for message deletion in delay tolerant network," in *Smart Spaces and Next Generation Wired/Wireless Networking*, ser. Lecture Notes in Computer Science, S. Balandin, R. Dunaytsev, and Y. Koucheryavy, Eds. Springer Berlin Heidelberg, 2010, vol. 6294, pp. 287–298.
46. E. Coe and C. Raghavendra, "Token based congestion control for DTNs," in *2010 IEEE Aerospace Conference*, Big Sky, MT, USA, March 2010, pp. 1–7.
47. Y. Cao, H. Cruickshank, and Z. Sun, "Active congestion control based routing for opportunistic delay tolerant networks," in *2011 IEEE 73rd Vehicular Technology Conference (VTC Spring)*, Budapest, Hungary, May 2011, pp. 1–5.
48. A. Grundy and M. Radenkovic, "Decongesting opportunistic social-based forwarding," in *2010 Seventh International Conference on Wireless On-demand Network Systems and Services (WONS)*, Piscataway, NJ, USA, Feb 2010, pp. 82–85.
49. T. Sturgeon, C. Allison, and A. Miller, "A WiFi virtual laboratory," in 7th *Annual Conference of the Higher Education Academy Subject Centre for Information and Computer Sciences*. Dublin, Ireland: HE Academy, 2006, pp. 207–211.

50. B. K. Saha, S. Misra, and M. S. Obaidat, "A web-based integrated environment for simulation and analysis with NS-2," *IEEE Wireless Communications*, vol. 20, no. 4, pp. 109–115, 2013.
51. A. Keränen, J. Ott, and T. Kärkkäinen, "The ONE simulator for DTN protocol evaluation," in *Proceedings of the 2nd International Conference on Simulation Tools and Techniques*, ser. Simutools '09. ICST, Brussels, Belgium, Belgium: ICST (Institute for Computer Sciences, Social-Informatics and Telecommunications Engineering), 2009, pp. 55:1–55:10.
52. Y. Cao and Z. Sun, "Routing in delay/disruption tolerant networks: A taxonomy, survey and challenges," *IEEE Communications Surveys Tutorials*, vol. 15, no. 2, pp. 654–677, 2013.
53. K. Wei, X. Liang, and K. Xu, "A survey of social-aware routing protocols in delay tolerant networks: Applications, taxonomy and design-related issues," *IEEE Communications Surveys Tutorials*, vol. 16, no. 1, pp. 556–578, 2014.
54. S. Tornell, C. Calafate, J. Cano, and P. Manzoni, "DTN protocols for vehicular networks: An application oriented overview," *IEEE Communications Surveys Tutorials*, vol. 17, no. 2, pp. 868–887, 2015.
55. E. P. C. Jones, L. Li, J. K. Schmidtke, and P. A. S. Ward, "Practical routing in delay-tolerant networks," *IEEE Transactions on Mobile Computing*, vol. 6, no. 8, pp. 943–959, Aug 2007.
56. M. Musolesi and C. Mascolo, "Car: Context-aware adaptive routing for delay-tolerant mobile networks," *Mobile Computing, IEEE Transactions on*, vol. 8, no. 2, pp. 246 –260, feb. 2009.
57. A. Balasubramanian, B. Levine, and A. Venkataramani, "DTN routing as a resource allocation problem," in *Proc. of the SIGCOMM '07*. New York, NY, USA: ACM, 2007, pp. 373–384.
58. Z. J. Haas and T. Small, "A new networking model for biological applications of ad hoc sensor networks," *IEEE/ACM Transactions on Networking*, vol. 14, no. 1, pp. 27–40, 2006.
59. X. Zhang, G. Neglia, J. Kurose, and D. Towsley, "Performance modeling of epidemic routing," *Comput. Netw.*, vol. 51, no. 10, pp. 2867–2891, Jul. 2007.
60. Y. Li, P. Hui, D. Jin, L. Su, and L. Zeng, "Evaluating the impact of social selfishness on the epidemic routing in delay tolerant networks," *IEEE Communications Letters*, vol. 14, no. 11, pp. 1026–1028, November 2010.
61. P. Mundur, M. Seligman, and G. Lee, "Epidemic routing with immunity in delay tolerant networks," in *Proc. of IEEE MILCOM*, San Diego, CA, USA, November 2008.
62. T. Matsuda and T. Takine, "(p,q)-epidemic routing for sparsely populated mobile ad hoc networks," *IEEE Journal on Selected Areas in Communications*, vol. 26, no. 5, pp. 783–793, 2008.
63. T. Spyropoulos, K. Psounis, and C. Raghavendra, "Spray and focus: Efficient mobility-assisted routing for heterogeneous and correlated mobility," in *Fifth Annual IEEE International Conference on Pervasive Computing and Communications Workshops, 2007. PerCom Workshops '07.*, White Plains, NY, March 2007, pp. 79–85.
64. T. Spyropoulos, K. Psounis, and C. S. Raghavendra, "Efficient routing in intermittently connected mobile networks: The single-copy case," *IEEE/ACM Transactions on Networking*, vol. 16, no. 1, pp. 63–76, Feb. 2008.
65. M. McNett and G. M. Voelker, "Access and mobility of wireless PDA users," *SIGMOBILE Mob. Comput. Commun. Rev.*, vol. 9, no. 2, pp. 40–55, Apr. 2005.
66. T. Spyropoulos, K. Psounis, and C. Raghavendra, "Efficient routing in intermittently connected mobile networks: The multiple-copy case," *IEEE/ACM Transactions on Networking*, vol. 16, no. 1, pp. 77–90, Feb 2008.
67. T.-K. Huang, C.-K. Lee, and L.-J. Chen, "Prophet+: An adaptive prophet-based routing protocol for opportunistic network," in *24th IEEE International Conference on Advanced Information Networking and Applications (AINA)*, April 2010, pp. 112–119.
68. N. Li and S. K. Das, "A trust-based framework for data forwarding in opportunistic networks," *Ad Hoc Networks*, vol. 11, no. 4, pp. 1497–1509, 2013.
69. P. Hui, J. Crowcroft, and E. Yoneki, "Bubble rap: Social-based forwarding in delay-tolerant networks," *IEEE Transactions on Mobile Computing*, vol. 10, no. 11, pp. 1576–1589, Nov 2011.
70. L. C. Freeman, "Centrality in social networks conceptual clarification," *Social Networks*, vol. 1, no. 3, pp. 215–239, 1978-1979.

71. P. Hage and F. Harary, "Eccentricity and centrality in networks," *Social Networks*, vol. 17, no. 1, pp. 57–63, 1995.

72. J. Scott, *Social Network Analysis*, 3rd ed. SAGE Publications, 2012.

73. S. C. Nelson, M. Bakht, and R. Kravets, "Encounter-based routing in DTNs," in *IEEE INFO-COM 2009*, Rio de Janeiro, April 2009, pp. 846–854.

74. S. Pal and S. Misra, "Contact-based routing in DTNs," in *Proceedings of the 9th International Conference on Ubiquitous Information Management and Communication*, ser. IMCOM '15. New York, NY, USA: ACM, 2015, pp. 3:1–3:6.

75. V. Erramilli, M. Crovella, A. Chaintreau, and C. Diot, "Delegation forwarding," in *Proceedings of the 9th ACM International Symposium on Mobile Ad Hoc Networking and Computing*, ser. MobiHoc '08. New York, NY, USA: ACM, 2008, pp. 251–260.

76. K. Massri, A. Vernata, and A. Vitaletti, "Routing protocols for delay tolerant networks: a quantitative evaluation," in *Proceedings of the 7th ACM Workshop on Performance Monitoring and Measurement of Heterogeneous Wireless and Wired Networks*. USA: ACM, 2012, pp. 107–114.

77. M. Kim, D. Kotz, and S. Kim, "Extracting a mobility model from real user traces," in *Proceedings of 25th IEEE International Conference on Computer Communications*, Barcelona, Spain, April 2006, pp. 1–13.

78. J. Burgess, J. Zahorjan, R. Mahajan, B. N. Levine, A. Balasubramanian, A. Venkataramani, Y. Zhou, B. Croft, N. Banerjee, M. Corner, and D. Towsley, "CRAWDAD data set umass/diesel (v. 2008-10-21)," Downloaded from http://crawdad.org/umass/diesel/, Oct. 2008, [Accessed: 22 Apr. 2015].

79. N. Eagle and A. S. Pentland, "CRAWDAD data set mit/reality (v. 2005-07-01)," Downloaded from http://crawdad.cs.dartmouth.edu/mit/reality, Jul. 2005, [Accessed: 19 Mar. 2014].

80. J. Scott, R. Gass, J. Crowcroft, P. Hui, C. Diot, and A. Chaintreau, "CRAWDAD data set cam-bridge/haggle (v. 2006-01-31)," Downloaded from http://crawdad.org/cambridge/haggle/, Jan. 2006, [Accessed: 22 Apr. 2015].

81. J. Leguay, P. Hui, J. Crowcroft, J. Scott, A. Lindgren, and T. Friedman, "CRAWDAD data set upmc/content (v. 2006-11-17)," Downloaded from http://crawdad.org/upmc/content/, Nov. 2006, [Accessed: 22 Apr. 2015].

82. B. Wietrzyk and M. Radenkovic, "CRAWDAD data set nottingham/cattle (v. 2007-12-20)," Downloaded from http://crawdad.org/nottingham/cattle/, Dec. 2007, [Accessed: 22 Apr. 2015].

83. P. Meroni, S. Gaito, E. Pagani, and G. P. Rossi, "CRAWDAD data set unimi/pmtr (v. 2008-12-01)," Downloaded from http://crawdad.cs.dartmouth.edu/unimi/pmtr, Dec 2008, [Accessed: 18 Mar. 2014].

84. J. Leguay and F. Benbadis, "CRAWDAD data set upmc/rollernet (v. 2009-02-02)," Down-loaded from http://crawdad.cs.dartmouth.edu/upmc/rollernet, Feb. 2009, [Accessed: 19 Mar. 2014].

85. I. Rhee, M. Shin, S. Hong, K. Lee, S. Kim, and S. Chong, "CRAWDAD trace ncsu/mobilitymodels/gps/kaist (v. 2009-07-23)," Downloaded from http://crawdad.cs.dartmouth.edu/ncsu/mobilitymodels/GPS/KAIST, Jul. 2009, [Accessed: 22 Apr. 2015].

86. G. Bigwood, D. Rehunathan, M. Bateman, T. Henderson, and S. Bhatti, "CRAWDAD data set standrews/sassy (v. 2011-06-03)," Downloaded from http://crawdad.cs.dartmouth.edu/standrews/sassy, Jun 2011, [Accessed: 19 Mar. 2014].

87. L. Bracciale, M. Bonola, P. Loreti, G. Bianchi, R. Amici, and A. Rabuffi, "CRAWDAD data set roma/taxi (v. 2014-07-17)," Downloaded from http://crawdad.org/roma/taxi/, Jul. 2014, [Accessed: 22 Apr. 2015].

88. S. Domancich, "Security in delay tolerant networks (DTN) for the android platform," Ph.D. dissertation, Royal Institute of Technology, KTH, Sweden, June 2010.

89. J. Scott, P. Hui, J. Crowcroft, and C. Diot, "Haggle: A networking architecture designed around mobile users," in *Proceedings of the Third Annual IFIP Conference on Wireless On-Demand Network Systems and Services (WONS 2006)*. France: IEEE, January 2006.

90. A. S. Pentland, R. Fletcher, and A. Hasson, "Daknet: Rethinking connectivity in developing nations," *IEEE Computer*, vol. 37, no. 1, pp. 78–83, Jan. 2004.

91. H. Ntareme, M. Zennaro, and B. Pehrson, "Delay tolerant network on smartphones: Applications for communication challenged areas," in *Proceedings of the 3rd Extreme Conference on Communication: The Amazon Expedition*, ser. ExtremeCom '11. New York, NY, USA: ACM, 2011, pp. 14:1–14:6.

92. B. Du and E. A. Brewer, "Dtwiki: A disconnection and intermittency tolerant wiki," in *Proceedings of the 17th International Conference on World Wide Web*, ser. WWW '08. New York, NY, USA: ACM, 2008, pp. 945–952.

93. M. Demmer, B. Du, and E. Brewer, "Tierstore: A distributed filesystem for challenged networks in developing regions," in *Proceedings of the 6th USENIX Conference on File and Storage Technologies*, ser. FAST'08. Berkeley, CA, USA: USENIX Association, 2008, pp. 3:1–3:14.

94. T. Islam, A. Turkulainen, and J. Ott, "Voice messaging for mobile delay-tolerant networks," in *2011 Third International Conference on Communication Systems and Networks (COMSNETS)*, Bangalore, India, Jan 2011, pp. 1–11.

95. P. Juang, H. Oki, Y. Wang, M. Martonosi, L. S. Peh, and D. Rubenstein, "Energy-efficient computing for wildlife tracking: Design tradeoffs and early experiences with zebranet," in *Proceedings of the 10th International Conference on Architectural Support for Programming Languages and Operating Systems*, ser. ASPLOS X. New York, NY, USA: ACM, 2002, pp. 96–107.

96. P. Zhang, C. M. Sadler, S. A. Lyon, and M. Martonosi, "Hardware design experiences in zebranet," in *Proceedings of the 2nd International Conference on Embedded Networked Sensor Systems*, ser. SenSys '04. New York, NY, USA: ACM, 2004, pp. 227–238.

97. T. Liu, C. M. Sadler, P. Zhang, and M. Martonosi, "Implementing software on resource-constrained mobile sensors: Experiences with Impala and ZebraNet," in *Proceedings of the 2nd International Conference on Mobile Systems, Applications, and Services*, ser. MobiSys '04. New York, NY, USA: ACM, 2004, pp. 256–269.

98. A. Balasubramanian, Y. Zhou, W. B. Croft, B. N. Levine, and A. Venkataramani, "Web search from a bus," in *Proceedings of the Second ACM Workshop on Challenged Networks*, ser. CHANTS '07. New York, NY, USA: ACM, 2007, pp. 59–66.

99. T. Hossmann, F. Legendre, P. Carta, P. Gunningberg, and C. Rohner, "Twitter in disaster mode: opportunistic communication and distribution of sensor data in emergencies," in *Proceedings of the 3rd Extreme Conference on Communication: The Amazon Expedition*, ser. ExtremeCom '11. New York, NY, USA: ACM, 2011, pp. 1:1–1:6.

100. A. Lindgren and A. Doria, "Experiences from deploying a real-life DTN system," in *4th IEEE Consumer Communications and Networking Conference*, Las Vegas, USA, Jan 2007, pp. 217–221.

101. A. Lindgren, A. Doria, J. Lindblom, and M. Ek, "Networking in the land of northern lights: Two years of experiences from DTN system deployments," in *Proceedings of the 2008 ACM Workshop on Wireless Networks and Systems for Developing Regions*, ser. WiNS-DR '08. New York, NY, USA: ACM, 2008, pp. 1–8.

102. Z. Li and H. Shen, "Utility-based distributed routing in intermittently connected networks," in *37th International Conference on Parallel Processing, 2008. ICPP '08.*, Portland, OR, Sept 2008, pp. 604–611.

103. A. Elwhishi and P. han Ho, "SARP - a novel multi-copy routing protocol for intermittently connected mobile networks," in *IEEE Global Telecommunications Conference, 2009. GLOBECOM 2009.*, Honolulu, HI, Nov 2009, pp. 1–7.

104. V. Prueksasri, C. Intanagonwiwat, and K. Rojviboonchai, "DNH-SaW: The different neighbor-history spray and wait routing scheme for delay tolerant networks," in *2011 International Symposium on Intelligent Signal Processing and Communications Systems (ISPACS)*, Chiang Mai, Thailand, Dec 2011, pp. 1–5.

105. J. Bloch, *Effective Java*, 2nd ed. Addison-Wesley, May 2008.

106. J. Burke, D. Estrin, M. Hansen, A. Parker, N. Ramanathan, S. Reddy, and M. B. Srivastava, "Participatory sensing," in *Workshop on World-Sensor-Web (WSW'06): Mobile Device Centric Sensor Networks and Applications*, Boulder, Colorado, USA, 2006, pp. 117–134.

107. A. Campbell, S. Eisenman, N. Lane, E. Miluzzo, R. Peterson, H. Lu, X. Zheng, M. Musolesi, K. Fodor, and G.-S. Ahn, "The rise of people-centric sensing," *IEEE Internet Computing*, vol. 12, no. 4, pp. 12–21, July-Aug. 2008.

108. M. Srivastava, T. Abdelzaher, and B. Szymanski, "Human-centric sensing," *Philosophical Transactions of the Royal Society A: Mathematical,Physical and Engineering Sciences*, vol. 370, no. 1958, pp. 176–197, Jan. 2012.

109. A. T. Campbell, S. B. Eisenman, N. D. Lane, E. Miluzzo, and R. A. Peterson, "People-centric urban sensing," in *Proceedings of the 2nd Annual International Workshop on Wireless Internet*, ser. WICON '06. New York, NY, USA: ACM, 2006.

110. R. Rao and G. Kesidis, "Purposeful mobility for relaying and surveillance in mobile ad hoc sensor networks," *IEEE Transactions on Mobile Computing*, vol. 3, no. 3, pp. 225–232, 2004.

111. S. Eswaran, A. Misra, F. Bergamaschi, and T. L. Porta, "Utility-based bandwidth adaptation in mission-oriented wireless sensor networks," *ACM Trans. Sen. Netw.*, vol. 8, no. 2, pp. 17:1–17:26, Mar. 2012.

112. C. H. Liu, K. K. Leung, C. Bisdikian, and J. W. Branch, "A new approach to architecture of sensor networks for mission-oriented applications," *Proc. SPIE*, vol. 7349, pp. 73 490L–73 490L–12, 2009. [Online]. Available: http://dx.doi.org/10.1117/12.820199

113. E. Pignaton de Freitas, T. Heimfarth, C. E. Pereira, A. Morado Ferreira, F. Rech Wagner, and T. Larsson, "Multi-agent support in a middleware for mission-driven heterogeneous sensor networks," *The Computer Journal*, vol. 54, no. 3, pp. 406–420, 2011.

114. H. M. Ammari and S. K. Das, "Mission-oriented k-coverage in mobile wireless sensor networks," in *ICDCN*, Kolkata, India, January 2010, pp. 92–103.

115. M. Grismer, "Field sensor networks and automated monitoring of soil-water sensors," *Soil Science*, vol. 154, no. 6, pp. 482–489, 1992.

116. Y. Kim, R. Evans, and W. Iversen, "Remote sensing and control of an irrigation system using a distributed wireless sensor network," *IEEE Transactions on Instrumentation and Measurement*, vol. 57, no. 7, pp. 1379–1387, July 2008.

117. J. K. Hart and K. Martinez, "Environmental sensor networks: A revolution in the earth system science?" *Earth-Science Reviews*, vol. 78, no. 34, pp. 177–191, 2006.

118. K. Martinez, J. Hart, and R. Ong, "Environmental sensor networks," *Computer*, vol. 37, no. 8, pp. 50–56, Aug 2004.

119. G. Werner-Allen, J. Johnson, M. Ruiz, J. Lees, and M. Welsh, "Monitoring volcanic eruptions with a wireless sensor network," in *Wireless Sensor Networks, 2005. Proceeedings of the Second European Workshop on*, Jan 2005, pp. 108–120.

120. R. Sherwood and S. Chien, "Sensor webs for science: New directions for the future," *AIAA Infotech@ Aerospace 2007*, pp. 7–10, May 2006.

121. A. G. Davies, S. Chien, R. Wright, A. Miklius, P. R. Kyle, M. Welsh, J. B. Johnson, D. Tran, S. R. Schaffer, and R. Sherwood, "Sensor web enables rapid response to volcanic activity," *Eos, Transactions American Geophysical Union*, vol. 87, no. 1, pp. 1–5, Jan. 2006.

122. J. B. Johnson, R. C. Aster, and P. R. Kyle, "Volcanic eruptions observed with infrasound," *Geophysical Research Letters*, vol. 31, no. 14, 2004. [Online]. Available: http://dx.doi.org/10.1029/2004GL020020

123. G. Werner-Allen, K. Lorincz, M. Ruiz, O. Marcillo, J. Johnson, J. Lees, and M. Welsh, "Deploying a wireless sensor network on an active volcano," *IEEE Internet Computing*, vol. 10, no. 2, pp. 18–25, March 2006.

124. M. V. Ramesh, S. Kumar, and P. V. Rangan, "Wireless sensor network for landslide detection," in *International Conference on Wireless Networks*, 2009, pp. 89–95.

125. M. Quaritsch, E. Stojanovski, C. Bettstetter, G. Friedrich, H. Hellwagner, B. Rinner, M. Hofbaur, and M. Shah, "Collaborative microdrones: Applications and research challenges," in *Proceedings of the 2nd International Conference on Autonomic Computing and Communication Systems*, ser. Autonomics '08. ICST, Brussels, Belgium, Belgium: ICST (Institute

for Computer Sciences, Social-Informatics and Telecommunications Engineering), 2008, pp. 38:1–38:7.

126. M. Quaritsch, K. Kruggl, D. Wischounig-Strucl, S. Bhattacharya, M. Shah, and B. Rinner, "Networked uavs as aerial sensor network for disaster management applications," *e & i Elektrotechnik und Informationstechnik*, vol. 127, no. 3, pp. 56–63, 2010. [Online]. Available: http://dx.doi.org/10.1007/s00502-010-0717-2

127. M. Castillo-Effer, D. Quintela, W. Moreno, R. Jordan, and W. Westhoff, "Wireless sensor networks for flash-flood alerting," in *Devices, Circuits and Systems, 2004. Proceedings of the Fifth IEEE International Caracas Conference on*, vol. 1, Nov 2004, pp. 142–146.

128. S. C. Nelson, A. F. Harris, III, and R. Kravets, "Event-driven, role-based mobility in disaster recovery networks," in *Proceedings of CHANTS '07*. New York, NY, USA: ACM, 2007, pp. 27–34.

129. M. Uddin, D. Nicol, T. Abdelzaher, and R. Kravets, "A post-disaster mobility model for delay tolerant networking," in *Proceedings of the 2009 Winter Simulation Conference (WSC)*, Austin, TX, USA, Dec. 2009, pp. 2785 –2796.

130. N. Aschenbruck, E. Gerhards-Padilla, and P. Martini, "Modeling mobility in disaster area scenarios," *Perform. Eval.*, vol. 66, no. 12, pp. 773–790, Dec. 2009.

131. L. Conceição and M. Curado, "Modelling mobility based on human behaviour in disaster areas," in *Wired/Wireless Internet Communication*, ser. Lecture Notes in Computer Science, V. Tsaoussidis, A. Kassler, Y. Koucheryavy, and A. Mellouk, Eds. Springer Berlin Heidelberg, 2013, vol. 7889, pp. 56–69.

132. ——, "Modelling mobility based on obstacle-aware human behaviour in disaster areas," *Wireless Personal Communications*, vol. 82, no. 1, pp. 451–472, 2015. [Online]. Available: http://dx.doi.org/10.1007/s11277-014-2235-8

133. T. Sakaki, F. Toriumi, and Y. Matsuo, "Tweet trend analysis in an emergency situation," in *Proceedings of the Special Workshop on Internet and Disasters*. New York, NY, USA: ACM, 2011, pp. 3:1–3:8. [Online]. Available: http://doi.acm.org/10.1145/2079360.2079363

134. T. Sakaki, M. Okazaki, and Y. Matsuo, "Earthquake shakes twitter users: Real-time event detection by social sensors," in *Proceedings of the 19th International Conference on World Wide Web*, ser. WWW '10. New York, NY, USA: ACM, 2010, pp. 851–860.

135. A. Crooks, A. Croitoru, A. Stefanidis, and J. Radzikowski, "Earthquake: Twitter as a distributed sensor system," *Transactions in GIS*, vol. 17, pp. 124–147, 2013.

136. K. Lorincz, D. Malan, T. Fulford-Jones, A. Nawoj, A. Clavel, V. Shnayder, G. Mainland, M. Welsh, and S. Moulton, "Sensor networks for emergency response: challenges and opportunities," *IEEE Pervasive Computing*, vol. 3, no. 4, pp. 16–23, Oct 2004.

137. S. George, W. Zhou, H. Chenji, M. Won, Y. O. Lee, A. Pazarloglou, R. Stoleru, and P. Barooah, "Distressnet: a wireless ad hoc and sensor network architecture for situation management in disaster response," *IEEE Communications Magazine*, vol. 48, no. 3, pp. 128–136, March 2010.

138. I. Sugino, "Disaster recovery and the R&D policy in Japan's telecommunication networks," in *Plenary talk at OFC/OFOEC2012*, Los Angeles, USA, Mar. 2012. [Online]. Available: http://www.ofcnfoec.org/osa.ofc/media/Default/Plenary/Sugino-Plenary-Final-Slides.pdf

139. X. Lu and P. Hui, "An energy-efficient n-epidemic routing protocol for delay tolerant networks," in *IEEE 5th International Conference on Networking, Architecture and Storage (NAS)*, Macau, China, 2010, pp. 341–347.

140. S. Saha, V. K. Shah, R. Verma, R. Mandal, and S. Nandi, "Is it worth taking a planned approach to design ad hoc infrastructure for post disaster communication?" in *Proceedings of the seventh ACM international workshop on Challenged networks (CHANTS '12)*. New York, NY, USA: ACM, 2012, pp. 87–90.

141. S. Russell and P. Norvig, *Artificial Intelligence: A Modern Approach*, 3rd ed. Upper Saddle River, NJ, USA: Prentice Hall Press, 2009.

142. C. M. Macal and M. J. North, "Tutorial on agent-based modelling and simulation," *Journal of Simulation*, vol. 4, pp. 151–162, 2010.

143. C. Skinner and S. Ramchurn, "The robocup rescue simulation platform," in *Proceedings of the 9th International Conference on Autonomous Agents and Multiagent Systems: Volume 1*, ser. AAMAS '10. Richland, SC: International Foundation for Autonomous Agents and Multiagent Systems, 2010, pp. 1647–1648.

144. F. Fiedrich and P. Burghardt, "Agent-based systems for disaster management," *Communications of the ACM*, vol. 50, no. 3, pp. 41–42, Mar. 2007.

145. D. Massaguer, V. Balasubramanian, S. Mehrotra, and N. Venkatasubramanian, "Multi-agent simulation of disaster response," in *ATDM Workshop in AAMAS 2006*, Hokkaido, Japan, May 2006.

146. N. Schurr and M. Tambe, "Using multi-agent teams to improve the training of incident commanders," in *Defence Industry Applications of Autonomous Agents and Multi-Agent Systems*, ser. Whitestein Series in Software Agent Technologies and Autonomic Computing, M. Pěchouček, S. Thompson, and H. Voos, Eds. Birkhäuser Basel, 2008, pp. 151–166. [Online]. Available: http://dx.doi.org/10.1007/978-3-7643-8571-2_9

147. M.-Y. Cheng and Y.-W. Wu, "Multi-agent-based data exchange platform for bridge disaster prevention: a case study in Taiwan," *Natural Hazards*, vol. 69, no. 1, pp. 311–326, 2013. [Online]. Available: http://dx.doi.org/10.1007/s11069-013-0708-9

148. X. Pan, C. Han, K. Dauber, and K. Law, "A multi-agent based framework for the simulation of human and social behaviors during emergency evacuations," *AI & SOCIETY*, vol. 22, no. 2, pp. 113–132, 2007.

149. A. T. Crooks and S. Wise, "GIS and agent-based models for humanitarian assistance," *Computers, Environment and Urban Systems*, vol. 41, pp. 100–111, 2013.

150. K. Mustapha, H. Mcheick, and S. Mellouli, "Modeling and simulation agent-based of natural disaster complex systems," *Procedia Computer Science*, vol. 21, pp. 148–155, 2013.

151. V.-M. Le, Y. Chevaleyre, and J.-D. Zucker, "Speeding up the evaluation of casualties in multi-agent simulations with linear programming application to optimization of sign placement for tsunami evacuation," in *2013 IEEE RIVF International Conference on Computing and Communication Technologies, Research, Innovation, and Vision for the Future (RIVF)*, Hanoi, Nov. 2013, pp. 215–220.

152. G.-H. Kim, S. Trimi, and J.-H. Chung, "Big-data applications in the government sector," *Communications of the ACM*, vol. 57, no. 3, pp. 78–85, Mar. 2014.

153. T. Shelton, A. Poorthuis, M. Graham, and M. Zook, "Mapping the data shadows of hurricane sandy: Uncovering the sociospatial dimensions of 'big data'," *Geoforum*, vol. 52, pp. 167–179, 2014.

154. S. Sagiroglu and D. Sinanc, "Big data: A review," in *2013 International Conference on Collaboration Technologies and Systems (CTS)*, San Diego, USA, May 2013, pp. 42–47.

155. P. Tin, T. T. Zin, T. Toriu, and H. Hama, "An integrated framework for disaster event analysis in big data environments," in *2013 Ninth International Conference on Intelligent Information Hiding and Multimedia Signal Processing*, Beijing, Oct. 2013, pp. 255–258.

156. S. Legg and M. Hutter, "A collection of definitions of intelligence," *CoRR*, vol. abs/0706.3639, 2007. [Online]. Available: http://arxiv.org/abs/0706.3639

157. M. R. Endsley, "Situation awareness global assessment technique (SAGAT)," in *Proceedings of the IEEE 1988 National Aerospace and Electronics Conference, 1988. NAECON 1988*, vol. 3, Dayton, OH, 1988, pp. 789–795.

158. ——, "Toward a theory of situation awareness in dynamic systems," *Human Factors*, vol. 37, no. 1, pp. 32–64, 1995.

159. ——, "Situation awareness," in *The Oxford Handbook of Cognitive Engineering*, ser. Oxford Library of Psychology, J. D. Lee and A. Kirlik, Eds. USA: Oxford University Press USA, March 2013, pp. 88–108.

160. M. M. Kokar, C. J. Matheus, and K. Baclawski, "Ontology-based situation awareness," *Information Fusion*, vol. 10, no. 1, pp. 83–98, Jan. 2009.

161. S. Vieweg, A. L. Hughes, K. Starbird, and L. Palen, "Microblogging during two natural hazards events: What Twitter may contribute to situational awareness," in *Proceedings of the SIGCHI Conference on Human Factors in Computing Systems*, ser. CHI '10. New York, NY, USA: ACM, 2010, pp. 1079–1088.

162. B. Tomaszewski, "Situation awareness and virtual globes: Applications for disaster management," *Computers & Geosciences*, vol. 37, no. 1, pp. 86–92, 2011.
163. S. Verma, S. Vieweg, W. Corvey, L. Palen, J. Martin, M. Palmer, A. Schram, and K. Anderson, "Natural language processing to the rescue? extracting "situational awareness" tweets during mass emergency," in *International AAAI Conference on Weblogs and Social Media*, 2011. [Online]. Available: https://www.aaai.org/ocs/index.php/ICWSM/ICWSM11/paper/view/2834
164. J. Yin, A. Lampert, M. Cameron, B. Robinson, and R. Power, "Using social media to enhance emergency situation awareness," *IEEE Intelligent Systems*, vol. 27, no. 6, pp. 52–59, 2012.
165. K. J. Bennett, J. M. Olsen, S. Harris, S. Mekaru, A. A. Livinski, and J. S. Brownstein, "The perfect storm of information: Combining traditional and non-traditional data sources for public health situational awareness during hurricane response," *PLoS Currents*, vol. 5, December 2013. [Online]. Available: http://www.ncbi.nlm.nih.gov/pmc/articles/PMC3871418/
166. J. Rogstadius, M. Vukovic, C. Teixeira, V. Kostakos, E. Karapanos, and J. Laredo, "Crisis-tracker: Crowdsourced social media curation for disaster awareness," *IBM Journal of Research and Development*, vol. 57, no. 5, pp. 4:1–4:13, September 2013.
167. D. Wang, M. Amin, S. Li, T. Abdelzaher, L. Kaplan, S. Gu, C. Pan, H. Liu, C. Aggarwal, R. Ganti, X. Wang, P. Mohapatra, B. Szymanski, and H. Le, "Using humans as sensors: An estimation-theoretic perspective," in *IPSN-14 Proceedings of the 13th International Symposium on Information Processing in Sensor Networks,* Berlin, April 2014, pp. 35–46.
168. A. Bruns and Y. Liang, "Tools and methods for capturing twitter data during natural disasters," *First Monday*, vol. 17, no. 4, 2012. [Online]. Available: http://firstmonday.org/ojs/index.php/fm/article/view/3937
169. B. Schilit and M. Theimer, "Disseminating active map information to mobile hosts," *IEEE Network*, vol. 8, no. 5, pp. 22–32, Sept 1994.
170. Y.-H. Feng, T.-H. Teng, and A.-H. Tan, "Modelling situation awareness for context-aware decision support," *Expert Systems with Applications*, vol. 36, no. 1, pp. 455–463, Jan 2009.
171. C. Alcaraz and J. Lopez, "Wide-area situational awareness for critical infrastructure protection," *Computer*, vol. 46, no. 4, pp. 30–37, April 2013.
172. W. Zhao, M. Ammar, and E. Zegura, "A message ferrying approach for data delivery in sparse mobile ad hoc networks," in *Proceedings of the 5th ACM International Symposium on Mobile Ad Hoc Networking and Computing*, ser. MobiHoc '04. New York, NY, USA: ACM, 2004, pp. 187–198.
173. N. A. Vien, N. H. Viet, S. Lee, and T. Chung, "Heuristic search based exploration in reinforcement learning," in *IWANN*, San Sebastian, Spain, 2007, pp. 110–118.
174. B. K. Saha, S. Misra, and S. Pal, "Utility-based exploration for performance enhancement in opportunistic mobile networks," *IEEE Transactions on Computers*, 2015, DOI: 10.1109/TC.2015.2441700.
175. G. Johnson. Theories of emotion. Last accessed: 18 February 2014. Drexel University, USA. [Online]. Available: http://www.iep.utm.edu/emotion/
176. P. Ekman and W. V. Friesen, *Unmasking the Face: A Guide to Recognizing Emotions From Facial Expressions*. Malor Books, September 2003.
177. P. Ekman and W. Friesen, "The repertoire of non-verbal behavior – categories, origins, usage and coding," *Semiotica*, vol. 1, pp. 49–98, 1969.
178. ——, "Constants across cultures in the face and emotion," *Journal of Personality and Social Psychology*, vol. 17, pp. 124–129, 1971.
179. P. Ekman, "Universal and cultural differences in facial expressions of emotion," in *Nebraska Symposium on Motivation*, J. Cole, Ed., vol. 19. Lincoln University of Nebraska Press, 1972.
180. P. Ekman and D. Cordaro, "What is meant by calling emotions basic," *Emotion Review*, vol. 3, no. 4, pp. 364–370, 2011.
181. A. W. Siegman and S. Feldstein, Eds., *Nonverbal behavior and communication*. L. Erlbaum Associates, 1978, vol. 1.
182. P. Ekman, W. V. Friesen, and P. Ellsworth, *Emotion in the Human Face: Guidelines for Research and an Integration of Findings*. New York: Pergamon Press, 1972.

183. T. Dalgleish and M. Power, Eds., *Handbook of Cognition and Emotion*, 1st ed. Wiley, March 1999.
184. R. Plutchik, *Emotion, a Psychoevolutionary Synthesis*. Harper and Row, 1980.
185. A. Mehrabian, "Framework for a comprehensive description and measurement of emotional states," *Genetic, social, and general psychology monographs*, vol. 121, no. 3, pp. 339–361, August 1995.
186. PAD emotional state model. Last accessed: 18 February 2014. [Online]. Available: https://en.wikipedia.org/wiki/PAD_emotional_state_model
187. A. Mehrabian, "Pleasure-arousal-dominance: A general framework for describing and measuring individual differences in temperament," *Current Psychology*, vol. 14, no. 4, pp. 261–292, 1996.
188. L. Peña, J.-M. Peña, and S. Ossowski, "Representing emotion and mood states for virtual agents," in *Multiagent System Technologies*, ser. Lecture Notes in Computer Science, F. Klgl and S. Ossowski, Eds. Springer Berlin Heidelberg, 2011, vol. 6973, pp. 181–188.
189. J. D. Velásquez and P. Maes, "Cathexis: A computational model of emotions," in *Proceedings of the First International Conference on Autonomous Agents*, ser. AGENTS '97. New York, NY, USA: ACM, 1997, pp. 518–519.
190. M. El-Nasr, J. Yen, and T. Ioerger, "Flame – fuzzy logic adaptive model of emotions," *Autonomous Agents and Multi-Agent Systems*, vol. 3, no. 3, pp. 219–257, 2000.
191. C. Pelachaud and M. Bilvi, "Computational model of believable conversational agents," in *Communication in Multiagent Systems*, ser. Lecture Notes in Computer Science, M.-P. Huget, Ed. Springer Berlin Heidelberg, 2003, vol. 2650, pp. 300–317.
192. J. Gratch and S. Marsella, "Evaluating a computational model of emotion," *Autonomous Agents and Multi-Agent Systems*, vol. 11, no. 1, pp. 23–43, 2005.
193. J. Dias and A. Paiva, "Feeling and reasoning: A computational model for emotional characters," in *Progress in Artificial Intelligence*, ser. Lecture Notes in Computer Science, C. Bento, A. Cardoso, and G. Dias, Eds. Springer Berlin Heidelberg, 2005, vol. 3808, pp. 127–140.
194. I. T. Meftah, N. L. Thanh, and C. B. Amar, "Towards an algebraic modeling of emotional states," in 5th *International Conference on Internet and Web Applications and Services (ICIW)*, Barcelona, Spain, May 2010, pp. 513–518.
195. S. Marsella and J. Gratch, "Computationally modeling human emotion," *Communications of the ACM*, vol. 57, no. 12, pp. 56–67, December 2014.
196. A. Chandra, "A computational architecture to model human emotions," in *Proceedings of Intelligent Information Systems, 1997*, Grand Bahama Island, Dec 1997, pp. 86–89.
197. K. Kühnlenz and M. Buss, "Towards an emotion core based on a hidden markov model," in 13th *IEEE International Workshop on Robot and Human Interactive Communication, 2004*, Japan, Sept 2004, pp. 119–124.
198. S. C. Banik, K. Watanabe, M. K. Habib, and K. Izumi, "Multi-robot team work with benevolent characters: The roles of emotions," in *Handbook of Research on Synthetic Emotions and Sociable Robotics: New Applications in Affective Computing and Artificial Intelligence*, J. Vallverdú and D. Casacuberta, Eds. IGI Global, 2009, pp. 57–73.
199. A. Moors, P. C. Ellsworth, K. R. Scherer, and N. H. Frijda, "Appraisal theories of emotion: State of the art and future development," *Emotion Review*, vol. 5, no. 2, pp. 119–124, 2013.
200. Appraisal theory. Last accessed: 18 May 2015. [Online]. Available: http://changingminds.org/explanations/theories/appraisal_theory.htm
201. J. Gratch, "Emile: Marshalling passions in training and education," in *Proceedings of the Fourth International Conference on Autonomous Agents*, ser. AGENTS '00. New York, NY, USA: ACM, 2000, pp. 325–332.
202. J. Gratch and S. Marsella, "A domain-independent framework for modeling emotion," *Cognitive Systems Research*, vol. 5, no. 4, pp. 269–306, 2004.
203. L. Li and J.-h. Chen, "Emotion recognition using physiological signals," in *Advances in Artificial Reality and Tele-Existence*, ser. Lecture Notes in Computer Science, Z. Pan, A. Cheok, M. Haller, R. Lau, H. Saito, and R. Liang, Eds. Springer Berlin Heidelberg, 2006, vol. 4282, pp. 437–446.

204. H. Gunes and M. Piccardi, "Bi-modal emotion recognition from expressive face and body gestures," *Journal of Network and Computer Applications*, vol. 30, no. 4, pp. 1334–1345, 2007, special issue on Information Technology.

205. N. Sebe, I. Cohen, T. Gevers, and T. S. Huang, "Multimodal approaches for emotion recognition: A survey," in *Proc. SPIE*, vol. 5670, San Jose, USA, January 2005, pp. 56–67. [Online]. Available: http://dx.doi.org/10.1117/12.600746

206. C. Stickel, M. Ebner, S. Steinbach-Nordmann, G. Searle, and A. Holzinger, "Emotion detection: Application of the valence arousal space for rapid biological usability testing to enhance universal access," in *Universal Access in Human-Computer Interaction. Addressing Diversity*, ser. Lecture Notes in Computer Science, C. Stephanidis, Ed. Springer Berlin Heidelberg, 2009, vol. 5614, pp. 615–624.

207. I. Maglogiannis, D. Vouyioukas, and C. Aggelopoulos, "Face detection and recognition of natural human emotion using Markov random fields," *Personal and Ubiquitous Computing*, vol. 13, no. 1, pp. 95–101, 2009.

208. G. Chanel, C. Rebetez, M. Bétrancourt, and T. Pun, "Emotion assessment from physiological signals for adaptation of game difficulty," *IEEE Transactions on Systems, Man and Cybernetics, Part A: Systems and Humans*, vol. 41, no. 6, pp. 1052–1063, Nov 2011.

209. S. Koelstra, C. Mühl, M. Soleymani, J.-S. Lee, A. Yazdani, T. Ebrahimi, T. Pun, A. Nijholt, and I. Patras, "DEAP: A database for emotion analysis using physiological signals," *IEEE Transactions on Affective Computing*, vol. 3, no. 1, pp. 18–31, Jan 2012.

210. S. Wioleta, "Using physiological signals for emotion recognition," in *2013 The 6th International Conference on Human System Interaction (HSI)*, June 2013, pp. 556–561.

211. M. Baig, E. Barakova, L. Marcenaro, M. Rauterberg, and C. Regazzoni, "Crowd emotion detection using dynamic probabilistic models," in *From Animals to Animats 13*, ser. Lecture Notes in Computer Science, A. del Pobil, E. Chinellato, E. Martinez-Martin, J. Hallam, E. Cervera, and A. Morales, Eds. Springer International Publishing, 2014, vol. 8575, pp. 328–337.

212. A. Kołakowska, A. Landowska, M. Szwoch, W. Szwoch, and M. Wróbel, "Emotion recognition and its applications," in *Human-Computer Systems Interaction: Backgrounds and Applications 3*, ser. Advances in Intelligent Systems and Computing, Z. S. Hippe, J. L. Kulikowski, T. Mroczek, and J. Wtorek, Eds. Springer International Publishing, 2014, vol. 300, pp. 51–62.

213. L. Buitinck, J. van Amerongen, E. Tan, and M. de Rijke, "Multi-emotion detection in user-generated reviews," in *Advances in Information Retrieval*, ser. Lecture Notes in Computer Science, A. Hanbury, G. Kazai, A. Rauber, and N. Fuhr, Eds. Springer International Publishing, 2015, vol. 9022, pp. 43–48.

214. G. Lindgaard, G. Fernandes, C. Dudek, and J. Brown, "Attention web designers: You have 50 miliseconds to make a good first impression," *Behaviour and Information Technology*, vol. 25, no. 2, pp. 115–126, 2006.

215. N. Bonnardel, A. Piolat, and L. L. Bigot, "The impact of colour on website appeal and users' cognitive processes," *Displays*, vol. 32, no. 2, pp. 69–80, 2011.

216. R. LiKamWa, Y. Liu, N. D. Lane, and L. Zhong, "Can your smartphone infer your mood?" in *2nd International Workshop on Sensing Applications on Mobile Phones (PhoneSense 2011)*, Seattle, WA, USA, 2011.

217. H. Lee, Y. S. Choi, S. Lee, and I. Park, "Towards unobtrusive emotion recognition for affective social communication," in *2012 IEEE Consumer Communications and Networking Conference (CCNC)*, Las Vegas, NV, USA, Jan. 2012, pp. 260–264.

218. J. Tang, Y. Zhang, J. Sun, J. Rao, W. Yu, Y. Chen, and A. C. M. Fong, "Quantitative study of individual emotional states in social networks," *IEEE Transactions on Affective Computing*, vol. 3, no. 2, pp. 132–144, 2012.

219. A. Ortigosa, J. M. Martín, and R. M. Carro, "Sentiment analysis in Facebook and its application to e-learning," *Computers in Human Behavior*, vol. 31, pp. 527–541, 2014.

220. M. Thelwall, D. Wilkinson, and S. Uppal, "Data mining emotion in social network communication: Gender differences in myspace," *Journal of the American Society for Information Science and Technology*, vol. 61, no. 1, pp. 190–199, 2010.

221. C. Gao and J. Liu, "Clustering-based media analysis for understanding human emotional reactions in an extreme event," in *Proceedings of the 20th International Conference on Foundations of Intelligent Systems*, China, 2012, pp. 125–135.

222. A. Beaudry and A. Pinsonneault, "The other side of acceptance: Studying the direct and indirect effects of emotions on information technology use," *MIS Quarterly*, vol. 34, no. 4, pp. 689–710, Dec. 2010.

223. R. Savolainen, "Emotions as motivators for information seeking: A conceptual analysis," *Library & Information Science Research*, vol. 36, no. 1, pp. 59–65, 2014.

224. R. Picard, *Affective Computing*. Cambridge: MIT Press, 1997.

225. I. Rhee, M. Shin, S. Hong, K. Lee, S. Kim, and S. Chong, "CRAWDAD data set ncsu/mobilitymodels (v. 2009-07-23)," Downloaded from http://crawdad.cs.dartmouth.edu/ncsu/mobilitymodels, Jul. 2009, [Accessed: 12 Dec. 2012].

226. W. Zhao, Y. Chen, M. Ammar, M. Corner, B. Levine, and E. Zegura, "Capacity enhancement using throwboxes in DTNs," in *IEEE International Conference on Mobile Adhoc and Sensor Systems (MASS)*, Canada, 2006, pp. 31–40.

227. S. Misra, S. Pal, and B. K. Saha, "Cooperation in delay tolerant networks," in *Next-Generation Wireless Technologies*, ser. Computer Communications and Networks, N. Chilamkurti, S. Zeadally, and H. Chaouchi, Eds. Springer London, 2013, pp. 15–35. [Online]. Available: http://dx.doi.org/10.1007/978-1-4471-5164-7_3

228. A. Brandenburger, "Cooperative game theory: Characteristic functions, allocations, marginal contribution," http://www.uib.cat/depart/deeweb/pdi/hdeelbm0/arxius_decisions_and_games/cooperative_game_theory-brandenburger.pdf, April 2007, [Accessed: 22 Apr. 2015].

229. S. Hart, "A comparison of non-transferable utility values," *Theory and Decision*, vol. 56, no. 1-2, pp. 35–46, 2004. [Online]. Available: http://dx.doi.org/10.1007/s11238-004-5633-7

230. Z. Han, D. Niyato, W. Saad, T. Baar, and A. Hjrungnes, *Game Theory in Wireless and Communication Networks: Theory, Models, and Applications*. Cambridge University Press, 2011.

231. T. S. Ferguson. (2014) Game theory. Online. University of California at Los Angeles. Last accessed: 15 October 2015. [Online]. Available: http://www.math.ucla.edu/~tom/Game_Theory/Contents.html

232. Y. Bachrach, E. Elkind, R. Meir, D. Pasechnik, M. Zuckerman, J. Rothe, and J. S. Rosenschein, "The cost of stability in coalitional games," in *Algorithmic Game Theory*, ser. Lecture Notes in Computer Science, M. Mavronicolas and V. G. Papadopoulou, Eds. Springer Berlin Heidelberg, 2009, vol. 5814, pp. 122–134.

233. W. Saad, Z. Han, M. Debbah, A. Hjorungnes, and T. Basar, "Coalitional game theory for communication networks," *IEEE Signal Processing Magazine*, vol. 26, no. 5, pp. 77–97, 2009.

234. G. Greco, E. Malizia, L. Palopoli, and F. Scarcello, "Non-transferable utility coalitional games via mixed-integer linear constraints," *Journal of Artificial Intelligence Research*, vol. 38, pp. 633–685, 2010.

235. J. M. Smith and G. R. Price, "The logic of animal conflict," *Nature*, vol. 246, no. 5427, pp. 15–18, 1973.

236. P. D. Taylor and L. B. Jonker, "Evolutionarily stable strategies and game dynamics," *Mathematical Biosciences*, vol. 40, pp. 145–156, 1978.

237. L. Fisher, *Rock, Paper, Scissors: Game Theory in Everyday Life*. Basic Books, 2008.

238. W. H. Sandholm, "Evolutionary game theory," in *Encyclopedia of Complexity and Systems Science*. Springer, 2009, pp. 3176–3205.

239. H. Gintis, "Classical versus evolutionary game theory," *Journal of Consciousness Studies*, vol. 7, no. 1-2, pp. 300–304, Jan. 2000. [Online]. Available: http://www.ingentaconnect.com/content/imp/jcs/2000/00000007/F0020001/1158

240. P. Hammerstein and R. Selten, *Game theory and evolutionary biology*, R. J. Aumann and S. Hart, Eds. Elsevier Science Publishers B. V., 1994, vol. 2.

241. W. D. Hamilton and R. M. May, "Dispersal in stable habitats," *Nature*, vol. 269, pp. 578–581, Oct. 1977.

242. S. Bowles, *Microeconomics: Behavior, Institutions, and Evolution*. Princeton University Press, 2004.
243. S. Bowles and H. Gintis, "Walrasian economics in retrospect," *Quarterly Journal of Economics*, vol. 115, no. 4, pp. 1411–1439, 2000.
244. A. Antoci, S. Borghesi, and M. Galeotti, "Environmental options and technological innovation: an evolutionary game model," *Journal of Evolutionary Economics*, vol. 23, no. 2, pp. 247–269, 2013.
245. S. Zeadally, R. Hunt, Y.-S. Chen, A. Irwin, and A. Hassan, "Vehicular ad hoc networks (VANETS): status, results, and challenges," *Telecommunication Systems*, vol. 50, pp. 217–241, 2012.
246. R. P. Barnwal and S. K. Ghosh, "Detection of misbehaving nodes in vehicular ad hoc network," in *Security for Multihop Wireless Networks*. CRC Press, 2014.
247. B. Das, S. Misra, and U. Roy, "Coalition formation for cooperative service-based message sharing in vehicular ad hoc networks," *IEEE Transactions on Parallel and Distributed Systems*, 2015.
248. S. Shivshankar and A. Jamalipour, "Spatio-temporal multicast grouping for content-based routing in vehicular networks: a distributed approach," *Journal of Network and Computer Applications*, vol. 39, no. 1, pp. 93–103, 2014.
249. ——, "Effect of node neighborhood on the evolution of cooperation using public goods game in vehicular networks," in *IEEE WCNC, Services, Applications, and Business*, 2014.
250. ——, "An evolutionary game theory-based approach to cooperation in vanets under different network conditions," *IEEE Transactions on Vehicular Technology*, vol. 64, no. 5, pp. 2015–2022, 2015.
251. D. Wu, Y. Ling, H. Zhu, , and J. Liang, "The RSU access problem based on evolutionary game theory for VANET," *International Journal of Distributed Sensor Networks*, vol. 2013, pp. 1–7, 2013.
252. A. Banerjee, V. Gauthier, H. Labiod, and H. Afifi, "Cooperation optimized design for information dissemination in vehicular networks using evolutionary game theory," *CoRR*, vol. abs/1301.1268, pp. 1–15, 2013.
253. J. Zhang, V. Gauthier, H. Labiod, A. Banerjee, and H. Afifi, "Information dissemination in vehicular networks via evolutionary game theory," in *IEEE International Conference on Communications*, 2014, pp. 124–129.
254. M. P. Anastasopoulos, P.-D. M. Arapoglou, R. Kannan, and P. G. Cottis, "Adaptive routing strategies in ieee 802.16 multi-hop wireless backhaul networks based on evolutionary game theory," *IEEE Journal on Selected Areas in Communications*, vol. 26, no. 7, pp. 1218–1225, 2008.
255. H. Tembine, E. Altman, R. El-Azouzi, and Y. Hayel, "Evolutionary games in wireless networks," *IEEE Transactions on Systems, Man, and Cybernetics*, vol. 40, no. 3, pp. 634–646, 2010.
256. C. Jiang, Y. Chen, and K. J. R. Liu, "Distributed adaptive networks: A graphical evolutionary game-theoretic view," *IEEE Transactions on Signal Processing*, vol. 61, no. 22, pp. 5675–5688, 2013.
257. I. F. Akyildiz, B. F. Lo, and R. Balakrishnan, "Cooperative spectrum sensing in cognitive radio networks: A survey," *Physical Communication*, vol. 4, no. 1, pp. 40–62, 2011.
258. B. Wang, K. J. R. Liu, and T. C. Clancy, "Evolutionary cooperative spectrum sensing game: How to collaborate?" *IEEE Transactions on Coommunications*, vol. 58, no. 3, pp. 890–900, 2010.
259. C. Hauert, S. D. Monte, J. Hofbauer, and K. Sigmund, "Volunteering as red queen mechanism for co-operation in public goods games," *Science*, vol. 296, no. 5570, pp. 1129–1132, 2002.
260. E. Palomar, A. Alcaide, A. Ribagorda, and Y. Zhang, "The peers dilemma: A general framework to examine cooperation in pure peer-to-peer systems," *Computer Networks*, vol. 56, no. 17, pp. 3756–3766, 2012.
261. E. H. Simpson, "The interpretation of interaction in contingency tables," *Journal of the Royal Statistical Society. Series B (Methodological)*, vol. 13, no. 2, pp. 238–241, 1951.

262. C. H. Wagner, "Simpson's paradox in real life," *The American Statistician*, vol. 36, no. 1, pp. 46–48, 1982.
263. P. J. Bickel, E. A. Hammel, and J. W. O'Connell, "Sex bias in graduate admissions: Data from berkeley," *Science*, vol. 187, no. 4175, pp. 398–404, Feb. 1975. [Online]. Available: http://www.jstor.org/stable/1739581
264. U. Shevade, H. H. Song, L. Qiu, and Y. Zhang, "Incentive-aware routing in dtns," in *Proceedings of the 16th annual IEEE International Conference on Network Protocols (ICNP 2008)*, 2008.
265. H. Zhu, X. Lin, R. Lu, and X. S. Shen, "A secure incentive scheme for delay tolerant networks," in *In Proc. 3rd International Conference on Communications and Networking in China (ChinaCom)*, 2008.
266. H. Zhu, X. Lin, R. Lu, Y. Fan, and X. S. Shen, "Smart: A secure multilayer credit-based incentive scheme for delay-tolerant networks," *IEEE Transactions on Vehicular Technology*, vol. 58, pp. 4628–4639, 2009.
267. B. B. Chen and M. C. Chan, "Mobicent: a credit-based incentive system for disruption tolerant network," in *Proceedings of INFOCOM*, San Diego, CA, 2010, pp. 1–9.
268. J. R. Douceur, "The sybil attack," in *Peer-to-Peer Systems*, ser. Lecture Notes in Computer Science, P. Druschel, F. Kaashoek, and A. Rowstron, Eds. Springer Berlin Heidelberg, 2002, vol. 2429, ch. The Sybil Attack, pp. 251–260. [Online]. Available: http://dx.doi.org/10.1007/3-540-45748-8_24
269. H. Zhu, S. Du, Z. Gao, M. Dong, and Z. Cao, "A probabilistic misbehavior detection scheme toward efficient trust establishment in delay-tolerant networks," *IEEE Transactions on Parallel and Distributed Systems*, vol. 25, no. 1, pp. 22–32, Jan 2014.
270. L. Wei, H. Zhu, Z. Cao, and X. S. Shen, "Mobiid: A user-centric and social-aware reputation based incentive scheme for delay/disruption tolerant networks," in *Ad-hoc, Mobile, and Wireless Networks*, ser. Lecture Notes in Computer Science, H. Frey, X. Li, and S. Ruehrup, Eds. Springer Berlin Heidelberg, 2011, vol. 6811, pp. 177–190.
271. X. Zhang, X. Wang, A. Liu, Q. Zhang, and C. Tang, "Pri: A practical reputation-based incentive scheme for delay tolerant networks," *TIIS*, vol. 6, no. 4, pp. 973–988, April 2012.
272. E. Hernández-Orallo, M. Serrat, J.-C. Cano, C. Calafate, and P. Manzoni, "Improving selfish node detection in manets using a collaborative watchdog," *Communications Letters, IEEE*, vol. 16, no. 5, pp. 642–645, May 2012.
273. M. Serrat-Olmos, E. Hernández-Orallo, J. Cano, C. Calafate, and P. Manzoni, "Accurate detection of black holes in manets using collaborative bayesian watchdogs," in *2012 IFIP Wireless Days (WD)*, Dublin, Nov 2012, pp. 1–6.
274. A. Al-Roubaiey, T. Sheltami, A. Mahmoud, E. Shakshuki, and H. Mouftah, "AACK: Adaptive acknowledgment intrusion detection for MANET with node detection enhancement," in *2010 24th IEEE International Conference on Advanced Information Networking and Applications (AINA)*, Perth, April 2010, pp. 634–640.
275. J. A. F. F. Dias, J. J. P. C. Rodrigues, C. Mavromoustakis, and F. Xia, "A cooperative watchdog system to detect misbehavior nodes in vehicular delay-tolerant networks," *IEEE Transactions on Industrial Electronics*, April 2015. [Online]. Available: DOI: 10.1109/TIE.2015.2425357
276. L. Buttyn, L. Dra, M. Flegyhzi, and I. Vajda, "Barter-based cooperation in delay-tolerant personal wireless networks," in *In Proceedings of the IEEE Workshop on Autonomic and Opportunistic Communications (AOC 2007)*, 2007.
277. L. Buttyan, L. Dora, M. Felegyhazi, and I. Vajda, "Barter trade improves message delivery in opportunistic networks," *Ad Hoc Networks*, vol. 8, pp. 1–14, 2010.
278. L. Yin, H. mei Lu, Y. da Cao, and J. min Gao, "Cooperation in delay tolerant networks," in *2nd International Conference on Signal Processing Systems (ICSPS)*, 2010.
279. T. Abdelkader, K. Naik, and W. Gad, "A game-theoretic approach to supporting fair cooperation in delay tolerant networks," in *2015 IEEE 81st Vehicular Technology Conference (VTC Spring)*, Glasgow, 2015, pp. 1–7.
280. A. Panagakis, A. Vaios, and I. Stavrakakis, "On the effects of cooperation in DTNs," in *2nd International Conference on Communication Systems Software and Middleware (COMSWARE)*, Bangalore, 2007, pp. 1–6.

281. G. Resta and P. Santi, "The effects of node cooperation level on routing performance in delay tolerant networks," in *6th Annual IEEE Communications Society Conference on Sensor, Mesh and Ad Hoc Communications and Networks*, Rome, June 2009, pp. 1–9.

282. ——, "A framework for routing performance analysis in delay tolerant networks with application to noncooperative networks," *IEEE Transactions on Parallel and Distributed Systems*, vol. 23, no. 1, pp. 2–10, Jan 2012.

283. A. Keränen, M. Pitkänen, M. Vuori, and J. Ott, "Effect of non-cooperative nodes in mobile DTNs," in *2011 IEEE International Symposium on a World of Wireless, Mobile and Multimedia Networks (WoWMoM)*, Lucca, June 2011, pp. 1–7.

284. R. B. Freeman, *The works of Charles Darwin: an annotated bibliographical handlist*, 2nd ed., F. Darwin, Ed. Dawsons, 1977.

285. S. Pal and S. Misra, "DISIDE: Distributed strategy identification in opportunistic mobile networks," *Computer Communications*, pp. –, September 2015. [Online]. Available: http://www.sciencedirect.com/science/article/pii/S0140366415003199

286. P. Jaccard, "Etude comparative de la distribution florale dans une portion des alpes et des jura," *Bulletin de la Societe Vaudoise des Sciences Naturelles*, vol. 37, pp. 547–579, 1901.

287. P. Hui and J. Crowcroft, "Predictability of human mobility and its impact on forwarding," in *3rd International Conference on Communications and Networking*, China, 2008, pp. 543–547.

288. J. M. Bahi, A. Makhoul, and M. Medlej, "Data aggregation for periodic sensor networks using sets similarity functions," in *Proceedings of the 7th International Conference on Wireless Communications and Mobile Computing*, 559–564, 2011, p. Turkey.

289. D. Hadj Sadok, T. Rodrigues, R. Amorim, and J. Kelner, "On the performance of heterogeneous MANETs," *Wireless Networks*, vol. 21, no. 1, pp. 139–160, 2015.

290. E. Freitas, R. Allgayer, M. Wehrmeister, C. Pereira, and T. Larsson, "Supporting platform for heterogeneous sensor network operation based on unmanned vehicles systems and wireless sensor nodes," in *2009 IEEE Intelligent Vehicles Symposium*, Xi'an, June 2009, pp. 786–791.

291. Z. Sanaei, S. Abolfazli, A. Gani, and R. Buyya, "Heterogeneity in mobile cloud computing: Taxonomy and open challenges," *IEEE Communications Surveys Tutorials*, vol. 16, no. 1, pp. 369–392, 2014.

292. E. De Poorter, I. Moerman, and P. Demeester, "Enabling direct connectivity between heterogeneous objects in the internet of things through a network-service-oriented architecture," *EURASIP Journal on Wireless Communications and Networking*, vol. 2011, no. 1, 2011.

293. R. C. A. da Rocha and M. Endler, "Evolutionary and efficient context management in heterogeneous environments," in *Proceedings of the 3rd International Workshop on Middleware for Pervasive and Ad-hoc Computing*. France: ACM, 2005, pp. 1–7.

294. R. Schmohl and U. Baumgarten, "Heterogeneity in mobile computing environmens," in *ICWN*, Las Vegas, Nevada, USA, 2008, pp. 461–467.

295. P. Stuedi and G. Alonso, "Transparent heterogeneous mobile ad hoc networks," in *the 2nd Annual International Conference on Mobile and Ubiquitous Systems: Networking and Services.*, San Diego, California, USA, Jul. 2005, pp. 237–246.

296. A. Petz, A. Bednarczyk, N. Paine, D. Stovall, and C. Julien, "MaDMAN: A middleware for delay-tolerant mobile ad-hoc networks," University of Texas at Austin, Tech. Rep. TR-UTEDGE-2010-010, 2010.

297. M. Le, J.-S. Park, and M. Gerla, "UAV assisted disruption tolerant routing," in *IEEE Military Communications Conference, 2006. MILCOM 2006.*, Washington, DC, Oct 2006, pp. 1–5.

298. V. Conan, J. Leguay, and T. Friedman, "Characterizing pairwise inter-contact patterns in delay tolerant networks," in *Proceedings of the 1st International Conference on Autonomic Computing and Communication Systems*, ser. Autonomics '07. ICST, Brussels, Belgium, Belgium: ICST (Institute for Computer Sciences, Social-Informatics and Telecommunications Engineering), 2007, pp. 19:1–19:9.

299. Y. Tian and J. Li, "Heterogeneity of device contact process in pocket switched networks," in *Proceedings of the 5th International Conference on Wireless Algorithms, Systems, and Applications (WASA'10)*. Berlin, Heidelberg: Springer-Verlag, 2010, pp. 157–166.

300. C. Lee and D. Eun, "Exploiting heterogeneity for improving forwarding performance in mobile opportunistic networks: An analytic approach," *IEEE Transactions on Mobile Computing*, 2015, DOI: 10.1109/TMC.2015.2407406.

301. V. Kostakos, "Temporal graphs," *Physica A: Statistical Mechanics and its Applications*, vol. 388, no. 6, pp. 1007–1023, 2009.

302. P. Erdős and A. Rényi, "On random graphs. I," *Publicationes Mathematicae*, vol. 6, pp. 290–297, 1959.

303. ——, "On the evolution of random graphs," *Publ. Math. Inst. Hungar. Acad. Sci*, vol. 5, pp. 17–61, 1960.

304. E. N. Gilbert, "Random graphs," *Ann. Math. Statist.*, vol. 30, no. 4, pp. 1141–1144, 12 1959.

305. A. Ferreira, "Building a reference combinatorial model for MANETs," *IEEE Network*, vol. 18, no. 5, pp. 24–29, Sept 2004.

306. A. Casteigts, P. Flocchini, W. Quattrociocchi, and N. Santoro, "Time-varying graphs and dynamic networks," *International Journal of Parallel, Emergent and Distributed Systems*, vol. 27, no. 5, pp. 387–408, 2012.

307. H. Alvestrand, "A mission statement for the IETF," Internet Requests for Comments, RFC Editor, RFC 3935, October 2004. [Online]. Available: https://www.rfc-editor.org/rfc/rfc3935.txt

308. W. Eddy and E. Davies, "Using self-delimiting numeric values in protocols," Internet Requests for Comments, RFC Editor, RFC 6256, May 2011. [Online]. Available: https://www.rfc-editor.org/rfc/rfc6256.txt

309. S. Symington, "Delay-tolerant networking metadata extension block," Internet Requests for Comments, RFC Editor, RFC 6258, May 2011. [Online]. Available: https://www.rfc-editor.org/rfc/rfc6258.txt

310. ——, "Delay-tolerant networking previous-hop insertion block," Internet Requests for Comments, RFC Editor, RFC 6259, May 2011. [Online]. Available: https://www.rfc-editor.org/rfc/rfc6259.txt

311. A. Lindgren, A. Doria, E. Davies, and S. Grasic, "Probabilistic routing protocol for intermittently connected networks," Internet Requests for Comments, RFC Editor, RFC 6693, August 2012. [Online]. Available: https://www.rfc-editor.org/rfc/rfc6693.txt

312. M. Yarvis, R. Shah, C. Wan, and Y. Wang, "Interface for a delay-tolerant network," Apr. 19 2011, US Patent 7,930,379. [Online]. Available: https://www.google.co.in/patents/US7930379

313. C. Karkanias, S. Hodges, and J. Scott, "Health-related opportunistic networking," Oct. 4 2011, US Patent 8,032,124. [Online]. Available: https://www.google.co.in/patents/US8032124

314. N. Fujita and M. Jibiki, "Routing method and node," Jan. 29 2013, US Patent 8,363,565. [Online]. Available: https://www.google.co.in/patents/US8363565

315. H. Lagar-Cavilla, K. Joshi, and A. Varshavsky, "Transmitting delay-tolerant data with other network traffic," Dec. 17 2013, US Patent 8,611,213. [Online]. Available: https://www.google.co.in/patents/US8611213

316. R. Rundquist, P. Bandera, Y. Yi, and H. Bai, "Multi-streaming multi-homing delay tolerant network protocol," Jul. 22 2014, US Patent 8,787,157. [Online]. Available: https://www.google.co.in/patents/US8787157

317. V. Conan, J. Leguay, and T. Friedman, "Routing method intended for intermittently connected networks," Sep. 23 2014, US Patent 8,842,659. [Online]. Available: https://www.google.co.in/patents/US8842659

318. E. Bathrick, "Automatic invocation of DTN bundle protocol," Apr. 21 2015, US Patent 9,015,822. [Online]. Available: https://www.google.co.in/patents/US9015822

319. M. Conti and M. Kumar, "Opportunities in opportunistic computing," *IEEE Computer*, vol. 43, no. 1, pp. 42–50, Jan 2010.

320. M. Conti, S. Giordano, M. May, and A. Passarella, "From opportunistic networks to opportunistic computing," *IEEE Communications Magazine*, vol. 48, no. 9, pp. 126–139, Sep. 2010.

321. D. Mascitti, M. Conti, A. Passarella, and L. Ricci, "Service provisioning through opportunistic computing in mobile clouds," *Procedia Computer Science*, vol. 40, pp. 143–150,

2014, Fourth International Conference on Selected Topics in Mobile & Wireless Networking (MoWNet2014).

322. M. Tehrani, M. Uysal, and H. Yanikomeroglu, "Device-to-device communication in 5G cellular networks: challenges, solutions, and future directions," *IEEE Communications Magazine*, vol. 52, no. 5, pp. 86–92, May 2014.
323. C. Mayer and O. Waldhorst, "Offloading infrastructure using delay tolerant networks and assurance of delivery," in *2011 IFIP Wireless Days (WD)*, Oct 2011, pp. 1–7.
324. Y. Li, M. Qian, D. Jin, P. Hui, Z. Wang, and S. Chen, "Multiple mobile data offloading through disruption tolerant networks," *IEEE Transactions on Mobile Computing*, vol. 13, no. 7, pp. 1579–1596, July 2014.
325. Y. Wu, S. Deng, and H. Huang, "Mobile cloud forwarding service in delay tolerant networks," in *Unifying Electrical Engineering and Electronics Engineering*, ser. Lecture Notes in Electrical Engineering, S. Xing, S. Chen, Z. Wei, and J. Xia, Eds. Springer New York, 2014, vol. 238, pp. 1511–1518.
326. O. Khalid, S. U. Khan, S. A. Madani, K. Hayat, L. Wang, D. Chen, and R. Ranjan, "Opportunistic databank: A context aware on-the-fly data center for mobile networks," in *Handbook on Data Centers*, S. U. Khan and A. Y. Zomaya, Eds. Springer New York, 2015, pp. 1077–1094.
327. D. Maggiorini, C. Quadri, and L. Ripamonti, "Opportunistic mobile games using public transportation systems: a deployability study," *Multimedia Systems*, vol. 20, no. 5, pp. 545–562, 2014.

Index

© Springer International Publishing Switzerland 2016
S. Misra et al., *Opportunistic Mobile Networks*,
Computer Communications and Networks, DOI 10.1007/978-3-319-29031-7